A Bastard Kind of Reasoning

SUNY series, Studies in the Long Nineteenth Century
———————
Pamela K. Gilbert, editor

A Bastard Kind of Reasoning
William Blake and Geometry

Andrew M. Cooper

Cover painting: William Blake, *Visions of the Daughters of Albion* frontispiece, copy I, 1793. Yale Center for British Art, Paul Mellon Collection.

Published by State University of New York Press, Albany

© 2023 State University of New York

All rights reserved

Printed in the United States of America

No part of this book may be used or reproduced in any manner whatsoever without written permission. No part of this book may be stored in a retrieval system or transmitted in any form or by any means including electronic, electrostatic, magnetic tape, mechanical, photocopying, recording, or otherwise without the prior permission in writing of the publisher.

For information, contact State University of New York Press, Albany, NY
www.sunypress.edu

Library of Congress Cataloging-in-Publication Data

Name: Cooper, Andrew M., 1953– author.
Title: A bastard kind of reasoning : William Blake and geometry / Andrew M. Cooper.
Description: Albany : State University of New York Press, 2023. | Series: SUNY series, studies in the long nineteenth century | Includes bibliographical references and index.
Identifiers: LCCN 2022041858 | ISBN 9781438493220 (hardcover : alk. paper) | ISBN 9781438493237 (ebook) | ISBN 9781438493213 (pbk. : alk. paper)
Subjects: LCSH: Blake, William, 1757–1827—Criticism and interpretation. | Literature and science—Great Britain—History—18th century. | Newton, Isaac, 1642–1727—Influence. | Geometry in literature. | Mathematics in literature. | Physics in literature. | Space and time in literature. | LCGFT: Literary criticism.
Classification: LCC PR4148.S35 C66 2023 | DDC 821/.7—dc23/eng/20221121
LC record available at https://lccn.loc.gov/2022041858

10 9 8 7 6 5 4 3 2 1

For Ellen

Contents

List of Illustrations		ix
Acknowledgments		xi
Abbreviations		xiii
Introduction	Geometry and Blake's *Newton* Print	1
Chapter 1	"Oh, but you're just analogizing . . ."	17
Chapter 2	Learning to Read in a Force Field: *Songs of Innocence*, Hartleyan Psychology, and the Physics of R. J. Boscovich	45
Chapter 3	*The Book of Urizen* as a Vortex of Perception	85
Chapter 4	A Brief Particular History of the Fourth Dimension of Space, with Special Reference to *Milton: A Poem*	105
Chapter 5	The Neoplatonism of Blake's Mundane Soul	147
Chapter 6	Berkeley: Very Close, but No Cigar	195
Conclusion	The Unified Space-Time of *The Vision of the Last Judgment*	219
Notes		237
Works Cited		297
Index		311

Illustrations

Figure I.1	William Blake, *Newton*, 1795/1805. Tate Collection.	1
Figure 1.1	William Blake, *Visions of the Daughters of Albion* frontispiece, copy I, 1793. Yale Center for British Art.	17
Figure 1.2	William Blake, *Jerusalem* pl. 62, copy E, c. 1821. Yale Center for British Art.	39
Figure 2.1	William Blake, *Milton's Mysterious Dream*, c. 1816–20. Morgan Library.	48
Figure 2.2	Roger Joseph Boscovich, *A Theory of Natural Philosophy* [*Theoria Philosophiae Naturalis*], 1758, figure 1.	50
Figure 2.3	William Blake, "A Cradle Song," *Songs of Innocence and of Experience*, copy L, 1789, 1794. Yale Center for British Art.	58
Figure 3.1	William Blake, *The Book of Urizen* pl. 13, copy A, 1794. Yale Center for British Art.	90
Figure 3.2	William Blake, *The Book of Urizen* pl. 9, copy C, 1794. Yale Center for British Art.	91
Figure 3.3	Roger Joseph Boscovich, *A Theory of Natural Philosophy* [*Theoria Philosophiae Naturalis*], 1758, figure 14.	93
Figure 3.4	William Blake, *The Book of Urizen* title page, copy C, 1794. Yale Center for British Art.	101
Figure 4.1	Klein bottle.	108
Figure 4.2	"Inside-out universe."	109
Figure 4.3	William Blake, *Jerusalem* pl. 25, copy E, c. 1821. Yale Center for British Art.	110

Figure 4.4	Center of a 3-sphere.	112
Figure 4.5	William Blake, *The Annunciation to the Shepherds*, c. 1815. Huntington Library.	115
Figure 4.6	William Blake, *The Last Judgment*, c. 1810. National Gallery of Art.	117
Figure 4.7	"Gluing" two spheres together.	121
Figure 4.8	Gustave Doré, *Canto 31: The Saintly Throng in the Form of a Rose*, 1868.	121
Figure 4.9	William Blake, *The Vision of the Deity from Whom Proceed the Nine Spheres*, 1824–27. Ashmolean Museum.	122
Figure 4.10	"Slicing" a sphere.	123
Figure 4.11	William Blake, *Milton* pl. 32, copy C, c. 1804–11. New York Public Library.	125
Figure 4.12	William Blake, *Milton* pl. 47, copy C, c. 1804–11. New York Public Library.	126
Figure 4.13	William Blake, *The Sun at His Eastern Gate*, c. 1816–20. Morgan Library.	129
Figure 4.14	William Blake, *Milton* pl. 40, copy C, c. 1804–11. New York Public Library.	131
Figure 4.15	Henry More, "Ogdoaz," *A Platonick Song of the Soul*, 1647.	139
Figure 6.1	William Hogarth, *Satire on False Perspective*, 1754.	211
Figure C.1	William Blake, *The Marriage of Heaven and Hell*, copy D, 1790, 1795. Library of Congress.	220
Figure C.2	James Barry, *Elysium and Tartarus or the State of Final Retribution*, 1791. Yale Center for British Art.	222
Figure C.3	William Blake, *The Vision of the Last Judgment*, 1808. National Trust.	224
Figure C.4	Different dimensions of Blake's *The Vision of the Last Judgment*.	230

Acknowledgments

Ellen Locy: light that gives heat, *sine qua non*, real space of mind in a body of substance, concentrated laughing gas. For being there when it matters, I'm also grateful to my siblings, Melissa Cooper and Thomas Van Cooper, and to my second family at Shambly Acres in North Truro: Rob DuToit, Janice Redman, and Alexander Redman-DuToit. Rob made the illustration of Blake's Vortex that appears in my conclusion as figure C.4. Big shout-out to Alex Webb and Becky Norris Webb, who heard out my early thoughts during a walk from Fisher Beach to the Pamet Harbor jetty in June 2019. Whatever its limitations, I hope this book's spirit stays true to my two very different teachers long ago, Jonathan Grandine and Karl Kroeber. Rebecca Colesworthy, my SUNY editor, has piloted the barge to port with great expertise and forbearance through multiple teapot tempests of revision. The manuscript readers Rebecca managed to obtain and their close, thoughtful suggestions improved this book beyond anything I anticipated or even wanted. I am deeply indebted to them. In solving technical issues, free-lance editor Matthew John Phillips was an absolute brick. I especially want to thank preproduction editor Jenn Bennett-Genthner for being so dead smack on the case all the time.

Chapter 1 borrows two or three pages from my article, "Small Room for Judgment: Geometry and Prolepsis in Blake's 'Infant Sorrow,'" *European Romantic Review* 31:2 (2020): 129–55. Here and there, I've repeated material amounting to maybe ten pages from my *William Blake and the Productions of Time* (Ashgate 2013/Routledge 2016).

Abbreviations

All Blake quotations are from *The Complete Poetry and Prose of William Blake*, ed. David V. Erdman, with commentary by Harold Bloom, 3rd ed. (Garden City, NJ: Doubleday, 1982). References to this text are designated as E. References to Blake's works are abbreviated as follows.

A	*America*
ARO	*All Religions Are One*
BU	*The Book of Urizen*
E	*Europe*
FR	*The French Revolution*
FZ	*The Four Zoas*
J	*Jerusalem*
M	*Milton*
MHH	*The Marriage of Heaven and Hell*
NNR	*There Is No Natural Religion*
PA	*A Public Address to the Chalcographic Society*
SI	*Songs of Innocence*
SE	*Songs of Experience*
VDA	*Visions of the Daughters of Albion*
VLJ	*A Vision of the Last Judgment*

Readers are strongly encouraged to visit Blake's designs and artwork on the Blake Archive website.

On one occasion Ma-tsu and Po-chang went out for a walk, when they saw some wild geese flying past.

"What are they?" asked Ma-tsu.

"They're wild geese," said Po-chang.

"Where are they going?" demanded Ma-tsu.

Po-chang replied, "They've already flown away."

Suddenly Ma-tsu grabbed Po-chang by the nose and twisted it so hard that he cried out in pain.

"How," shouted Ma-tsu, "could the wild geese ever have flown away?"

This was the moment of Po-chang's awakening.

—A. W. Watts, *The Way of Zen* (1959)

Space is nothing other than the finest light.

—Proclus, *Elements of Physics*, 142a

But cloudy, cloudy is the stuff of stone.

—Richard Wilbur, "Epistemology"

Introduction

Geometry and Blake's *Newton* Print

Figure I.1. William Blake, *Newton*, 1795/1805. Tate Collection.

For half a century or more, Blake's color print *Newton* (1795/1805) was regarded as a savage rejection of Newton, mathematics, and *l'esprit géometrique* (figure I.1). Did not Blake fulminate all his life, "God forbid that Truth should be Confined to Mathematical Demonstration" (E 659)? Newton is shown in splendid profile similarly to the two-dimensional picture of "The Tyger" a couple of years earlier, which associates such flattening with the argument of divine Design while implying that, in reality, good and evil intertwine in antinomian fashion without right-left Manichaean "symmetry." Then came Donald Ault's *Visionary Physics* (1974), which opens by showing how Blake's print exposes a profound contradiction in Newton's thought—evidence, Ault argued, of Blake's close knowledge of Newtonian mechanics, the calculus, and the problem of gravitation's physical cause. Ault says that in Blake's print "the human figure is constructing a limited, fixed, and unchanging model of his fundamental bodily experiences to stave off the sense of the dissolving quality of the outer world. Yet . . . it is the very act of constructing the model that separates the world into inner and outer, definite and indefinite, action and background, symmetry and asymmetry. The background is both the cause and the effect of the central action."[1]

Much as Ault's argument would suggest, my claim here is that Blake did not simply reject Newton, geometry, and science. Quite the opposite—Blake's way of representing perspective, geometric figures, and nested relations between objects builds on Newton's physics through insights and intuitions which we today ascribe to Einstein and relativity's tendency to suspend cause and effect by dissolving objects into their background "field." He was not anti- but rather post-Newtonian. Contrary to what generations of Blake critics have supposed, the outlook of Romantic-period physics and chemistry was far from materialist.[2] By redefining the idea of material substance, these sciences foregrounded a deep (and, today, well-known) Neoplatonic tendency in Newton's thought that until the middle-late twentieth century had been obscured by his emphasis on contact mechanics—an emanationist tendency that Blake recognized and embraced.

Ault's point that Blake collapses cause and effect has since been noticed in a variety of contexts. Marxist critics have deemed it part of the poet's dialectical-materialist critique of how mystified social reality reconstitutes the past within the present.[3] On the other hand, Steven Goldsmith has argued, against liberal academic criticism's sentimental adoption of Blakean radical "enthusiasm," that Blake subscribed to Paine's liberal-democratic assumption "that power can be collapsed into indeterminate signs, that freedom corresponds to the capacity for perpetual subversion in and by language," such that change and difference become institutionalized within democratic discourse as an endless play of competing representations: Derridean deferral made real.[4] Thus, Blakean prophecy tends to conflate speech and action, event and discourse, without necessarily changing anything. More positively, for Angela Esterhammer, Blake's poetry transforms the force of speech acts based on social convention into "the phenomenological performative," whose force is metaleptic and derives from "an author's ability to 'create' reality through poetic or fictional utterance": "Prophecy and performativity interconstitute one another; what the poet predicts will happen *is* happening in and through his writing, and vice versa."[5] One concludes that if "Let there be light" (Gen. 1:3) was the original speech act that "does what it says"—unlike human speech acts grounded in conventional social agreement about which kinds of syntax signal performance in the world ("hereby," "henceforth," "it is decreed that," etc.)—then prophetic poetry operates as mankind's conditionalized mortal reiteration of God's command. By contrast, Robert Essick has examined the dark downside of such prophesizing. He shows how *Newton* and the other great color prints of 1795 merge graphic media with pictorial content and themes of fallenness, thereby making the medium the message. Says Essick, the sometimes deliberately blotchy tactility of these prints, so contrary to Blake's celebrations of radiant "Florentine" fresco and determinate outline, instantiates corporeal fears that were beginning to occlude the artist's vision[6]—perhaps because the government's November 1794 clampdown on dissent was driving him into the complicity of self-censorship, as I'll argue in chapter 5.

Most recently, Sarah Haggarty has developed these paradoxes in relation to geometry, based on her claim that Blake was "intrigued by diagrams" because of their proximity to line drawing. Blake's "engagement with—and fascination by—geometry as such, or more precisely, with both Euclidean and practical geometry as they were taught and theorized by his contemporaries," leads Haggarty to conclude that the *Newton* print "temporize[s] geometry's very origin, exhibiting demonstration as practical intelligence rather than act of pure thinking," and so "allows geometry to coexist with artistry."[7] In other words, the print's fusion of cause and effect no longer conveys Newton's entrapment by mathematic formalism but rather his redemption through the materialized self-consciousness of Blake's art. It seems that the closer critics look, the more complicated, sympathetic, and even self-projected Blake's image of Newton becomes.

The source of these various critical observations may be seen to lie in Blake's substitution of a creative principle of immanency for Newtonian mechanism's transcendental first cause.[8] In a physics context, *Newton*'s underwater background—less than transparent, seemingly oozy, and suffocatingly silent, "both the cause and the effect of the central action" as Ault says—resembles the aether, an entity Newton had boldly conjectured in the General Scholium to his *Principia Mathematica* as constituting gravity's "physical seat" (as he often phrased it), and that he described at length in the *Opticks*. The aether, frequently deemed to be a liquid or "subtle fluid"—only by the later nineteenth century did it become "fixed" and "luminiferous"—was dualist mechanism's acknowledgment that some medium was needed by which to connect mind and matter: a *tertium quid* such as Coleridge was always calling for.[9] This "subtle matter" served to ground gravity by enabling it to act not just mechanically on the surfaces of bodies but on all their parts. By flowing through masses with different degrees of density, so highlighting the distance between them, the aether could supply a physical basis for Newton's inverse-square law. The aether's ubiquity provided a platform for measurements and established a uniform observational perspective on all objects, as required by Newton's idea of "absolute" space.[10]

Following the work of David Hartley, eighteenth-century investigations of this semimetaphysical "third kind" were increasingly undertaken by medical scientists and anatomists, who located it in the human brain and nerves. Their physiological approach had the sanction of Newton himself. As "a certain most subtle Spirit which pervades and lies hid in all gross bodies," the aether might transmit the force of gravitation across planetary space and along the nerves to the brain.[11] It filled in the pore between hard particles while remaining, itself, real and atomic, and not just a physical property of space (as classical aether physicists would argue during the 1910s in a last-ditch attempt to defuse Einsteinian Relativity). Perhaps, Newton wondered at the very end of the *Principia*, "all sensation is excited, and the members of animal bodies move at the command of the will, namely, by the vibrations of this Spirit, mutually propagated along the solid filaments of the nerves, from the outward organs of sense to the brain, and from the brain into the muscles." Hitherto, the *Principia* had relegated all bodily interaction to the level of accidental changes in the relations of masses. Here, Newton readmits substantial contact, potentially restoring the place of the experimental observer within his system because "this electric and elastic Spirit" is an implicitly anthropomorphic one. In his open letter to Henry Oldenburg, secretary of the Royal Society, Newton calls it a "Mediator of Sociablenes"[12] by means of which we become acquainted with objects metacognitively (we not only perceive objects but are reflexively aware of it), as opposed to accessing them through independent mental representations as Locke subsequently seemed to suggest.

In Blake's *Newton*, the glowing background murk—illumined, perhaps, by inner light emanating from the human figure—portrays aetherous "Sociablenes" in its alienated materialist form. Hence, the curious drapery hanging over Newton's left shoulder. If this represents the skintight bodysuit or tunic with which Blake typically clothes his spirits, then Newton's having shuffled it off (though it apparently remains attached by a neck strap below his jaw) signifies the ascetic side of his dualism. He seems oblivious to the huge undifferentiated reef of matter—rocky corral coated with algae and bits of ectoplasm—whose bench supports his muscular buttock and thigh. The denseness of the aether's all-surrounding invisible medium is suggested by a pair of anemones below him, their tentacles drifting in a current. Hunched over to form a series of triangles in imitation of his compass, Newton is bending his body into another measuring instrument in a kind of parody of Vitruvian Man. (If he fell forward, however, his posture would resemble the similarly triangulated Nebuchadnezzar crawling on all fours like a beast along the floor of *his* cavern in another of Blake's large color prints.)

One wonders if Blake's design alludes to Newton's famous statement, supposedly made "a little before he died," that to himself he seemed "only like a boy playing on the sea shore, and diverting myself in now and then finding a smoother pebble or a prettier shell than ordinary, whilst the great ocean of truth lay all undiscovered before me."[13] These words were well known. Wordsworth appears to recall them when his Immortality Ode tells how aged adults, "Though inland far we be," still "have sight of that immortal sea / . . . / And see the Children sport upon the shore, / And hear the mighty waters rolling evermore."[14] Less reverently than Wordsworth, Blake's print takes Newton at his (reported) word. The "great ocean of truth"—the world aether whose created human form is Albion the Divine Humanity, as we'll see—"[lies] all undiscovered" before the mathematician fixated upon his geometrical abstractions. The print's 1795 version, which shows light entering a subterranean cavern through a rift in the rock above Newton's back, makes the allusion to Plato's Cave more explicit, perhaps linking it with the unspecified "dark chamber" where Newton says he performed the optical experiments described in his *Opticks*.[15] Turned away from the light, the figure in Blake's design is preoccupied with shadow representations, ironically those of so-called "divine" geometry itself. In both versions, Newton bending to his task recalls the antihero of Blake's *Book of Urizen*, published less than a year earlier, who "formed golden compasses / And began to explore the Abyss" (*BU* 20:39–40, E 81)—the cause-effect transposition being, here, that the conditions of measurement are established through Urizen's own acts of creation by division, by which Eternity's space-time of pure duration is reduced to metric space and chronological time (itself measured spatially by the movement of the sun or clock hands). Blake knew it was by means of geometry that the ancient astronomers measured time. They divided the 360 degrees of a circle or the sphere of Earth into sixty parts or "minutes" and then divided each minute into sixty "seconds." Time as a form of movement is invisible, but geometry serves to arrest and reify it. Like the supreme Blakean reifier, Urizen, Newton is evidently constructing a model of Creation similarly to the Demiurge in Plato's *Timaeus*, the ultimate source of the triangles in Blake's print. There, Plato portrays the requisite intermediate "third thing" between mind and matter as the barely real Receptacle, a "room" (*khora*) that is the "place" of things without containing them in a definite "space," and that is said to contain molecular particles of matter configured as regular geometrical solids made up of various arrangements of atoms in the form of elementary right triangles (53c–55c).

Let me suggest that Blake's engagements with geometry were not incidental but key to how he understood the workings of the universe. It's no accident that his first books in illuminated printing, the little tractates of 1788, *All Religions Are One* and *There Is No Natural Religion* [a] and [b], are sets of axioms or postulates in the form of Euclidean proofs. Elsewhere, I've argued they are not direct satires of Reason so much as incremental skewings of rational argumentation that dramatize its dependence upon rhetoric, persuasion, performance, and feeling, following the example of Hume's dramatically emotional *Dialogues concerning Natural Religion*.[16] That Blake should have launched his career this way is hardly surprising, given geometry's enormous cultural prestige ever since Plato's *Meno* linked it with eternal truth. The "Euclidean method" of deducing propositions from theorems became the very paradigm of knowledge. In Blake's time, geometry was regarded not as a waystation between algebra and calculus, as it is today, but as a philosophy of the world, the soul of mathematics, and a representation of space in its purest form independent of limited human perception. In 1805, Wordsworth called it "An independent world, / Created out of pure intelligence," perhaps echoing his Kantian friend, Coleridge, who later observed: "the Circle in [a] diagram is only a picture or *remembrancer* of the Circle, on which the mathematician is *reasoning*."[17] Coleridge here recalls the *Meno*'s Doctrine of Recollection, set forth by means of an ignorant slave boy's ostensibly intuitive knowledge of geometry. Kant even claimed Euclidean geometry was true a priori, as the necessary form of cognition that structures human experience of external objects. Similarly, fifty years before Kant, Hume asserted from an opposing empirical perspective that geometrical propositions "are discoverable by the mere operation of thought" instead of being "matters of fact": "Though there were never a circle or triangle in nature, the truths demonstrated by Euclid would for ever retain their certainty and evidence."[18] So, when Hume wanted to contrast his "experimental method of reasoning" on "moral" subjects with demonstrative reasoning from propositions, he *attacked* the Parallel Postulate for exhibiting "the fallacy of geometrical demonstrations, when carry'd beyond a certain degree of minuteness": "How can [a mathematician] prove to me . . . that two right lines cannot have one common segment? . . . [S]upposing these two lines to approach at the rate of an inch in twenty leagues, I perceive no absurdity in asserting, that upon their contact they become one. . . . The original standard of a right line is in reality nothing but a certain general appearance."[19] And when Thomas Reid then wanted to preserve practical "common sense" against Hume's broader argument that mere facts of experience alone daily suffice to subvert reason, he devised a geometrical thought experiment, albeit a non-Euclidean one. To an eye placed in the center of a sphere, all "great" lines traced across the sphere's surface will "return to themselves" and appear straight even though they curve. Visible straight lines therefore differ from the tangible straight lines of Euclidean geometry, which if projected will never return to their starting point. This suggested the possibility of a spatial fourth dimension; though, as a recent critic points out, Reid's paradigm remained "notionally Euclidean."[20] What all these different positions share is a view of Euclidean geometry as the universal basis for reasoning about truth claims and an accurate representation of the space of thought itself.

Newton's tendency to absolutize three-dimensional space even led him, in the early *De gravitatione et aequipondio fluidorum*, to insist that mathematical shapes and figures are already actually contained in spatial extension while they remain beyond human sense: "There are everywhere all kinds of figures, everywhere spheres, cubes, triangles, straight lines, everywhere circular, elliptical, parabolical, and all other kinds of figures, and those of all shapes and sizes, even though they are not disclosed to sight. . . . We firmly believe that space was spherical before the sphere occupied it, so that it could contain the sphere. . . . And so on of the other figures."[21] Space's preexisting dimensions thus made it a receptacle for correspondingly configured bodies, but in a way exactly opposite to the potentiality of sensible body represented by Plato's virtual Receptacle. This is much the same logic as Newton used to devise his theory of fluxions, the infinitesimal straight lines that exist in a space outside of time as the differential of the curve.

Blake abhorred Newton's nonsensible fluxions: "A Thing that does not Exist" (E 783). But he could not have read *De gravitatione*, which remained unpublished until 1962. Still, he satirized Newton's ideas through intuition and inference. About the time of the *Newton* print, *The Four Zoas* Night Two strikingly anticipates the preceding Newton passage when Urizen the Workmaster builds the Mundane Shell as a "weighd & orderd" Euclidean solid space within which the ordered ranks of his Sons and Daughters travel underwater "In right lined paths . . . / And measure mathematic motion . . .":

> Others triangular right angled course maintain. others obtuse
> Acute Scalene, in simple paths. but others move
> In intricate ways biquadricate. Trapeziums Rhombs Rhomboids
> Parallelograms. triple & quadruple. polygonic
> In their amazing hard subdued course in the vast deep.
> (*FZ* 33:32–36; E 322)

Newton-Urizen's descendants form a cadre of dehumanized corpuscles (L. *corpusculum*: small body), their forward march driven not by desire but disciplined obedience to the force of logic. Anybody who doubts Blake's interest in geometry will need to explain his complex attitude toward these baroque anthropomorphisms whose pompous self-importance seems freighted with Gillrayan comedy and pathos. That they appear to allude to a passage in *Paradise Lost* comparing the dance of angels to the motions of planets and stars—"mazes intricate, / Eccentric, intervolv'd, yet regular / Then most, when most irregular they seem"—extends Blake's little satire of self-delusion beyond Newton to Milton, theodicy, and the theory of divine Design in general.[22] What Blake rejects here is not Bacon's "advancement of learning" or the Enlightenment's "grand march of intellect," as Keats later called it,[23] but rather the idea that human progress and forward movement can be reduced to mathematics and the determinations of logical reason.

So, Newton-Urizen's nonsensible absolute space appears as a parody of Plato's creation myth in the *Timaeus*. Urizenic Creation *is* the Fall. Blake recognized the fallacy of misplaced concreteness by which Newton purported to detect substantial forms within the chaotic potentiality of Plato's divine Receptacle as it existed even before heaven was made. Tellingly, the Receptacle—described by Francis Cornford as "nothing yet but a flux of shifting qualities, appearing and vanishing," and by A. E. Taylor, sounding more like a particle physicist, as a matrix "agitated everywhere by irregular disturbances, random vibratory movements, and exhibiting in various regions mere rude incipient 'traces' . . . of the definite structure we know as characteristic of the various forms of body"[24]—is viewed by Urizen as nothing but "the draught of Voidness to draw Existence in" (*FZ* 24:1; E 314). But this also implies that the object of Blake's satire of Urizenic architecture in the rest of the *Four Zoas* passage above is not the *Timaeus*, as Ault seems to suppose (132–33). Rather, Blake satirizes Newtonianism's inability to see Plato's Receptacle as *meta*physical: as a dynamic and mediatory precondition of visibility, unlike the sheer material "Voidness" which is all Urizen sees.[25] Newton's failure to recognize nature's potentiality to produce something more, new, and different results in a circular and self-reinforcing materialism, a "ratio of all we have already known" (*NNR* [b]; E 2, also E 659).[26]

The aetherous background to Blake's *Newton* looks, then, like an effect of Newtonianism's materialist reduction of the Platonic Receptacle to absolute space, the physical container of objects. Notice Newton is touching the straight line in the diagram with his forefinger. Like Hume and Berkeley, Blake here points to the basis of geometric lines in sense, not mathematic calculation. Geometry is indeed, as Plato described his own attempt to tease out the virtual space of the Receptacle, an illicit, "bastard kind of reasoning."[27] Descartes's opponents frequently objected it is impossible to imagine a point not situated in a space occupied by that point. Accordingly, in Blake's design, the difference between the equilateral triangle formed by the upright physical divider and the triangle Newton draws within his diagram calls attention to the diagram's perspectival foreshortening. Blake was likely familiar with research showing the "sphere" of human vision to be a function of the roundness of the eyes themselves, as Reid explained in his realist "geometry of visibles" (examined in the next chapter). No matter how it is geometrized, visual perspective remains an organic experience. As Blake stressed, scientific instruments such as "The Microscope" and "Telescope" can assist the eye but only as prostheses whose data needs to be seen, judged, and interpreted in its turn (*M* 29:17–18; E 127). In fact, Roger Joseph Boscovich had already imagined a non-Newtonian calculus based on curves rather than straight lines:

> A straight line seems to our human mind to be the simplest of all lines . . . But really all lines that are continuous & of uniform nature are just as simple as one another. Another kind of mind

> which might form an equally clear mental perception of some property of any one of these curves, as we do the congruence of a straight line, might believe these curves to be the simplest of all & from that property of these curves build up the elements of a far different geometry, referring all other curves to that one, just as we compare them with a straight line. Indeed, these minds, if they noticed & formed an extremely clear perception of some property of, say, the parabola, would not seek, as our geometricians do, to rectify the parabola; they would endeavour, if one may use the words, to *parabolify a straight line*.[28]

Boscovich's alertness to the possibility of "another kind of mind" reflects his appreciation of geometry's dependence upon appearances. In Blake's print, Newton can be seen as attempting, precisely, to "*parabolify a straight line*" by means of projected "conic sections" formed by the intersection of a plane with a cone.[29] One is not surprised that Niels Bohr and Werner Heisenberg both celebrated Boscovich's importance for the curved spaces of relativity and subatomic particle physics.[30] By extending Plato's barely real Receptacle into a curved four-dimensional space-time, Blake, and later Einstein and Bohr, approached closer to the non-Euclidean cosmogenesis Plato had pursued in *Timaeus*.

What therefore sets Blake apart from his contemporaries is his much more far-reaching and systematic investigation of non-Euclidean geometry, the ground of his most startling insights into the temporal nature of space and matter. Thomas Young never succeeded in changing the prevailing corpuscular view but, beginning in 1801, his experimental single- and then double-slit demonstrations of interference effects indicating light to be a wave, like sound, helped to dematerialize Newton's absolute space and laid a basis for Faraday's early field theory.[31] The post-Newtonian redefinition of "that calld Body" (*MHH* 4; E 34) formed part of the Enlightenment's broader reconception of a panoply of received ideas such as God, Heaven and Hell, "Earth," man, "globes of attraction" including the human eye (*BU* 3:36; E 71), and even substance itself.

The aether would live on for another century until Einstein finally exploded it with his special relativity paper of 1905, but already in Blake's time its days were numbered. Based on Boscovich's mathematics, there were mounting efforts to supplement Newton's contact mechanics with a theory of field conceived increasingly in electromagnetic terms. Accordingly, Blake's print depicts Newton from an implicitly revolutionary historicist viewpoint, similarly to Joseph Priestley in his *History and Present State of Electricity* when he says of an esteemed predecessor: "Though we know much more than he did, we, at the same time, know how much more is unknown better than he could."[32] In concluding his book with a series of "Queries and Hints" in imitation of Newton's Queries at the end of the *Opticks*, Priestley goes so far as to remark that, to future electricians "in a more advanced state of the science," many of his ideas "will probably appear idle, frivolous, or extravagant ones. . . . But if this chapter be a means of . . . accelerating the progress of electrical knowledge, I am very willing that it should, ever after, stand as a monument of my present ignorance." Indeed, obsolescence is an unavoidable entailment of Newton's famous remark to his rival Robert Hooke that if he saw further, it was by standing on the shoulders of giants. Far from being naïvely "Whiggish," the age's Baconian confidence that great discoveries lay to hand spawned an appreciation of the contingent nature of present knowledge in relation to past and future times that was fully as sophisticated as the Academy's recent new historicism. Look again at Blake's print. Insofar as it portrays a classical hero, does it not carry a tragic hint that Newton is fated to be superseded by his very success, not just despite the concentrated intensity of his gaze but because it is so narrowly focused?

Indeed, Blake's confounding of cause and effect in "Newton"—and everywhere in his poetry through a vast array of puns, ambiguous prepositions, dangling modifiers, associative rather than grammatical punctuation, two-way syntax, and recursive subnarratives, an array which far transcends the conventionalized performativity of social speech acts—can be seen to anticipate relativity's replacement of Newtonian mechanics with a more phenomenal and descriptivistic kind of science epitomized, notoriously, by Heisenberg's uncertainty principle. This holds that the observer is included in the scene of observation because the act of measuring not only disturbs what is measured (as anthropologists and psychologists were already beginning to suspect) but even defines it since measuring always occurs within a wider, indefinite set of interrelationships whose ongoing flow of information it arrests at the local level. Two tiny particles or "minute particulars" at the limit of observability are so deeply embedded in their local areas of space-time that they don't really exist as objects in a field; thus, there is no metric backdrop by which to compare them. (As has been noted, Heisenberg's empirical term, *unschärfe*, blurry, is much more apt than Bohr's public relabeling of the principle as sheer epistemological "uncertainty.") Through the resulting process of approximation, observers "become what they behold" (J 39:32; E 187); in Enlightened ideological terms, we are structurally implicated in "the system." But this gorgonic principle doesn't just imply a criticism of Newton in relation to the objects of his science. It holds true of Blake himself looking back at Newton across a century, and viewers today looking at *Newton* across another. From a 2023 standpoint, we can say Blake's color print critiques Newton's reduction of time to space, in disregard of the unified space-time of events that included him as a historically limited observer and contributor to Bacon's overarching program of scientific progress.

So, the present study differs from Ault's *Visionary Physics* in offering a less binary, more interinvolved, yin-yang or Blakean-Contrary view of the relation of Eternity to three-dimensional existence. Ault's position rests on the claim, "Blake's Eternity is constructed in such a way that the concept of measurement as we have characterized it [i.e., as Newtonian and Euclidean] could never come into existence." Lacking any "standard unit" or metric, "an 'Eternal' would never be concerned with comparing the 'lengths' of any two objects, since length would be dependent on his own perception."[33] Or as Blake puts it, "Every thing in Eternity shines by its own internal light" (*M* 10:16; E 104). Says Ault, it is the imposition of "some additional uniform limiting conditions" on individual perception that produces "the emergence of an 'external' world peopled by 'objects' whose existence is independent of the individual" (129). Thus, "Blake's Eternals could never derive the idea of rigid bodies fixed in space" (128).

Granted, Eternity as portrayed at *Jerusalem*'s close is nonmetric in the quasi-Kantian or Coleridgean sense of being the infinite, universal, absolute space of all imaginable spaces—not a place, object, or thing but the very condition for imagining things. It constitutes the antecedent realm of continuous topological shapeshifting that supplies the basis of the various turnings inside-out and outside-in within three-dimensional space that pervade Blake's work. More than Kant, whose writings he evidently didn't know, Blake's Eternity is close kin to the indefinite, all-but-unperceivable Receptacle in Plato's *Timaeus*. Yet Plato's Receptacle is only the ground of creation. It isn't, itself, anything created. So, one struggles to see how the utopian, uncreated no-place of Aultian Blake's Eternity could be habitable even by "Eternals"—unless, of course, they are simply nonhuman.

After all, when Blake's Eternals do look back at the Newtonian Urizen in *The Book of Urizen* and *Milton*, they perceive to their horror just the same fixed and outward world as Ault insists it is impossible for them to imagine. In other words, what Eternity's unrestricted, nonextensive becoming becomes, in Blake's cosmogenesis, is a place standing in some definite relation to Urizen's arrested world of measurable, externalized, substantial being. Thereby, Los and the other Eternals in *The Book of Urizen* become what they behold. Walling themselves off from Urizen, they establish a fearful symmetry that eventually turns Eternity into the traditional otherworldly Heaven. Causality is shown to be a two-way street, as distance and isolation generate perversely sturdy forms of relationship.[34] After all, Urizen doesn't somehow cease to be Eternal—an impossibility. His fall "out of" Eternity therefore establishes causal connection with the other unfallen Eternals no matter what they do. Consequently, Eternity, too, begins to acquire extension relatively to his geometric universe. The reader reflects that physical reality must have involved, from the outset, some causal interlocking of the Contraries, being and becoming. In Blake's cosmology—which I'll later characterize as Platonic-realist, like Whitehead's—the very form of fallenness exists in Eternity even before its instantiation on earth. Urizen allows us to discover it and, thence, ourselves. Blake's myth thus occupies a middle ground between "discovery" and "invention." Call it, revelation. Relativistic laws of nature exist independent of the observer, but their mathematics remains descriptive and acausal ("kinematic") until they are imagined and translated into earthly sense.

On the one hand, then, I want to agree with Ault when he insists, *à la* Henri Bergson, "*the rise of temporal succession is the response of Eternal energy to the intrusion of Urizenic spatialization into the causally independent interaction of Contraries in Eternity*" (173; his italics). Ault's claim seems based on an analogy to the way relativity theory conceives of events outside the light paths between two different light cones as constituting an "absolute elsewhere": a set of world lines unknown to occupants within those two light cones, barring arrival of some other event able to provide linkage—as, for instance, Milton's return in *Milton* to "this earth of vegetation on which now I write" (*M* 14:41; E 109), namely "1804" (*M* pl. i; E 95), activates in the living Blake poet the alienated potential of his dead precursor's utopian Christian vision. Aultian Blake's Eternity represents "elsewhere," and the Urizenic "intrusion" brings the independent Contraries down to a warring marriage of heaven and hell on earth, where *Milton*'s divine comedy finally takes place.[35]

On the other hand, the either-or opposition Ault draws between Newton's three-dimensional geometry and Blake's ostensibly nondimensional, noncausal, symbolical Eternity seems, itself, a "Newtonian" abstraction from Eternity's underlying energetic becoming, which in Blake's myth is what sustains calcified three-dimensional existence in the first place. Paradoxically, Ault makes Eternity into the same kind of unimaginable idealization as

Hume's *Dialogues concerning Natural Religion* (1779) satirizes in the character Demea, whose Calvinism premises a deity so remote as to constitute a kind of vanishing point of relevancy. Though the existence of orthodoxy's omnipotent God is never disproved in the *Dialogues*, he is seen to amount to nothing in real human terms. Conversely, the purpose of Priestley's equally rationalist *Disquisitions Relating to Matter and Spirit* (1777)—an important early influence on Blake as we'll see—is to *defend* Scripture and divine revelation against the "modern" Cartesian view of spirit as an immaterial substance without extension like Demea's all-transcendent God. According to the Cartesian view, says Priestley, "it is even improper to say that an immaterial being *exists in space*, or that it *resides* in one place more than in another; for, properly speaking, it is *no where*."[36] When orthodox dualists speak of "the omnipresence of the Deity, . . . they mean his power of *acting every where*, though he exists *no where*": the metaphysical form of Newton's gravitational force at a distance, which however is also the form of Aultian Blake's immaterial Eternity. Priestley goes on to point out "that if nothing but immaterial substances, or pure intelligences, had existed, the very idea of *place*, or *space* could not have occurred to us" (56). Recall Ault's claim that Blake's Eternals "could never derive the idea of rigid bodies fixed in space" (128), and you can see the appeal for Blake of Priestley's paradoxical "immateriality of matter." Priestley promised a dynamic idea of body as occupying three-dimensional space not as substance, whether solid or nonextended, but as a psychophysiological process wherein heaven's immaterial space extends down or "falls" to earth and becomes available to human sense (namely, via the Blakean space-time Vortex, as I later try to demonstrate). In contrast to Enlightened Cartesians, Priestley claimed "the vulgar [his democratic term for ancient Christians unfettered by Church doctrine] who consider *spirit* as a *thin aerial substance*," would regard "the modern idea of a proper *immaterial being* . . . to be only a *negation of properties*, though disguised under the positive appellation of *spirit*" (73). The fact that Hume and Priestley alike regarded the Cartesian idea of immaterial, unextended spirit as empty, Hume arguing the point in favor of atheism or at least agnosticism, and Priestley in favor of natural religion which Blake equally rejected, shows how urgently Blake needed to develop an alternative conception.

But it took an Einsteinian revolution for the physical implications of Blake's perspective to snap into focus. It might be objected that the "geometry" of my title is misleading and should be replaced by "cosmology," the usual term in Blake criticism. I prefer "geometry" precisely because it defamiliarizes

Blake's cosmology, which wasn't given to him via the authority of Einstein and others but had to be built up with difficulty from insights and perceptions that appear incoherent—"bastard"—from a Euclidean perspective. Through the lens of what came later, we can understand what Blake thought was at stake and why it aligned so awkwardly with the Newtonian physics of the day. Partly, too, my aim is to appreciate how old Newtonian debates over the nature of matter, time, motion, and change implied four-dimensional interpretations of space long before their explicit development. What Blake like Einstein grasped is that all these concepts are abstracted for purposes of measurement from the extensive, "thick," partially sensible relations existing between events. Relations between the abstractions themselves are therefore essentially analogical. Indeed, analogy, as it was beginning to be understood in the early nineteenth century through research in chemistry and electricity, included not only relations between different physical phenomena but also the relations of those phenomena to their visual representations. According to Andrea Henderson, with the rise of field theory, analogy as "a generally applicable principle of equivalence . . . facilitated a rapprochement between a reality understood as fundamentally comprised of consistent formal features and representations of that reality."[37] Arguably, it isn't only Newtonianism Blake parodies in *Newton* and other noticeably flat designs but pictorialism and its supposed direct correspondence with a mechanical universe of solid matter in space whose underlying force, gravity, nevertheless remained notoriously uncharacterized except in nonsensible mathematical terms. But even if "expanded sense perception" was suppressed by Newtonian science, there still were ways to foster it through visualizations based in field theory's powerful abstractive analogies and heightened awareness of relationality.

CHAPTER ONE

"Oh, but you're just analogizing . . ."

Whatever else, viewers of Blake's frontispiece to *Visions of the Daughters of Albion* (figure 1.1) can agree the design exhibits quite a complicated geometry. The overall skull shape looks like a morbid conflation of Locke's caverned mind with Plato's Cave.

Figure 1.1. William Blake, *Visions of the Daughters of Albion* frontispiece, copy I, 1793. Yale Center for British Art, Paul Mellon Collection.

In fact, the metaphysical cast of Locke's inaccessible "material substratum" did not escape eighteenth-century interpreters. A few years before Blake's poem, Thomas Reid anchored his "commonsense" critique of representative realism in the observation that "Plato's subterranean cave, and Mr. Locke's dark closet [of the mind], may be applied with ease to all the systems of perception that have been invented; for they all suppose that we perceive not external objects immediately, and that the immediate objects of perception are only certain shadows of the external objects."[1]

Examine, then, the *Daughters of Albion* frontispiece more closely. Unlike what is often assumed, it portrays not a cavern but rather the mistaken idea of one. Water appears not only behind the unhappy tableau but in front; thus, the setting must be a promontory along a coastal point. In other words, the "cavern" is just an open arch, even a proscenium whose theatricality casts ironic light on the melodrama of confinement taking place below. Several critics have seen in Blake's question mark-shaped stony archway the outline of a head in profile, the threesome inside forming "constituent parts of a single fragmented personality."[2] If viewers do indeed look through the arch of a skull shown in cutaway, what they confront in back is, apparently, the arresting sun-eye of the jealous God who has sanctioned all this alienation, a reflected image of the morality Calvinistic Deists would have brought to the poem (for example, Demea in Hume's *Dialogues concerning Natural Religion*). In this way, the design can be seen to represent Earth's "intrinsic geometry." The viewer is made to occupy the unbounded yet finite surface of a sphere, endlessly traversable without any point of exit. As opposed to flat Euclidean geometry's "extrinsic" view of Earth "As of a Globe rolling thro Voidness" (*M* 29:16; E 127), Blake portrays a local inhabitant's experience of its curvature as "one infinite plane" (15:32; E 109) wherein all "straight" or "great" lines return to their starting point after traversing the length of the equator, contrary to Euclid's Fifth ("Parallel") Postulate. So considered, Blake's distant eye-sun mirrors and mocks the viewer's own gaze.

But there is a wrinkle. The sun is low on the horizon, yet foreground and background are evenly lit. This is possible only if the shoal in front curves inward *through the viewer* to meet the open ocean in back. So, the eye-sun is not a far-off, flat reflection of our own eye but consubstantial with it. We seem to be looking through a hole in the distinctly thickish side of a container of three-dimensional existence, much as the sun is peeping through it from the opposite side. More precisely, the viewer occupies—like Theotormon, the poem's compromised man-in-the-middle—a Möbius-strip "threshold" (*VDA* 2:6, 2:41; E 47) between inside and outside. We stand between the spacious freedom of the sun and the apparently caverned realm of the three humans. Whereas the cave wall's vertical dark top half tends to separate us from the depicted scene, its bottom half shelves out into a plateau (not necessarily a cave floor), with the viewer stationed a few feet away. At this perceptual threshold, we face an impossible choice between solidarity with the ideologically deluded humans inside the cavern and a visionary but solipsistic transcendence of social connection. This is the same choice Oothoon faces within the poem: stand by her man, or give him up and uphold her personal truth alone, unheard?

The design's wrinkled threshold resembles the space of the Klein bottle, a higher-dimensional Möbius strip that effectively contains itself (figure 4.1). In topology, the Klein bottle is usually described as a two-dimensional shape "embedded" in four-dimensional space. However, when we consider how the bottle's inside and outside flow together continuously, we can see that, as Steven M. Rosen stresses, "the Kleinian object and its spatial context are of the *same* dimension."[3] The bottle as object runs back through its hole into the subject. Despite the hypostatizations of classical-modern mathematics, this shape is not strictly two, three, or four dimensional but rather, in Maurice Merleau-Ponty's phrase, "a coiling-over of the visible" that discloses space itself as "the formative medium of the object and the subject." Says Merleau-Ponty of the Kleinian shape he senses when he holds his right hand with the left: "This circle which I do not form . . . can traverse, animate other bodies as well as my own. . . . I appear to myself completely turned inside out under my own eyes."[4] The fractal lattices of *Songs of Innocence*, which my next chapter shows to exist indefinitely in space *between* different dimensions (analogously to how the mathematical curvature of a fractal surface measures between two and three dimensions), extend forward in the *Daughters of Albion* frontispiece directly into the viewer. As elsewhere in Blake's designs, we see through the frontispiece's illusory three-dimensional scene in the double sense of seeing it *as* illusion while seeing *by means of* it. Metaphysically speaking, the illusory image is a representation of how the natural order exceeds all human conceptual systems for imagining it, so allowing new concepts to emerge, inexhaustibly.

Indeed, the rocky outcropping in the *Daughters of Albion* frontispiece looks surprisingly like an instantaneous three-dimensional slice of curved four-dimensional space-time, as in a Minkowski diagram. Like Bergson, Blake criticizes "the cinematographic mechanism of thought" which reduces cosmic becoming to a succession of worlds-at-an-instant.[5] Why is brutal Bromion looking off at right angles toward something he finds perfectly hair-raising?

If this images his mortified reaction to his tentative spiritual awakening to "other" wars, sorrows, and joys than literal, corporeal ones, as dramatized in the text at 4:13–24, then it seems what Bromion beholds here is Leviathan. He is looking past the point where the outcropping and its cavern illusion end—a point beyond three-dimensional perspective where foreground and background, inside and outside, meet in "fleshly," Kleinian union. From his materialist perspective, the egalitarian *jouissance* of Oothoon's copulatory visions of the morning sun and erotic, glowing Marygold constitutes an apocalyptic splitting-apart of his solid world.[6]

Your own perspective on three-dimensional reality is probably closer to Bromion's than to Blake's. Accordingly, my aim in this book is to render Blake's vision of a fourth dimension visualizable—even though time, change, and motion, which lie at its center, cannot be seen as such but only their effects. Granted, visualization and vision are two different things, like ratiocination and reason. That is why we each must see Blake's vision for ourselves, or not at all. Unlike spontaneous natural seeing, visualization often makes use of analogy, and visualization of a fourth dimension necessarily so. At the core of this book lies a set of analogies between Blake's cosmology and twentieth-century relativity, particularly as interpreted by its most thoughtful metaphysician, Alfred North Whitehead. Whitehead features so prominently in this study because he was the first to map the traditional history of British empiricism from Bacon to Locke, Berkeley, and Hume as a series of false starts and confused prefigurings of physical field theory and relativistic space-time.

My study mostly bypasses social-political issues that have occupied Blake criticism for the last twenty or thirty years concerning gender, feminism and the rights of women, slavery and the slave trade, English imperialism and colonialism, embryology, madness, and the social construction of the body. Efforts to specify Blake's relation to historical topics often ignore his work's essential multivalency, including its occasional tendency to degenerate into what Morris Eaves aptly calls "an antihistorical network of self-duplicating analogies" that "begins to deny the possibility of real change."[7] The displaced analogical equivalences that Eaves from a social-historical perspective considers disturbingly "antihistorical" are what Oothoon, *Daughters of Albion*'s Cassandra-like prophet *manqué*, celebrates as the fugitive bliss of vision. I am not advocating a return to the structuralism of Northrop Frye and the systematization of Blake's myth, which culminated in S. Foster Damon's once useful *Blake Dictionary* (1965). Rather, I hope to unite structural and historical perspectives through Blake's idea of space-time as a physical field of ongoing events, perceptual as well as historical. Even as he scathingly confronts the main social and political issues of his time and place, Blake also leads reader-viewers to look beyond them toward an underlying reality whose relativistic geometry remains contemporary for us today.

The point here is not that Blake's transcendental vision necessarily eclipses the late eighteenth-century injustices he condemns or renders them incidental and parochial. Far from it. Blake hoped that by imparting his vision as current, he might enable his audience—including "Children of the future Age, / Reading this indignant Page" which was written "in a former time" (*SE*; E 29)—to see how his own, fairly typical Enlightened liberal-radical politics form the surface tip of a universal moral crisis of apocalyptic urgency. The local historical positivities addressed in his work are the obverse of an immanent four-dimensional space-time from which they derive their material visibility (somewhat as the *Marriage of Heaven and Hell* title page, examined in my conclusion, shows level Earth as giving way to a swirling Hell below [figure C.1]). For present-day Anglo-American readers, the Atlantic slave trade and women's suffrage belong to the past. Less so, their persisting long-range effects. Arguably, the ongoing moral authority of Blake's work—its appeal to conscience registered, one way or another, by over a century of Blake criticism—springs from his prophetic power to make us bear witness to the inheritors and analogues in our own day of injustice in his. That authority, in turn, depends on his vision of a space-time continuum that fuses past and present, both there and then and here and now, and so makes painfully real to the imagination what has been happening beyond our horizon "way back then and over there." In other words, Blake's space-time logically precedes, and provides a physical basis for, both his own contemporary social-historical concerns and those of his critics. For, the simple truth of relativity is that, to adopt Whitehead's phrase, "everything is everywhere at all times."[8] Thereby, Blake demonstrates that the judgment of conscience is categorical and transcends local time and place. You can't have a conscience more or less; either you've got one or you don't. "Conscience in those that have it is unequivocal" (E 613). Of course, conscience can be ignored, its small voice is sometimes difficult to make out amid competing claims, and different consciences may differ (for example, many if not most people reasonably reject Kant's absolute prohibition on lying). Notwithstanding, Blake's evident position is that the judgment of conscience being essentially first-personal and final—a Last Judgment—there can be no further adjudication of that judgment without temporizing and falsehood or second-guessing based on mere probability and "Prudence" (*MHH* 7:4; E 35).

Visions of the Daughters of Albion, perhaps the most topical of all Blake's Illuminated Books, illustrates the intersection in his work of geometry with urgent social criticism and shows how he strives to render it a crisis of conscience for readers now as well as then. Far from being a cohesive dramatic character, the heroine, Oothoon, is at once an African woman raped and enslaved, a Wollstonecraft-like English feminist, a virgin awakening to the world of Experience, and "the soft soul of America" whose ideal of liberty has been violated by the English slave trade. To be sure, these different historical positions are all compatible, combining as they do to make Oothoon a victim of oppression (though her self-comparison to an enslaved African, common among 1790s feminists, would nowadays be considered an "appropriation"). But how to explain that she is, further, a nymph who plucks a flower only to become deflowered? "I see thee now a flower; / Now a nymph!" (*VDA* 1:6–7; E 46), she tells the Marygold before she plucks it from its "golden shrine" and places it "here to glow between my breasts" (1:10–12; E 46), thus becoming its new shrine, herself another flower-nymph and hence, through rape, a poignantly self-reflexive figure like Milton's Proserpine who "gath'ring flow'rs, / Herself a fairer Flow'r by gloomy Dis / Was gather'd."[9] The initial transferal of flower power signifies the reciprocal creation of subject and object in each other's image, by contrast with the assignment of objectified social identity as when Bromion boasts of his conquest, "Behold this harlot" (*VDA* 1:18; E 46). Oothoon's status as a betrothed or unmarried "bride" to her straitlaced beloved, Theotormon, further dissolves into symbolism once we explore that word in an etymological sense that Blake likely understood from his readings about "Jacobin plants" in Erasmus Darwin's anthropomorphizing *Loves of the Plants* (1789): nymph, "an insect at that stage of development which intervenes between the larva and the imago" (*OED* sense 3; L., after Gr. *nympha*: bride).[10] Unlike the view of her shared alike by both males, Bromion and Theotormon, Oothoon upholds her transitional, nondualist, "middle" identity as a "bride," a role Blake links with romance and not marriage, auratic dazzle not solid representation, polymorphic dalliance not institutionalized sex (like Wollstonecraft, Oothoon emphasizes that marriage is just sanctioned prostitution). Thereby, Blake's multiform "nymph" harkens back to the sylphs of Pope's *Rape of the Lock* (its symbolical rape an overlooked source for *Daughters of Albion*, along with Milton's *Comus*), the patently metaphorical nature of angelic substance in *Paradise Lost*, and as we'll see in later chapters, the aetherous "aery vehicles" of Henry More and late seventeenth-century Neoplatonism.

Indeed, when we consider the Marygold's dawn flashes of light shown in the design to the Argument plate,[11] the reflective "smile" of the "clear spring" at 2:18, the erotic union Oothoon celebrates with "the morning sun" at 6:22, and the presence of three maidens dancing in the ring of a rainbow on the title page, it is not farfetched to regard her as iridescent, perhaps a damselfly like the lepidopterous fairy seen at the top of *Jerusalem*'s title page.¹² The apparent sadomasochism and scopophilia of Oothoon's proposal to pander nymphs to Theotormon—

> silken nets and traps of adamant will Oothoon spread,
> And catch for thee girls of mild silver, or of furious gold;
> I'll lie beside thee on a bank & view their wanton play
> In lovely copulation bliss on bliss with Theotormon
> (*VDA* 7:23–26; E 50)—

can be seen to disclose nothing more abject than mutual childlike pleasure in the frottage of sunlight gleaming on water skimmed by shiny metallic insects, as in her professed solar "copulation" through the "eyes" (*VDA* 6:23–7:1; E 50). Not only are "silken nets" a standard implement for catching insects on the wing (and the wings of damselflies are, themselves, distinctively silken and netlike, more so than their cousin dragonflies), but the retina of the eye is, etymologically, a "net" for catching representations.¹³ The lines' syntactic inversions join Innocent and Experienced perspectives. If the nets and traps will spread Oothoon herself, shown in one design as an erotically splayed Prometheus, then the *jouissance* she anticipates seems purely aesthetic, contrary to the suggestion of "adamant will" and the voyeurism of her proposal as viewed in descriptive-realistic terms. The Experience viewpoint, which interprets the girls' play and copulation as shared exclusively with Theotormon while Oothoon looks on, opens onto an Innocent vision where Oothoon lies stretched beside Theotormon "on a bank" and enjoys, together with him, an experience of bright prismatic shimmering as the nymphic "girls" play amongst themselves, the water, and the light.¹⁴ Such a vision would constitute a prototype of the one Blake describes in his epistolary verses of 2 October 1800:

> In particles bright
> The jewels of Light
> Distinct shone & clear—
> Amazd & in fear
> I each particle gazed
> Astonishd Amazed
> For each was a Man
> Human formd. (E 712)

Indeed, *Daughters of Albion*'s Argument plate shows no horizon or perspective but only the genesis of vision from a sweeping field of light. Oothoon appears bathed in a vortex of dawn splendor that skies sidelong across the earth, transforming nearby Marygold flowers into small suns (presumably, through "copulation"), from one of which a nymph leaps forth to embrace her. Her speech invites Theotormon to enjoy a similar vision.

The somewhat disappointing way the poem develops further reduces character to voice. *Visions of the Daughters of Albion* opens as a lyrical drama in the aftermath of violence, but by the end of plate 4 it abandons drama and unspools into a long soliloquy or complaint. Thereby, the reader's classical expectation of a heroic or mock-heroic narrative of realistic outward action is denied. A literalistic interpretation of the rape is associated with brutal Bromion, an action figure straight out of the *Iliad* (Blake modeled his contempt of pagan epic shame culture after Milton, and it is just as severe). Likewise, Theotormon's self-absorbed "reflections of desire," "lamplike eyes," and jealous "watching" (*VDA* 7:9, 7:22; E 50) are associated with the spectatorship induced by conventional three-dimensional representation. Rather than depicting realistic dramatic characters, the fractured, wobbly voices heard in *Daughters of Albion* imply the interdependence of its three characters beyond individual personality. The frontispiece shows as much, no matter whether its image is viewed as a three-dimensional tableau of mutual enslavement or imagined, Contrariwise, as pointing left toward a post-Newtonian, four-dimensional sea of joy such as Oothoon struggles to describe.

In fact, for all their shared anxiety of contamination, Bromion and Theotormon both intuit utopian possibilities largely overlooked in criticism of this poem. Bromion's adventurous sensualism leads him into much the same childlike state of open-eyed contemplation as Oothoon, his victim:

> Thou knowest that the ancient trees seen by thine eyes have fruit;
> But knowest thou that trees and fruits flourish upon the earth
> To gratify senses unknown? trees beasts and birds unknown:
> Not unpercievd, spread in the infinite microscope,
> In places yet unvisited by the voyager. and in worlds
> Over another kind of seas, and in atmospheres unknown:
> Ah! are there other wars, beside the wars of sword and fire!
> And are there other sorrows, beside the sorrows of poverty!
> And are there other joys, beside the joys of riches and ease?
> And is there not one law for both the lion and the ox?

And is there not eternal fire, and eternal chains?
To bind the phantoms of existence from eternal life?
(*VDA* 4:13–24; E 48)

This is the very no-place to which Bromion directs the viewer's eye in the frontispiece: an "other" realm where Eden's "ancient trees" bear new fruit, and past, present, and future are united in the perpetual dawn of Oothoon's "morning sun": "Eternity's sun rise" (E 493, E 470). Bromion's "Ah!" carries the full force of wonder: "the beginning of philosophy" for Plato and Aristotle, Einstein and Whitehead alike.[15] One recalls the "awe and wonder" of Coleridge's Hobbesian "man of trade" in the "Essay on Method," whose mind has been "raised . . . to the consideration of EXISTENCE, in and by itself, as the mere act of existing."[16] Yet, Bromion's awe reflects less calm wonder before what Coleridge religiously calls "the presence of a mystery" than amazement and horror at reality's extension into an ungrounded modal realm of possibility and potentiality, namely, the imagination. Thus, his questions turn rhetorical in the last three lines as he remembers with a start that he is under surveillance by a jealous God who punishes transgression, namely, via Bromion himself, sin's "scourge" (*VDA* 1:22; E 46). Still, Bromion, like Oothoon, craves further erotic contact. His "stamp[ing]" (1:21; E 46) of his "signet" upon the world represents a totalitarian reduction of the old Renaissance idea of material-body touching as a meeting of sensible surfaces: a reduction, scientifically, to the mechanical view of matter as atomic billiard balls in motion; commercially, to the mass printing of duplicate texts; and, philosophically, to the vulgar Lockean notion of sense impressions as mind's imprinting by unknown substance. The impress of his "signet" travesties Blake's lavish, labor-intensive art of relief etching, which purports to unify copy and original, invention and execution, idea and embodiment.[17] Bromion's questing after new worlds to conquer expresses an atavistic instinct to overcome separateness by getting hold of things—a primitive expression of the desire to know. Theotormon's speech evinces a similar receptiveness despite its nostalgia and Job-like self-pity:

Tell me where dwell the joys of old! & where the ancient loves?
And when will they renew again & the night of oblivion past?
That I might traverse times & spaces far remote and bring
Comforts into a present sorrow and a night of pain
Where goest thou O thought? to what remote land is thy flight?
(*VDA* 4:3–8; E 48)

Only a slight change of tone would turn the caverned Lockean epistemology of these doubts into the soaring speculation of *The Marriage of Heaven and Hell*'s "How do you know but ev'ry Bird that cuts the airy way, / Is an immense world of delight, clos'd by your senses five" (*MHH* 7; E 35).[18]

So, if Oothoon exhibits a satirist's insight into critical agency available through parody, irony, and modes of imitation beyond the mirroring of reified social realities, as Fredrick Hoerner argues, then it appears Bromion and Theotormon are not entirely beyond her reach.[19] Oothoon's pandering proposal is an extreme version of her previous summoning of Theotormon's "Eagles" of thought. "Rend away this defiled bosom that I may reflect. / The image of Theotormon on my pure transparent breast (*VDA* 2:15–16; E 46), she commands; for "How can I be defild when I reflect thy image pure?" (3:16; E 47). Thus, she gently mocks her lover's puritanism. In reply, "Theotormon severely smiles. her soul reflects the smile; / As the clear spring mudded with feet of beasts grows pure & smiles" (2:18–19; E 46). Significantly, "As" here not only sets up an analogy but submits it to temporal process. Oothoon turns the other cheek toward Theotormon's grim self-satisfaction similarly to how many Songs of Innocence parody received ideas of purity when they "[stain] the water clear" (*SI*; E 7). She goes on to emphasize that bathing, far from negating the flesh, constitutes a tactual bodily act:

> The new wash'd lamb ting'd with the village smoke & the
>   bright swan
> By the red earth of our immortal river: I bathe my wings.
> And I am white and pure to hover around Theotormon's breast.
> (*VDA* 3:17–20; E 47)

This seems a loose allegory of the impure bodily spirituality of Blakean relief etching, no less than the standardly identified passages in the *Innocence* "Introduction" and *The Marriage* plate 15. The "white" paper is "bathe[d]" and softened with water to receive an imprint "ting'd" with color, that reader-viewers might then "hover around" it.

Notwithstanding, the desperation behind Oothoon's attempts to get inside Theotormon's head is illustrated in the design to plate 7, where she pleads for mercy from a wave-like thought balloon beneath which Theotormon "sits / Upon the margind ocean conversing with shadows dire" (*VDA* 8:11–12; E 51). His brooding self-absorption casts the image as a lurid male fantasy of bondage. Oothoon renews her bliss of solar copulation "every morning" (8:11; E 51), yet the poem leads one to ask what good this does under the circumstances, for her or anybody else. Does her private bliss merely sublimate the masturbation she deems an effect of sexual repression at 7:3–7? Satire is, notoriously, a matter of perspective—just consider how many have taken Swift's *Modest Proposal* straight-up, or the longstanding uncertainties over Gothic self-parody (whether campy or morally serious) in Matthew Lewis, Charles Maturin, and Coleridge's *Christabel*. It remains debatable whether Blake's poem upholds radical intellectual irony against existing social reality, or rather affirms irony's ineffectuality in the face of a status quo that cannot be bothered to notice.

What seems clear is that Blake deliberately saddles his audience with this problem. Theotormon's "hypocrite modesty" (*VDA* 6:16; E 49) belongs also to the *hypocrite lecteur*. His "cold floods of abstraction" (5:19; E 49) and "margin'd ocean" (8:12; E 51) project outward into three dimensions so soon as readers perceive the poem and its frontispiece pictorially in detachment from beyond the page margins.[20] *Visions of the Daughters of Albion* was written in 1793 after hysterical newspaper reports of the September Massacres of 1792 (the "First Terror"), which were soon followed by grisly Gillray caricature prints. One might therefore interpret the fellow traveling of Theotormon and Bromion as allegorizing the reconsolidation of the Establishment. *Bien-pensant* English Girondists suddenly reevaluating their support for the Revolution (for instance, Wordsworth and Coleridge) had closed the gap between themselves and reactionary home-and-hearth English nationalists like the recent Association for Preserving Liberty and Property against Republicans and Levellers. It is this paralysis of liberal reformism that Oothoon sees to seal the triumph of Urizen, here identified for the first time in Blake's work as the "mistaken Demon of heaven" (5:3; E 49)—that is, not just a once-heavenly but now self-deceived demon like Milton's Satan; nor, from the Blake devil's inverted viewpoint, the tyrant demon whom the orthodox take for God; but the source of these unhappy twins, the metalepsis or self-projection by which misprision and hypocrisy license moral accusation and scapegoating in the first place. She deconstructs the self-serving reversal behind Bromion's "behold this harlot" (1:18; E 46). "Urizen" names the overarching dualism formed by the unlikely complicity between Bromion's antinomian sensualism and Theotormon's ascetic piety. By arresting time and change, Urizen's morality stands to flatten Oothoon's flexible, pacifistic compassion into mere ratification of the status quo. Despite that an unnamed, all-encompassing Newtonian-Euclidean "they" has "inclos'd [her] infinite brain into a narrow circle, / . . . a red round globe hot burning" (2:32–33; E 47), she affirms the experiential reality of "the morning sun" conjoined, in all its fugitive glory, with "the moment of desire" (6:22; E 50). She affirms it—but is anybody listening?

What, then, do the Daughters of Albion sound like to Blake's real-life audience at the poem's periphery as the Daughters "hear her woes, & eccho back her sighs" (*VDA* 2:20; E 46, 5:2; E 48, 8:13; E 51)? How do *we* hear this chorus and also its source, Oothoon? Evidently, we can't be sure. Blake remains enough of a skeptical utopian to indicate we can't be sure because it is still too early to tell, because the future may be shaped by our own efforts, now, to audit these voices attentively within the mind's ear. Oothoon's "wail" or "lamentation" articulates a broad range of moods including anguish, delight, indignation, exultance, and wonder, even as it occasionally lapses into melodrama and histrionics, the emotive shadows cast by innocence in a darkening world.[21] Are the Daughters' echoes a ragged welter of solitary moans, or do they swell apocalyptically with solidarity and protest like the street cries in "London"? The question brings the moment of reading to its crisis. The answer constitutes the passing of a Last Judgment.

I have to agree with Steven Goldsmith's insight into "a distinctly modern, nonactivist Blake whose orientation toward the world was both public and radically aesthetic—not in terms of a retreat into fantasy but in terms of a critical engagement with history intense enough to have felt like participation in history."[22] By externalizing the essential ambivalence of his politics—its own Urizenically arrested self-reflexivity—Blake brings contemporary reader-viewers to the brink of moral decision while refusing to tip the scales. The unbearable pressure to view the present as a moment that demands conclusive action, exactly what he cannot say, defines Blake's negative ironies as modern. The doubtful urgency of his message is why he still speaks to us today.

The preceding analysis of *Visions of the Daughters of Albion* and its frontispiece will perhaps serve to show I am not "merely" analogizing between Blake's visionary physics and relativity, in a woozy attempt to prove the poet's relevance or garnish some lingering New Age glamor from "the fourth dimension." As we've just seen, Blake portrays his prophet Oothoon's space-time vision of Eternity's sunrise as a revolutionary threat to a society whose alienation and dishonesty are founded on the assumed separateness between three-dimensional objects and perceivers ("subjects," in every sense of the word). Close behind the other charge I can hear the derisive snort, "*Of course*, Blakean vision was four-dimensional"—as if the idea explained itself. However, this is not an ahistorical study, and the devil is in the details, most especially in Blake's efforts to develop a spatiotemporal fourth dimension from a purely spatial one and then sustain the self-reflexivity of that idea against the temptations of system and allegory. Field theory, which culminated in Einsteinian Relativity but arose from Newton's aetherous *Opticks* speculations following his (mostly) consistent refusal to locate the power of gravitation in matter, had already been given a formal mathematical basis in Blake's day by Roger Joseph Boscovich. With illicit materialist bias, Joseph Priestley then extended Boscovich's doctrine of "atom-points" into physics and Hartleyan sensory-motor physiology.[23] As we shall see, Priestley's doctrine of "the immateriality of matter" was a lifelong influence on Blake, and its underlying Neoplatonism ensured that Blake's engagements with Newton's already somewhat Neoplatonic natural science were intimate as well as critical.

So, this study examines Blake in relation to a specific history of ideas, namely, the British idea of four-dimensional space-time. Blakean space-time is not some formal end-all existing, itself, beyond space and time, like traditional Heaven or as Einstein's universe was popularly deemed in the 1940s and 50s. The reason I focus throughout on contemporary sources is to indicate where his cosmology stands in relation to Britain's wider history of the idea as it emerged before and during his lifetime. My chapter sequence further shows how Blake developed his space-time in the course of his work. It seems worthwhile to examine the paradox that a mythology whose complexities appear to us unreasonably arcane and intricate is one Blake himself felt driven to create by the force of contemporary intellectual history. As he told his patron Butts, probably with reference to *Milton*, he was inspired "without Premeditation & even against my Will" (E 729). I want to recover the context that made his unlikely choices appear, to him, as necessary and unavoidable despite their difficulty and risk. If this single-author study foregrounds contextual materials to such an extent that recent Blake scholarship is overshadowed by the philosophical tradition surrounding Blake's putative immaterialism on the one hand, and the more recent body of thought around twentieth-century relativity on the other—thus inverting the figure-ground relation between a writer and his age—then I will not be sorry for my book to be read as a history of ideas with Blake as their conduit.

For, my assumption is that ideas have primordial life and agency of their own. This simply does justice to the fact that Blake came of age during the American and French Revolutions, culminating events of the European Enlightenment and its tremendous commitment to the power of ideas.

Even today, some internal coherency and integrity are presumed intrinsic to what it means to "have an idea," as opposed to a vague "notion" or fleeting "thought." In Blake's extreme view, it isn't the thinker who has an idea, it is the idea that seizes a thinking mind and renders it its "vehicle." Objective ideas are not invented or made up, they are uncovered or disclosed, albeit their mortal apprehension is partial, conditioned, and subject to improvement. Thus, Jesus the Imagination, the contact-point between Heaven and Earth, is a *living* truth that must be *revealed*. My approach thus runs parallel to Blake's "ontologizing," eerily Humean view of perceptions as freestanding substances independent of the self, which is an illusion (see chapter 2). Blake inverts Hume's reductivism and, instead, raises up ephemeral perceptions to the status of divine Ideas. His position resembles but isn't quite the same as the realism of mathematicians such as Boscovich (and Kurt Gödel and, sometimes, his late friend Einstein), who considered concepts as existing within the formal space of their logical interrelations, a "concept space." For Blake, ideas are different from abstract concepts because ideas arrive freighted with contingency and history, not only the history of the errors they purport to correct but also their modal relations with merely possible ideas or "notions" whose realization their own reality necessarily supplants. Unlike the mathematicians, for Blake, ideas must win their way in the world. This is to say that Blake was enough of a Hartleyan empiricist to believe that ideas, like perceptions, have a "vectoral" aspect or "impetus" because they take place as unfolding events along a physical "extensive continuum," as Whitehead terms it (see chapter 5).

Nevertheless, this book is not a source history. I don't mind leaving unresolved the question of how exactly Blake came to know certain of the works his writings seem to echo. Let us accord Blakean prophecy the hermeneutic power to anticipate other writers' ideas without his having read them in a book, as in the Oulipovian theory of "anticipatory plagiarism." Indeed, Muireann Maguire and Timothy Langen point out that the entire tradition of Christian typological interpretation is built on "reading backwards" from New Testament history to its putative prefiguring in the Hebrew Bible or (retroactively so-called) "Old" Testament.[24] The same might be said of the whole linguistic concept of "intention," which often seems less an immediate psychological reality about how speech is chosen and manipulated in real time than a way of explaining utterances after the fact, especially if they have been deemed mistaken or confusing. To be sure, the dialectic of reading backwards easily degenerates into allegory and reading-into, as much Christian exegesis and verificationist theorizing of verbal behavior both have gone to show.

Consider, nonetheless, the strenuously self-conscious words of J. L. Borges: "each writer *creates* his precursors. His work modifies our conception of the past, as it will modify the future."[25] One might compare with this T. S. Eliot's less dramatic, more Burkean idea of the literary canon as "an ideal order" whose shape is continually shifting to absorb new arrivals, whose inclusion then modifies the relation in which all the previous works stood toward one another.[26] The element of relativity in all these paradoxes boils down to the simple fact that writers possessed of sensitive noses can turn them to the wind and sense what is coming or, more subtly, discover within "poetic form itself the privilege of a future perspective on the present," as one critic has described Shelley's core claim in *A Defence of Poetry*.[27] The new historicist's travel "back to the future"[28] has a history of its own, and finally reduces to the truth that inspired writers express more than they know—much as critics prove when they belatedly update a writer's meaning. So, Blake's vehicle of time travel, "the Fiery Chariot of . . . Contemplative Thought" (*VLJ*; E 560), doesn't *transcend* time. Rather, time supplies traction for the vehicle to move across the different "places" of fixed Urizenic space—most especially the space between the illuminated page and the reader-viewer's head, as seen in *Daughters of Albion*'s distance-dissolving frontispiece.

That said, let me state plainly that while Blake does imagine a geometry of four-dimensional space-time and enables his reader-viewers to imagine it, he does not demonstrate space-time's reality in the physical world. Mythology is not proof. Blake's radically empiricist (and Christian and British Romantic) criterion of proof—namely, self-evidence grounded in visionary witnessing through testimony of the senses, an experience he strove to extend to his audience by means of illuminated printing—falls far short of scientific demonstration, as he knew. So, Einstein did not "*create*" Blake, nor Blake Einstein, if only because Blake was a "writer" not a "scientist"—the latter a word commonly traced to William Whewell in 1834, that is, the distinguishing idea emerged *after* Blake though well in time for Einstein to exemplify it, often quite deliberately. (By contrast, Humphry Davy, who tried much harder than Einstein to shape his public image until his death in 1829, portrayed himself as an all-round Romantic "genius.") Even so, Einstein as a realist concurs with Blake that objective scientific facts only exist relatively to human existence and the further fact of subjective human perception. *Visions of the Daughters of Albion*'s argument that different animal species analogous to different human classes inhabit unique but overlapping worlds or *umwelten*, owing to the different ways their minds receive and interpret the report of the same five senses, is not just an allegory of Romantic humanism or ecologism; it is also thoroughly scientific for its day, as Erasmus Darwin shows below. At any rate, I would contend that Blake read every one of the major predecessors whose writings I examine. The possible exception is Boscovich's *Theoria Naturalis Philosophiae* (English translation 1763), which, however, he could have read about in several of Priestley's writings. Indeed, if Blake read Priestley, as seems all but certain, he could scarcely have avoided learning about Boscovich, so pervasive was the mathematician's influence on Priestley's thought.

To return to analogy. No question, some analogies are bad, and for reasons just beginning to be investigated during the Romantic period. Late Enlightenment scientists strove hard to distinguish their rejection of Humean epistemological skepticism from a simple return to the Renaissance doctrine of occult correspondences based on sympathetic magic.[29] Eventually, Coleridge even extolled Francis Bacon, godfather of the empirical method, as "the British Plato," the first to see that science, poetry, metaphysics, even cognition itself would not exist without analogizing since something truly *sui generis* is beyond meaningful description. Bacon's concern for "truths which have their signature in nature, and which . . . may indeed be revealed to us through and with, but never by the senses" is shown, says Coleridge, by the way he "names the laws of nature, ideas; and represents . . . facts of science . . . as signatures, impressions, and symbols of ideas." Bacon's *lumen siccum* (reason's "dry light") thus supplies "the indispensable conditions of all science, and scientific research, whether meditative, contemplative, or experimental."[30] Or as Whitehead puts it in less Kantian terms: "The claim to the unity of the soul is analogous to the claim to the unity of the body, and is analogous to the claim to the unity of body and soul, and is analogous to the claim to the community of the body with an external nature."[31]

I am suggesting that Blake considered himself a late-Enlightened participant in the overarching Baconian project of "the advancement of learning." My history-of-ideas approach is Bacon's distant little heir. From the present vantage, then, what arguably rescued Blake from the condescension of posterity—from being turned into the Carlylean artist hero of his Victorian biographer Alexander Gilchrist, or into the Jungian Neoplatonic quester of Mona Wilson's fine 1927 *Life*, a figure shaped in part by *fin-de-siècle* theosophists like Yeats who imagined a purely spatial fourth dimension—was Anglo-American culture's absorption of relativity theory.[32] Whitehead's metaphysics of space-time, together with its Buddhist ramifications, had wrought upon the mind of Northrop Frye long before he published his groundbreaking *Fearful Symmetry* in 1947.[33]

Erasmus Darwin, another widely acknowledged influence on early Blake, agreed with Bacon that analogy is a cognitive necessity. Darwin saw that because analogy abstains from causal explanations, it could render irrelevant, or circumvent, Newtonian mechanism's disturbing mystery of "action at a distance." Mid-nineteenth-century research into magnetism, electricity, and chemistry relied increasingly on analogy because analogy allowed for the construction of physical models based on proportional relations and algebra rather than contact mechanics.[34] In anticipation of this development, Darwin's preface to *Zoonomia* affirms analogy can also help resolve Hume's skeptical problem of induction—that is, thought's inability to prove the rationality of its habitual progression from known particulars to unknown generalities. In Darwin's words: "The great CREATOR of all things has infinitely diversified the works of his hands, but has at the same time stamped a certain similitude on the features of nature, that demonstrates to us, that *the whole is one family of one parent*. On this similitude is founded all rational analogy; which, so long as it is concerned in comparing the essential properties of bodies, leads us to many and important discoveries."[35]

Appearances to the contrary, Darwin is not here appealing presumptively to the theory of divine Design, any more than Blake is when he asks if he who made the Lamb made the Tyger. Rather, both men emphasize, like other vitalists, that "the laws of organic life" will never be understood except by comparing "the properties belonging to animated nature with each other" rather than "those of mechanism and chemistry" (*Zoonomia* 1.1–2). Based on the "Penetrability of Matter"—the idea of "two bodies existing together in the same space," as propounded, he notes, by Boscovich, and "lately published by Dr. Priestley" (1.113)—Darwin developed a field-like theory of a general sensorial power able both to propel motion in the form of vibrations and transmit it as physical stimuli to organic bodies and minds. On these grounds, he also postulated a second, nonrational and synthetic kind of analogy, defined as "an act of reasoning of which we are unconscious except from its effects in preserving the congruity of our ideas, and [that] bears the same relation to the sensorial power of volition, that irritative ideas, of which we are unconscious except by their effects, do to the sensorial power of irritation" (1.196). "Intuitive analogy" serves to maintain the consistency of experience in the face of Humean skepticism about the objective existence of personal identity. Devin S. Griffiths says Darwin "naturalized analogy as part of the sensational mechanism, rather than as a trope or form of argument."[36] Operating in tandem, Darwin's two types of analogy bridge the epistemological gap between scientific deduction and physical sensation.

For, as Whitehead observes, a fully stand-alone substance such as Hume's skepticism requires—Aristotle's independent "matter" combined with Cartesian dualism—must be incapable of interaction, so it "can tell no tale as to the survival of order in its environment."[37] The collapse of Blake's Urizen into unmeaning and Eternal Death following his withdrawal from Eternity illustrates this point. Urizen's Democritean hell of discrete particles in flux lacks all differentiation. Only by interacting with prophetic Blake-Los does he acquire a relation to past and future states: a story, however horrible. In fact, *any* forecasted future—whether a religious prophecy or a scientific prediction—is necessarily grounded in a presumed continuity between that future and some aspect or feature of the present. Since the present is nothing more or less than the universe objectivized as data from the standpoint of a local observer, there always already exists an analogy in the assumption that current data can be applied to reveal something about the universe in a different state. Says Whitehead, "Induction presupposes metaphysics. . . . You cannot have a rational justification for your appeal to history till your metaphysics has assured you that there is a history to appeal to; and likewise your conjectures as to the future presuppose some basis of knowledge that there is a future already subjected to some determinations."[38] Jonathan Bain comments, "It is clear then that Whitehead's argument is not so much a justification of induction as it is a clarification of the requirements necessary for the concept of induction to be meaningful."[39] Science and prophecy run on parallel metaphysical tracks, and both need to "bracket out" the ineliminable possibility of a self-defeating empty or demotivated skepticism like Hume's.

Analogy thus promises the *tertium quid* or "third kind" able to conjoin or reconcile, within the mind, forms of judgment or measurement with observed empirical phenomena like change and motion. It is the conceptual form of Plato's virtual cosmological Receptacle. The idea that analogy's "middle" nature suits it to investigate Zenonian paradoxes of displacement raised by the physical realities of temporal passage and motion goes back at least to the *Timaeus*. The Neoplatonist Proclus argues in his commentary as translated by Blake's friend, Thomas Taylor:

> For it is this ["the middle"] through which all analogy consists, collecting the extremes according to ratio, from one power to the other. For analogy is that which is principally and properly a bond. But it is a bond as that *through which*, and the middle. For through the middle analogy binds the extremes. . . . [S]ameness is the end of all this analogy . . . the ascent is through

> sameness to union. For analogy indeed is suspended from equality, being a habitude ingenerated in the boundaries of equality. But equality is suspended from sameness, and sameness from union.[40]

Proclus here implies a theory of analogy as an emanative "bond" of "middles," a set of perceived equivalencies between "contraries" or "extremes" whose source and object is nonperceptual "union" with higher truth. As I argue in chapters 4 and 5, it was the Cambridge Platonist Henry More who transmitted this middle to Blake via his elaborate doctrine of mediatory spiritual "vehicles." Not that Blake's work is necessarily immune to appearing, in retrospect, as a mere allegorical precursor to Whitehead and Einstein—and also depth psychology, Marxism, deconstruction, hypertext, ecocriticism, animal rights theory, and the body electric, to name a few recent candidates. My claim—rather more old-fashioned and less ideological than these—is that the striking similarities between Blake's cosmology and relativity theory are genuine because he knowingly chose to participate in the same intellectual history as led from Newton and Boscovich to Einstein and Whitehead. This history began with Newton and Newtonian Neoplatonism, carried on through Boscovich and Priestley, and then continued beyond Blake's purview with the work of Davy, Faraday, Maxwell, and Einstein.[41] That name sequence forms the main through-line of the history of field theory—albeit the middle, Boscovich-Priestley portion, crucial for Blake, is not so well known.

To reiterate, Blake was no scientist. It is his metaphysical concern with how to interpret sensory experience in the light of late eighteenth-, early nineteenth-century science research that generates the correspondences with relativity and especially Whitehead. My goal is not to confer scientific authority on Blake's mythmaking, but the opposite. I want to demonstrate that the universe Blake imagined over two hundred years ago was a four-dimensional space-time no less consistent than relativity's, which it resembles. Blake's chief means of imagining space-time was analogy, which by serving to extend one pattern of experience through its observed correspondence with another leads to higher levels of abstraction and a bigger picture. It helps that analogy as a mental tool is, in many ways, more effective—more visualizable—in the "beautiful flexible hands" (*M* 28:13; E 126) of a poet-artist than it is in the cuticle-chewed digits of a nominalistic modern quantum scientist. Blake's exploitation of impure metaphors, long chains of analogy, self-reflexive similes, inconsistent syntax, variable verb moods, the dialogism of multiple voices and, especially, topological inversions of perspective—all deployed alongside provoking multidimensional designs and disturbingly restrictive three-dimensional illustrations like his *Newton* print (figure I.1)—serves to push the reader-viewer's experience to extremes where Newton's laws break down and a more unified model of space and time dawns into view. It is not difficult to discern a Priestleyan idea of "the immateriality of matter" in the Romantic-neoclassical style of Blake's artist friend John Flaxman, whose airy geometric abstractions, says Robert Rosenblum, "exploit the dissolution of normative human scale" and "elude rational spatial treatment," and whose Dante illustrations appear "exempt from the gravitational and perspectival laws that generally dominated post-medieval art."[42] After all, Newton himself did not believe attraction was a real physical force, even as he further recognized the futility of devising a mechanical theory of gravity. On the other hand, the reason Blake's much more solid, Michelangelesque, figures fall or float in space is not because they irrationally seek liberation from the laws of physics and optics like Flaxman's, but because they reflect a coherent theory of the existence a higher spatiotemporal manifold of passage, movement, and change, of which each image is a quantum snapshot. Together, Priestley and Boscovich pointed the way.

In Blake's day, the problem of induction Hume had raised in the *Enquiry concerning Human Understanding* (1748) and its essay "Of Miracles" compelled scientists like Erasmus Darwin and Priestley to reassert analogy's power to synthesize the unwieldy heaps of information that had been generated, for over a century, by indiscriminating Royal Society amateurs in their pursuit of Bacon's program for the advancement of learning.[43] Griffiths dubs it "the age of analogy."[44] Priestley claimed, with excusable exaggeration: "Analogy is our best guide in all philosophical investigations; and all discoveries, which were not made by mere accident, have been made by the help of it."[45] With similar holism, Erasmus Darwin emphasized that his open-ended empiricism often led him "into conjectures . . . not . . . supported by accurate investigation or conclusive experiments. Extravagant theories however . . . are not without their use; as they encourage the execution of laborious experiments, or the investigation of ingenious deductions . . . And since natural objects are allied to each other by many affinities, every kind of theoretic distribution of them adds to our knowledge by developing some of their analogies."[46]

Hume himself stresses, similarly to Whitehead above, that the basis for reasoning about probability in the world is "analogy" rooted in resemblance: "We transfer our experience in past instances to objects which are resembling, but are not exactly the same with those concerning which we have had experience."[47] "All our reasonings concerning matter of fact are founded on a species of ANALOGY, which leads us to expect from any cause the same events, which we have observed to result from similar causes."[48] Blake perhaps detected this line of skeptical empiricism behind Bishop Watson's miracle mongering in *An Apology for the Bible*, Hume's reversed religious image. In agreement with Hume, Blake commented in the margin: "Prophets in the modern sense of the word have never existed . . . Every Honest Man is a Prophet he utters his opinion both of private & public matters/Thus/If you go on So/the result is So/ He never says such a thing shall happen let you do what you will. a Prophet is a Seer not an Arbitrary Dictator" (E 617). The remark is sometimes taken as evidence of Blake's earthy, plain-man empiricism. It can also be regarded as an anthropic protest that if we do not envision the future analogically in human terms, then the future remaining to us will necessarily turn out to be a contradictory one from which human agency has been excluded, leaving in its place the rule of Laplace's Demon of absolute determinism. "Honest" forecasting acknowledges the presence of the forecaster embedded in all the thick particularity of his or her limited perspective. "Seeing into the future" takes place when the future is inferred inductively by analogy from present signs and portents.

Humphry Davy offered an overview of analogy's scientific role in his *Elements of Chemical Philosophy* (1812), published just after Blake's *Milton* and six years before his good friend Coleridge published the "Essay on Method," quoted earlier: "By observation, facts are distinctly and minutely impressed on the mind. By analogy, similar facts are connected. By experiment, new facts are discovered; and, in the progression of knowledge, observation, guided by analogy, leads to experiment, and analogy confirmed by experiment, becomes scientific truth."[49] Previously, Davy's "Introductory Lecture to the Chemistry of Nature" had asserted a Blakean-Contrary, rather than a dualistic or oppositional, relation between knowing facts and doing science: "The body of natural science, then, consists of facts; its governing spirit is analogy,—the relation or resemblance of facts by which its different parts are connected, arranged, and employed, either for popular use, or for new speculative improvements."[50] Together, "body" and "spirit" form a human whole. That said, the key question, as Robert Oppenheimer later insisted, is whether the correspondences picked out by an analogy are detailed and logical enough to lead to fruitful *dis*analogizing through closer structural analysis of the target object or whether they're just superficial metaphorical embellishments.[51] The latter possibility was not only a weapon critics brandished against Erasmus Darwin and Davy but a constant source of inner unease for both men.

Blake was similarly concerned to disanalogize his analogies lest they harden into pleasant, self-reinforcing allegories. His understanding of pictorial representation as a two-dimensional formalism whose structure is no more than isomorphic to its three-dimensional objects was fully as self-conscious as that of an art theorist like Erwin Panofsky or a modern particle physicist like Niels Bohr. When his pictures do "map" as analogies of physical objects and events, they are usually disconcerting and ironical. Blake made a number of diorama-like designs whose overt pictorialism appears as no more than analogical to a higher-dimensional vision encompassing time and change. Examples include the designs to "A Cradle Song" and "Infant Sorrow" (both discussed in chapter 2), "The Chimney Sweeper" (*SE*), "The Tyger," and *Milton* plate 40 (discussed in chapter 4). The crudely flat look of these images can be seen to satirize the rectilinear perspective boxes of Hogarth's Rake's Progress and Harlot's Progress sequences, whose dense graphic realism reinforces a moral allegory where the meaning of even incidental details has been monitored and worked out beforehand. Blake deploys the modernized neoclassical style of his artist friends Flaxman and Cumberland, characterized by its "severe planarity," "lack of modeling," and "willful two-dimensionality,"[52] to lay bare the rational, Euclidean perspective that underpins Hogarth's moralizing realism. Similarly, the text in Blake's later Illuminated Books often appears as a formidable "wall of words" perpendicularly opposed to the spread-out three-dimensional design.[53] One sees how entrance into the world of the narrative requires adoption of a viewpoint quite unlike—even at right angles to—the viewpoint by which one perceives a static, synchronic three-dimensional representation. The necessity of this shift is brought home on *Jerusalem* plate 62 (copy E), which satirically reduces the temporal processes of reading, interpreting, and imagining the text to the passive beholding of its graven image. Albion, the Great Humanity Divine, stands aghast at the text placard his body has become during an age of Lockean nominalism, while a small human figure—presumably the poet-artist, perhaps the reader-viewer—witnesses this monumental horror from ground level (figure 1.2).

Blake's mockery of nominalism here anticipates the worrisome question posed by James Clerk Maxwell in a youthful essay written for the Cambridge Apostles in 1855: "Are we to conclude that these various departments of nature in which analogous laws exist, have a real interdependence; or that their relation is only apparent and owing to the necessary conditions of human thought?"[54] Some forty years later, at the outset of science's biggest watershed since Newton, the influential "empiriocriticism" of Ernst Mach elevated Maxwell's question to crisis proportions. Notoriously, Mach not only denied the existence of Newton's absolute space—as did Einstein and other aether doubters—but also atoms and anything else not detectable by sensory impressions or observable quantities, or not reducible to a thermodynamic law based directly on some combination of empirical observations.

"Oh, but you're just analogizing . . ." / 39

Figure 1.2. William Blake, *Jerusalem* pl. 62, copy E, c. 1821. Yale Center for British Art, Paul Mellon Collection.

Maxwell and Boltzmann's statistical treatment of thermal processes relied upon analogy-dependent physical models to extend the scope of theory and thus explain evidence whose sources, though invisible, could nevertheless be detected by their effects on manmade instruments. In contrast to Einstein,

Mach regarded their evidence as little more than gimmicks by which to make up for insufficient data. For Mach, the true business of science was controlled laboratory observation subject to confirmation, not the use of analogies to perform wide-ranging thought experiments that might or might not be verifiable *ex post facto*.

Mach's radical sensationist critique targeted the dualist metaphysics of eighteenth-century mechanism, specifically the idea that all phenomena are appearances produced by matter moving through space, the subjective experience of which then becomes known without mediation by an abstract human spirit. The irony is that his monist reduction of the physical and mental worlds to the status of sensed "events," and of knowledge to man's concrete historical activity of organizing these events with ever greater economy and accuracy, was misconstrued as a positivistic kind of subjective idealism. Einstein said of the quantum mechanics that Mach's polemics helped launch: "The basic positivistic attitude . . . seems to me to come to the same thing as Berkeley's principle, *Esse est percipi*."[55] Kant's "Copernican Revolution" consisted in the claim that knowledge must conform to the knower—a doctrine of which Stanley Cavell memorably remarked, "You don't . . . have to be a romantic to feel sometimes about his settlement: Thanks for nothing."[56] For Kantian science, the proper objects of inquiry are not "things in themselves" but "the appearances." But if the reality of the physical world reduces to nothing but our own sensations, which are private and inaccessible, then everything turns into swarms of anthropocentric ideas: light rays, nerve fibers, and the whole shared, independently existing world that it was the original aim of empiricism to investigate. Indeed, this is the very paradox that befell Lockean empiricism two centuries before Mach and that he was attempting to critique.

Locke's emphasis on mechanical causes drove him to postulate that material substance was not anything sensible but rather "the constitution of the insensible parts of a body, upon which its powers depend," that is, its atomic structure.[57] Like the atom (which no one had ever observed until Einstein's 1905 paper on Brownian motion and light quanta effectively measured its size), this substrate matter was purely a logical implication of thought. Locke calls it "an unknown X" (1.230), "I know not what" (1.392). Thus, Berkeley's radical idealist solution lay just around the corner. For Blake, as for Berkeley, Locke's mysterious material substratum merely hypostatizes his mystification. He parodies it in Urizen's vainglorious "[search] for a solid without fluctuation" (*BU* 4:11; E 72). In this satire, the materialist can only justify his atheistic science by positing a metaphysical entity "created in his own image" (Gen. 1:27), namely "matter," a substance no less inaccessible than Calvin's supreme God.

Blake himself, like the nineteenth-century field theorists, wanted to exploit the imaginative reach of formal abstractions to generate a model of how the physical world extends beyond ordinary human sense. His critique of reason thus remains less idealist than Berkeley's, and less "transcendental" than Kant's. Far from seeking reconciliation with the world through an acceptance of necessary limits like Kant, Blake seeks to approach and touch limits *within* the empirical world, thereby glimpsing the infinite for a brief cosmic Moment. His mythology with its multiple different allegorical levels can be seen as a Platonic "concept space" grounded in the notion that the mind tracks a concept the way the eye tracks a visual object, shifting angles and perspective the better to see it whole. If this work is demanding, it is because the concepts are real and exist independently of the limited human perceiver. A conceptual model is, therefore, an approximation of physical reality that can be improved. It is a *weltbild* or "physical world image" in Max Planck's sense, which Einstein embraced.[58] Similarly, Blake's sequence of Prophetic Books builds up his mythology, bringing his fundamental concepts into progressively sharper focus. As a "free invention of the human spirit" (to quote one of Einstein's favorite, semi-Kantian phrases), the mythological model always remains formal and incomplete with respect to the metaphysically real world it represents. To put it differently, Blake, like the later Einstein and unlike Mach, came to feel that concepts exist in a space logically independent from sense experience, a space from which they cannot be derived inductively without giving rise to a damaging circularity. Nevertheless, the concepts can be intuited and even visualized.

On this view, physical models always contain an element of anthropomorphism through the very presence of an observer inferring analogies between different laws in nature. So, the answer to Maxwell's previous either-or question can only be yes and yes. The human observer participates in the conditions that make possible scientific observation of an analogy. One reason why Blake's mythology consists not of concrete symbols but rather, as Leopold Damrosch, Jr. has emphasized, displaced "symbols *about* symbols,"[59] is that the objects symbolized exist not in nature but within science's shared concept space. They are at once things and ideas. Likely thinking of Leibniz's principle of the identity of indiscernibles, Wallace Stevens famously declared, "Both in nature and in metaphor identity is the vanishing point of resemblance."[60] Blake adds volume and dimension to the thought. For him, converging perspective lines meet at the point where the externalized three-dimensional world of imperfect copies and analogies leads back to an invisible four-dimensional world of self-identical Forms beyond resemblance which constitute the true identities of things. This is the realm of Blake's concept space.

When early twentieth-century scientists employed the so-called "hypothetico-deductive method" to develop a theoretical model for observed data, they were building an analogy between relations and proportions among human sensations on the one hand, and corresponding relations and proportions in physical nature on the other. Likewise, Blake as a figurative artist constructs an analogy between his visual perceptions and the images he draws on the page or incises on the copperplate—an analogy the more powerful when his perceptions refer to things imagined rather than seen. This suggests why Blake, who greatly disliked naturalism in art, nevertheless still considered himself an imitative artist. "Men think they can Copy Nature as Correctly as I copy Imagination this they will find Impossible" (*PA*; E 574). "To learn the Language of Art Copy for Ever. is My Rule" (E 636).[61] If it seems paradoxical to "copy Imagination," that is because a copy, for Blake, is not an empirical duplicate but a conceptual, that is, Platonic analogy brought down to earth and rendered visible in all its concrete spatiotemporal displacement. Accordingly, he ensured no two iterations of his Illuminated Books are exactly alike. The copy discloses, not nature, but an object of perception anthropomorphized by the hand and eye of the artist engaged, like Michelangelo, in an unrepeatable discovery process of bringing forth the true form hidden inside, as if immediate visual sensations were uncut blocks of raw material. Blake's "romantic neoclassicism" holds the neoclassical mirror up to nature in the romantic sense of creating a fractal hall of mirrors by which to reveal imitation itself as an imaginative process at the horizon. For Blake, imitation chips away at sensory perceptions and pares them down to model an idea at the very limit of what can be humanly apprehended. *The Book of Urizen* and *Milton* portray three-dimensional visual perception and its literary correlative, descriptive realism, as copyings whose seeming givenness is produced dynamically through recursive series of analogies. In these poems, the dominant concept space equates to the overall structure formed by repeating, roughly isomorphic (but never quite allegorical) subnarratives that serve to bring a conceptual object into prophetic perspective: respectively, "Earth" (*BU* 3:36; E 71) and Milton. The concept space remains formal, yet it can be apprehended by means of a metaphysical "Vortex" (*M* 15:21–35; E 109) linking four-dimensional Eternity at the open end to three-dimensional earthly existence, and ourselves, at the apex.

In a late essay, Heisenberg recounts how, as a young man, he beseeched his mentor Bohr: "If the inner structure of the atom is as closed to the descriptive accounts as you say, if we really lack a language for dealing with it, how can we ever hope to understand atoms?" Ever enigmatic, Bohr hesitated a moment before replying: "I think we may yet be able to do so. But in the process we may have to learn what the word 'understanding' really means."[62] Understanding understanding differently, even differentially, as an act of mind conditioned by the instruments and material technology that render the object perceptible by its traces, whether these occur in cloud chambers and silver-bromide photographic plates or in the bespoke processes of relief etching: that is what Blake's cosmology invites. Consider, in this light, Einstein's bold eclecticism in a letter of 1908, before the dominance of quantum mechanics drove him into a conservative stance of hard realism:

> Science without epistemology is—insofar as it is thinkable at all—primitive and muddled. . . . However [the scientist] must appear to the systematic epistemologist as a type of unscrupulous opportunist: he appears as *realist* insofar as he seeks to describe a world independent of the acts of perception; as *idealist* insofar as he looks upon the concepts and theories as the free inventions of the human spirit (not logically derivable from what is empirically given); as *positivist* insofar as he considers his concepts and theories justified *only* to the extent to which they furnish a logical representation of relations among sensory experiences. He may even appear as a *Platonist* or *Pythagorean* insofar as he considers the viewpoint of logical simplicity as an indispensible and effective tool of his research.[63]

Blake's metaphysics—like the early Einstein's, like Whitehead's—is an attempt to recognize and unite such multiple perspectives in the face of a physical world that continues to transcend them all. It is because reality demands a multitiered approach that Blake continually disanalogizes his analogies, the better to introduce new ones lest the old decay into allegory. One hears a lot, these days, about the cognitive value of literature and "thinking through poetry." Blake's "visionary physics" rooted, my next chapter shows, in David Hartley's psychophysiological account of reading and learning to read offers an astonishing example of what such thinking can be like.

CHAPTER TWO

Learning to Read in a Force Field

Songs of Innocence, Hartleyan Psychology, and the Physics of R. J. Boscovich

Elusive presences sweep across Blake's *Songs of Innocence*, overlapping, interpenetrating, evanescing. The poems blend voices, moods, and echoes built partly out of grammatical ironies such as two-way syntax and changes in tense, partly out of recurring images and analogies that give rise to a kind of ghostly pentimento as one moves through the volume. This chapter traces these impressions to a certain narrow current in eighteenth-century materialism, a current that floated Blake's spiritual boat even though as a fellow traveler he was always pulling in a different direction.

Songs of Innocence marries the associationist psychology of David Hartley with R. J. Boscovich's atomic theory based on alternately attracting and repelling force points, as it was likely transmitted to Blake by Boscovich's English champion, Joseph Priestley.[1] In the world of *Songs*, associations of ideas materialize as independent entities or "minute particulars" of experience immediately visible and audible to the reader-viewer, in and as themselves. Hartley had noticed that "in passing over Words with our Eye, in viewing Objects, in Thinking, and particularly in Writing and Speaking, faint Miniatures of the Sounds of Words pass over the Ear."[2] Blake embraces Hartley's model of mental growth, particularly his account of how children learn to read. *Songs* presents reading as a synesthesia of seeing with subaudible vocalization and hearing, its physiological basis deriving from Hartley's influential *Observations on Man* (1749). In Blake's volume, the reader-viewer's mind travels, almost shamanically, to Hartley's tiny "miniatures" of perception in a place of vision where the abandoned macroscopic world, including one's own three-dimensional body, recedes to a distant "bound or outward circumference" (*MHH* 3; E 34). And the physics of this process, a cognitive version of Coleridge's psychological suspension of disbelief, lies in Boscovich's theory of alternating attractive and repulsive shells or spheres of force.

This is to say Blake, like many eighteenth-century way-of-ideas empiricists, tends to "ontologize" perceptions as objects existing independent of the self.[3] Unlike them, he exploits British empiricism for the visionary purpose of turning the Lockean self inside-out, so exposing it to the cosmos to which it belongs. Berkeley thought his *esse est percipi aut percipere* (to be is to be perceived or to perceive) showed the idea of mind-independent material substance to be a Scholastic fallacy of misplaced concreteness, while at the same time he took pains to emphasize that his seemingly outlandish immaterialism alters nothing in actual empirical experience. Inasmuch as substantial matter's traditional dearth or badness arises from misunderstanding of God's given "language of nature" rooted in sense perception, Berkeley's philosophy constitutes a kind of theodicy. But then Hume's atheistic skepticism stood Berkeley on his head. If all that we know is our own perceptions, Berkeley's phenomenalism turns knowledge of causation into mere subjective, folk ideas of "cause and effect." Notoriously, Hume compares the mind to "a kind of theatre, where several perceptions successively make their appearance; pass, re-pass, glide away, and mingle in an infinite variety of postures and situations." Then, having reduced perception to the spectatorial watching of some private inner lava lamp, he coolly lands the final blow: "nor have we the most distant notion of the place, where these scenes are represented, of the materials, of which it is compos'd."[4] Hume thus established the conditions for an alarming "problem of knowledge" that nevertheless he, too, cheerfully concedes to be merely academic, inasmuch as it disappears so soon as you doubt your doubt in Pyrrhonian fashion and get on with the business of living life according to "feeling" and your proper "human nature": "I dine, I play a game of back-gammon, I converse, and am merry with my friends; and when after three or four hour's amusement I wou'd return to these speculations, they appear so cold, and strain'd, and ridiculous, that I cannot find in my heart to enter into them any further" (269).

In contrast to these opposing dualisms, Blake's "bastard" metaphysics—his dualistic monism—or is it monistic dualism?—develops within the context of the auditory and visual impressions produced directly by his own poems and paintings. Hence, the importance to him of Hartley's physiological theory of ideas based not in "Reason" but "the Body" (*MHH* 3; E 32). Hume's estranging collapse of the punctual Lockean self into "nothing but a bundle or collection of different perceptions" (*Treatise* 252) becomes an opportunity to release those hitherto caverned private perceptions into the open, indefinite space of reading and viewing. There, they cease to be hard, atom-like idealizations and take on life as anthropomorphic entities flexibly responsive to the demands of the moment—namely, the reader-viewer's ongoing associations of ideas. In Blake, Hartleyan physiology serves to extend Berkeley's immaterialism by granting perceived ideas tangible, physical existence. Since no concrete realities can exist that are not experiences, the result is a species of panexperientialism or panpsychism.[5] (And Berkeley rolled in his grave.) Hence, in several Blake paintings, for example *Milton's Mysterious Dream* (figure 2.1), mystical eyes abound in a way that seems directly indebted to Hartley. The eyes—mini-Vortexes of vision shown in the process of coalescing out of the landscape—work to heighten viewers' awareness of how their own eyes interact with the physical image and form it into a representational object within the mind.

These eyes—*our* eyes spiraling into physical space through the temporal process of vision—are located monistically within the landscape as real objects of perception. Blake thus pursues the very implication Richard Price feared in his debate with Priestley, published as *A Free Discussion of the Doctrines of Materialism* (1778). Priestley argued that "the term, *thing*, or *substance*, signifies nothing more than that to which properties are ascribed," namely, those of attraction and repulsion. Price remonstrated, *à la* Newton, that attraction and repulsion must be effects of some further action upon matter. For "if the effects of such action [be] as that of ideas and motives on conscious and thinking beings, then since all matter attracts and repels, all matter must be conscious and intelligent. . . . Is it only powers that circulate in our veins, vibrate in the nerves, revolve around the sun, &c."[6] Yes, Priestley replied, all is "powers." In what Robert Schofield calls a "surprisingly modern declaration,"[7] Priestley affirms it is best simply to drop the whole metaphysical notion of material substance:

> Solid *atoms*, or *monads of matter*, can only be hypothetical things; and till we can either touch them, or come at them, some way or other, by actual experiment, I cannot be obliged to admit their existence. . . . A definition of any particular *thing*, *substance*, or *being* (call it what you will) cannot be anything

48 / A Bastard Kind of Reasoning

Figure 2.1. William Blake, *Milton's Mysterious Dream*, Illustrations to Milton's "l'Allegro" and "Il Penseroso," c. 1816–20. Morgan Library & Museum.

more than an enumeration of its known *properties* . . . the terms *substance, thing, essence* &c. being . . . nothing more than a help to expression, . . . a convenience in speech. (45–46)

Boscovich's Proto-field Theory and Priestley's "Matter in a Nutshell"

Behind Priestley's emphasis on phenomenally observed "properties," as opposed to independent "substances," lies a principle very like the one the mathematician Boscovich borrowed from Leibniz and placed at the core of his physics of nested attractive and repulsive forces: "The Law of Continuity, as we here deal with it, consists in the idea that . . . any quantity, in passing from one magnitude to another, must pass through all intermediate magnitudes of the same class."[8] Boscovich was a figure of international renown.[9] He met Priestley at least once when he visited England in 1760, and was made a member of the Royal Society in January 1761. Boscovich thought he had finally offered a unified theory of force. His introduction claims to have fulfilled Newton's wish in the *Opticks* that, as Boscovich puts it, "important principles [beyond Newton's stipulated "two or three"], such as impenetrability & impulsive force, be reduced to a single principle" (15). Rejecting mechanical aether theories, he postulated a highly rarified universe that fed directly into Priestley's doctrine of "the immateriality of matter."

For, Boscovich's atoms are not hard corpuscles but unextended mathematical "points of force" or *puncta*, possessed nonetheless of inertia, and wrapped in asymptotically alternating spheres of attraction and repulsion in space (figure 2.2). As opposed to explaining the transfer of motion between material bodies by contact action, Boscovich's theory is kinematic and based entirely on parameters of force and distance. The force of gravity is derived from acceleration at distances from the infinite resistance at an atom's center. At very small distances, the forces between particles become strongly repulsive, which explains why the universe does not contract to a single point. In this way, Boscovich solved the longstanding puzzle of how bodies in collision could produce seemingly instantaneous changes in velocity and position.

Simply put, communication between bodies does not require impact. Dugald Stewart recognized that the longstanding metaphysical assumption that bodies cannot act where they are not was finally overthrown: "It must appear a very curious circumstance in the history of science, that philosophers have been so long occupied in attempting to trace all the phenomena of matter and even some of the phenomena of mind, to a general fact, which, upon accurate examination, is found to have no existence."[10] But if "matter" was now available for redefinition, then so was mind—as Priestley

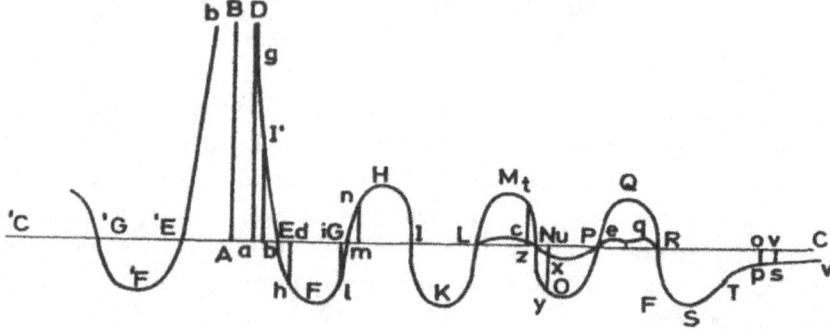

Figure 2.2. Roger Joseph Boscovich, *A Theory of Natural Philosophy* [*Theoria Philosophiae Naturalis*] (1758), fig. 1, p. 41. At axis AB, the force of repulsion is infinite and "will destroy the relative velocities [of particles], no matter how great they may be" (145). Toward C, Newton's weak force of gravitational attraction takes over. For a particle at distances E, G, I, L, N, P, and R from A, repulsive and attractive forces are in stable equilibrium. At points E, I, N, and R, increasing the distance increases the attractive force, and reducing the distance increases the repulsive force. In these instances, the particle will return to its previous distance, a state Boscovich terms a "limit-point of cohesion." At points G, L, and P, particles are in an unstable equilibrium moving either away or toward each other: a "limit-point of non-cohesion" (143). As particles pass their limits of cohesion, they gain or lose energy quanta corresponding to the areas delimited by the arcs of repulsion and attraction.

was one of the first to see, even though, in his haste to propound matter's immateriality, Priestley dropped Boscovich's requirement of *vis inertia*, leaving it unclear how the atoms could still obey Newton's laws of motion, which Boscovich still accepted. Priestley's failure to grasp Boscovich's system is evident from his claim that absolute space would remain in the absence of matter (*Disquisitions* 299–300). Thereby, he sought to rescue the world aether from mathematics by regarding it as a physical property of space. Notwithstanding, Priestley adequately summarizes Boscovich in *History and Present State of Discoveries in Relation to Vision, Light, and Colours* (1772), which adopts

> the hypothesis of M. Boscovich, who supposes that matter is not impenetrable . . . but that it consists of physical points only, endued with powers of attraction and repulsion, taking place at different distances; that is, surrounded with various spheres of attraction and repulsion, in the same manner as solid matter

is generally supposed to be: Provided, therefore, that any body move with a sufficient degree of velocity, or have sufficient *momentum*, to overcome any powers of repulsion that it may meet with, it will find no difficulty in making its way through any body whatever; for nothing will interfere, or penetrate one another, but *powers*, such as we know, do, in fact, exist in the same place, and counterbalance, or overrule one another."[11]

Can the doctrine of Contraries be far behind?

The really striking feature of the new system—which explains its influence on Humphry Davy, Michael Faraday, and James Clerk Maxwell, and why many scientists have considered it to anticipate relativity field theory and even Bohr's quantum mechanics[12]—is that Boscovich's forces do not act between atomic mass points but *constitute* them. As L. L. Whyte wrote in 1957, Boscovich "substituted a continuous range of density for the sharp dualism of matter and empty space, setting no upper limit to density, and predicting the penetration of bodies by high-speed particles confirmed 150 years later." His theory "was essentially relational, indeed almost 'relativistic,' giving the first analysis of space and time measurements in which the spatial and temporal relations of a system are in some degree affected by its state of motion relative to all other particles in the universe (a century before Mach)."[13] In Boscovich's words, "since things that are contiguous as well as non-extended must compenetrate," his nonextended points are therefore separated from one another and so possess "distance relations" (273). "Now what kind of extension can that be," he asks, "which is formed out of non-extended points & imaginary space, i.e., out of pure nothing? How can Geometry be upheld if no thing is considered to be actually continuously extended? Will not groups of points, floating in an empty space of this sort be like a cloud, dissolving at a single breath?" (131). Contra Newton, he delivers his stunning answer like a true mathematician: "Space has no actual existence. It is only something that is possible, indefinitely imagined by us; that is to say, it is the possibility of real local modes of existence, pictured by us after we have mentally excluded every gap." Descartes's dense, etherous space has struck some commentators as "relativistic" in the loose sense that its vortexes wrap around objects agglomeratively, unlike Newton's void. Boscovichean space is relativistic in the truly modern sense of forming part of a physical field while conveying, well before Einstein—who never did give up on the idea of a physical world in-itself with properties different from the observed phenomena—a deeply discomfiting epistemological awareness that ideas of substance cannot be detached from our own organic capacity for "picturing."

Accordingly, for Priestley, the solidity and impenetrability of material bodies are simply effects of the human body's inability to overcome those bodies' exerted force of repulsion. Matter therefore exists in Priestley's system only as, in Blake's terms, "a Limit of Opakeness, and a limit of Contraction; / In every Individual Man" (*J* 42:29–30; E 189), much as the righthand side of Boscovich's graph shows (figure 2.2). As the astronomer Sir William Rowan Hamilton wrote to Coleridge in 1832, by way of emphasizing his adherence both to Boscovichean atomism and Berkeleyan immaterialism:

> I regard a certain atomistic theory as having subjective truth, and as being a fit medium between our understanding and certain phenomena: although objectively, and in the truth of things, the powers attributed to atoms belong not to them but to God. The atomistic theory of which I speak . . . consists in representing all phenomena of motion as produced by the action of localized energies of attraction or repulsion, each energy having a centre in space; and this centre, which is supposed to be a mathematical point, without any figure or dimension, being called an *atom* instead of a point, merely to mark its conceived possession of, or connexion with, physical properties and relations.[14]

"Atom," in other words, is a mere concession to the weakness of human sense. Like Berkeley, like Blake, Hamilton was both idealist and empiricist. As he put it elsewhere, in a remark Coleridge and Blake as symbolists would both have appreciated: "I am becoming more disposed to value facts . . . the more I regard them as but passive states of our own being."[15] Hard corpuscles having dissolved into a continuous field of interpenetrating forces, the power of environment to condition perception is emphasized as never before. Boscovich's field physics thus complements Blake's Hartleyan focus on psychology and education. Since no two environments are alike, each minute particular is absolutely unique, "particular" signifying here not an atom, much less a person or self, but an electron-like succession of extremely brief events or "perceptions."[16] To quote the philosopher of science Milič Čapek, "one never observes the *same* particle a second time."[17] Identity is a repeating spatiotemporal occurrence. It is our own body's restriction to a biological "zone of the middle dimensions" between the microscopic and the astronomical that leads us to substantialize these different occasions as if they were enduring atomic corpuscles.[18] That is why Niels Bohr insisted on the principle of "complementarity," without which the discontinuous quantum world of "leaps" or "jumps" in orbit could never become known within the broader classical world of everyday perception, or be joined to the language of classical physics based on the smooth transitions of the differential calculus.

Newton himself had already conjectured a highly rarified universe with his popular "nutshell" theory of matter in the *Opticks*. In contrast to Cartesian billiard-ball matter in motion based on the impact of two moving bodies, Newton's was a universe of impenetrable particles floating in a void pervaded by forces of attraction and repulsion. These forces were manifest in vibrations of light and sound carried into the nerves (he speculated) by means of the aether. Thereby, gravity's seemingly occult power to act at a distance might be explained. The lengthy Question 31 of the *Opticks* (4th ed. 1730) begins: "Have not the small Particles of Bodies certain Powers, Virtues, or Forces, by which they act at a distance, not only upon the Rays of Light for reflecting, refracting, and inflecting them, but also upon one another for producing a great Part of the Phaenomena of Nature?"[19] Matter's property of impenetrability was an effect of these widespread forces, while material substance itself Newton deemed to be extremely rare and porous. As Priestley put it in *Disquisitions Relating to Matter and Spirit* (1777), "For any thing we know to the contrary, all the matter in the solar system might be contained within a nut-shell."[20]

The nutshell theory was widespread. Nevertheless, when Priestley developed Newton's conjecture by merging it with his materialist appropriation of Boscovich's mathematical proto-field theory of continuously opposed forces or "powers" of attraction and repulsion, the result was political dynamite because Priestley further grafted onto it a progressive doctrine of mind, education, and cultural enlightenment. Boscovich, a conservative Jesuit, was appalled. It was Priestley—known as "Hartley's bulldog"—who first propounded the link between Hartley's psychology and Boscovich's physics: "The relation that attractions and repulsions bear to several modes of thought, may be seen in *Hartley's Observations on Man*" (*Free Discussion* 20). By conjoining Hartley's physiology with Newtonian physics as revised by Boscovich's mathematical geometry, Priestley claimed to dissolve the dualism inherent in Newton's model. He opens the *Disquisitions* by remarking that "the notion of two substances that have no *common property*, and yet are capable of *intimate connection* and *mutual action* is both absurd and modern," that is, Cartesian (xxxviii). In *Songs of Innocence*, this missing medium between mind and matter—the "third kind" or *tertium quid* sought by Newtonians and Cartesians alike, and generally hypothesized as the corpuscular aether—is found to be human sense perception itself. Blake's subsequent, seemingly retrograde embracing of Henry More's Neoplatonic doctrine of "aetherial vehicles" followed naturally. As Hartley explained it, the forces of attraction and repulsion that pervade the universe also penetrate the human nerves and cause their atomic particles to vibrate, in the process attuning the senses to specific frequencies that correspond to the arousal in the mind of perceptual ideas. Echoing certain well-known speculations of Newton and Locke discussed in my next chapter, Priestley added that a chief advantage of his theory's restricting of matter's properties to attraction and repulsion is that it "greatly relieves the difficulty which attends the supposition of the *creation of it* [matter] *out of nothing* . . . For, according to this hypothesis, both the creating mind, and the created substance are equally destitute of *solidity* or *impenetrability*; so there can be no difficulty whatever in supposing that the latter [matter] may have been the offspring of the former [mind]" (*Disquisitions* 18). And so Priestley's purported demolition of Christian dualism was complete: "Matter, destitute of what has hitherto been called *solidity*, [is] no more incompatible with sensation and thought, than that substance, which, without knowing any thing farther about it, we have been used to call *immaterial*." Such is the Blakean "space of reading": a self-consciously creative process of cognition where sensibilia form into perceptual objects. Glance again at Boscovich's recursive force-curve in figure 2.2. My later chapters will suggest Blake regarded it as a sidelong, cutaway two-dimensional representation of a vortex, itself a three-dimensional "concept space" of the temporal, four-dimensional process by which visual perception non-instantaneously spirals into and constructs its focal object.

While Boscovich derived his law of continuity from Leibniz, he could also have derived it inductively from phenomena in Newtonian fashion, and in that case, his atoms would not have had to be unextended as required by his mathematics. Likely, this was Priestley's rationale for interpreting the theory mechanistically and thinking it could be synthesized with Hartley's vibratiuncles.[21] Boscovich's graph shows that as the alternating forces approach ordinary macroscopic distances and become three-dimensional objects of perception, they settle down to Newton's inverse-square law of weak gravitational attraction. Like Einstein, he treats Newton's law of gravitation as an approximation, a "classical limit" that suffices over great distances but breaks down at the atomic scale where increasing oscillation between attractive and repulsive forces takes over.[22]

This implicitly Cambridge Platonist theme of the interpenetrability of the mind and material creation is taken to its radical extreme in one of Blake's very first works in illuminated printing, which concludes in a phrase I will repeat as a mantra throughout this book: "Therefore God becomes as we are, that we may be as he is" (*NNR* [b]; E 3). The emanative divine spirit generously materializes itself not once upon a time, as in Genesis, but all the time—dynamically, through the human senses—in order that mankind, created after God's image, may reciprocate him by recreating themselves and their world in the image of God and Eternity. The human form divine is an unstable composite of attractive and repulsive forces, ever in danger of falling back into the solid, three-dimensional world of matter that it projects outward due to bodily weakness, even as it remains capable of expanding beyond material objects into virtual, utopian realms of becoming whose immanency transcends perspectival representation. To put it differently, Blake opens an Enlightened dialectic between the two claims Berkeley largely conflates together, *esse est percipi* (to be is to be perceived) and *esse est percipere* (to be is to perceive). On one hand, he insists, "As a man is So he Sees" (E 702) and "As the Eye—Such the Object (E 645): we are each responsible for the world we see, and for taking control of the mind's chronic propensity toward what Berkeley calls three-dimensional "outness." On the other hand, Blake also recognizes, like many Revolutionary thinkers, that people tend to "become what they behold" (J 39:32; E 187). When the externalized perceptual world we each construct assumes collective, social form, *it* takes *us* for its object and conditions us according to its own ends—and rarely for the best.

In other words, there remains a historicist dualism at the base of Blake's idealism that makes it more dynamic and more empirical than Berkeley's. If miracles exist in the eye of the beholder as the infinity of Jesus, who "pass[es] the limits of possibility, as it appears / To individual perception" (J 62:18–19; E 213), so conversely "the all tremendous unfathomable Non Ens / Of Death" (J 98:33–34; E 258) represents perception's zero-limit of terrified self-delusion. For, "What seems to Be: Is: To those to whom / It seems to Be, & is productive of the most dreadful / Consequences to those to whom it seems to Be: even of / Torments, Despair, Eternal Death" (J 32:50–54; E 179). As in Blake's late painting, *The Vision of the Last Judgment* (figure C.3), Jesus and Eternal Death constitute invisible Contraries with humankind in-between. Blake agrees with Berkeley that the mind creates its own reality (as aging New Agists still say). But he also recognizes that this notion alone can hardly liberate us from the historical effects of injustices arising from the "imposture" of corrupt human institutions, much as Oothoon's plight goes to demonstrate in *Visions of the Daughters of Albion* (see chapter 1).

Fractal Vision, Miracles, Bayesian Probability, and Hume in *Innocence*

Boscovich's nested shells of alternating attraction and repulsion can be seen to underlie the thick, interdependent affectivity of the Contraries parent and child, sheep and shepherd, man and God in *Songs of Innocence* no less than they do *The Marriage*'s pronouncement that "Attraction and Repulsion . . . are necessary to human Existence" (*MHH* 3; E 34). The resultant Blakean "space of interpretation" or "allegory of reading" is no mere critical metaphor as it was in much deconstructive hermeneutic theory of the 1980s and 90s. For example, in "A Cradle Song," to which I will return, a mother gazes intently at her child's face and muses, almost as if reading a Bible: "Sweet babe, in thy face, / Holy image I can trace" (*SI*; E 12). Elsewhere, I've argued this baby symbolizes an incunabulus (from L. *cunae*: cradle): a book made using metal type before 1500 when printing was in its "swaddling clothes," and before Blake's *bête noire* of commercial engraving for letterpress existed.[23] However, the mother is less a point-for-point allegory of what the poem's reader is doing than she is the form of that activity projected, via the lullaby's sleepy syntax and shifting hypnogogic imagery, into the physical space between the eyes in our head and the book held in our hands or lap (not unlike a baby).

Throughout *Songs of Innocence*, such dissolution of the subject-object dichotomy that provides the basis for rational knowledge allows the shifting thoughts and feelings expressed by the imaginary humans in the poem "out there" to mingle intersubjectively with the reader-viewer's actual responses. The latter become, in great measure, performative, even as the feelings alleged within the poem are self-reflexively seen and heard to become real and efficacious. As a result, it is very hard, if not impossible, to tell the difference between the two. Are we inside the poem with the lambs and flowers and children, or are they inside us? How can we know the singer from the song, the blossom from the bole? This expansion of sense perception within a space of "mind wandering" understood to be coextensive with the imaginary heterocosm of the poem is well described by Blake himself in the catalog to his painting of the Last Judgment: "If the Spectator could Enter into these Images . . . or could make a Friend & Companion of one of these Images . . . then would he meet the Lord in the Air & then he would be happy" (*VLJ*; E 560).[24] Compare Blake's lifelong hero James Barry's 1784 *Account of a Series of Pictures, in the Great Room of the Society of Arts* (a couple of its phrases echoed in Blake's catalog to his painting of the Last Judgment, perhaps via Barry's 1809 *Works*): "It is difficult, and would lead us into the depths of philosophy, to say, what is the difference between the actual thing and its image in glass; and yet . . . they are both equally actual to the sense of seeing in a third spectator . . . the painted object is equally permanent with what we may call the actual or real one. The truth is, that they are all pictures alike, painted equally on the retina or optical sensorium."[25] *Songs of Innocence* occupies the physiologic middle ground between copy and original similarly to Barry's visual phenomenology of "the painted object" where, a recent critic comments, "painting is . . . not the communication or mediation of the idea but rather its miraculous instantiation."[26]

Priestley's materialist account of Boscovich's nested shells of attraction and repulsion, in *A Free Discussion of the Doctrines of Materialism* and *History and Present State of Discoveries Relating to Vision, Light, and Colours*, also helps to explain the notably "fractal" aspect of Blake's human form divine. The key feature of fractals is that the measurable amount of space they fill lies *in between* different topological dimensions. A fractal object possesses self-similarity: it exhibits the same pattern across different scales. Standard examples include snowflakes, crystals, tree branches, and the coast of Britain, in all of which the length that is measured depends on the length of the ruler used or the observational instrument's degree of resolution. "As the Eye—Such the Object" (E 645). The distance between any two points on a fractal line or surface is infinite, the object being too complex—curved at every point—to reduce to strict Euclidean geometry, which offers only approximations. Likewise, though Blakean God's infinite degree of detail can be apprehended through expanded sense perception locally, in small, "smoothed-out" portions, the whole exceeds the three-dimensional space in which we represent objects to ourselves. *Songs of Innocence* reminds us that a lower-dimensional version of this also holds true for the less sublime objects we perceive daily in three-dimensional perspective. These lie between two and three dimensions, as we can never actually see their front and back at the same time in consequence of our own location in space.

As Erwin Panofsky recognized, the goal of Renaissance rational perspective was to assimilate this empirical limitation inherent in any specific point of view to the homogeneous, totalizing idealizations of mathematics and geometry, so creating the illusion of an impersonal, panoptical field of vision that disappears the experience of space. In a realistic, single-point perspective, objects are still seen between two and three dimensions; what the viewer grasps, however, is a complete, stable, fully focused visual field available for scanning as if frozen in time. In *Songs of Innocence,* Blake extends his fractal awareness of perspective's artificiality into a temporal fourth dimension whose avatars—infants, lambs, flowers, and birds—"expand inward." Panofsky himself points out the paradox, already inherent in Euclidean perspective, that reducing objects to a single set of points and relations further implies "that from every point in space it must be possible to draw similar figures in all directions and magnitudes," thereby rendering visible the relational structure of the perceptual field itself and restoring perspective to its etymological sense of "seeing-through" a flat surface into an illusion of depth.[27]

58 / A Bastard Kind of Reasoning

Similar to Panofsky's critique, numerous pages of *Songs* convey their visual imagery as in-between. On the title page, for example, the letters, the vegetation of the gothic letter flourishes, the branches and leaves of the tree, and the small human figures nestled among its branches together form a material continuum from one to two and from two to three dimensions, and from symbolic ciphers (alphabet letters) to representations, while at the same time they call attention to the convention whereby some two-dimensional shapes are recognized as 3-d. In "A Cradle Song," the picture on the second page (figure 2.3) is literally pegged to the upper text plane, as if the latter formed a solid wall.

Figure 2.3. William Blake, "A Cradle Song," *Songs of Innocence and of Experience* (1789, 1794), Copy L. Yale Center for British Art, Paul Mellon Collection.

Yet, the image only literalizes what we habitually overlook when reading, namely, that printed letters would be illegible if they did not stand forth against their ground on the page. Over and over in *Songs*, tree branches appearing in a vignette at the page bottom gradually shed their depth and flatten into purely ornamental dividers between the poem's different stanzas. If the branches thus instantiate the process by which, reading downward, we construe their two-dimensional lines as three dimensional, is it too much to regard them self-reflexively as symbols of the Hartleyan nerve fibers by which the page becomes transmitted to the reader's brain for perception?

Blake's fractal approach to miniaturization in the Illuminated Books thus serves to heighten proprioceptive awareness of reading and viewing as mental processes of introjection. Where the infinitely tender particulars of the *Songs* take place is evidently inside the human body, starting in the head. As Priestley wrote, no doubt recalling Hartley's medical researches, "the necessary seat of thought"—the soul—"is a property of the *nervous system*, or rather of the *brain*" (*Disquisitions* 27). Blake came to term the concentering process by which visual focus spirals inward toward the object a "Vortex," a largely four-dimensional image-concept. As Northrop Frye pointed out, the Blakean Vortex symbolizes the cone of vision, its apex pointing to the object, through which the mind takes in the two eyes' shared point of focus.[28] To take a small sidestep, one could say *Songs* reconceives the artificial stasis of Deism's *concordia discors*, the God-given checkerboard of black and white by which evil becomes aestheticized, as a natural fractal process based on a dialectic of interlocking Contraries. In Blake, aesthetic experience is not a reflection upon the world, it is immanent in the world through our own bodies as precognitive awareness of the "upwelling" pattern of feelings between actual entities already, themselves, in process of changing symbiotically within space-time.[29] And the form of this upwelling or revelation of aesthetic feeling is the Vortex-Moment through which Eternity floods the world of ordinary three-dimensional experience.

An important source for much of Blake's thinking about vision seems to have been Priestley's compendious *History and Present State of Discoveries Relating to Vision, Light, and Colours*. There, Priestley relays the discovery by William Porterfield of quick eye movements or saccades, without which vision would be reduced to sequences of infinitesimally small, atom-like focus points whose overall meanings could only be deduced, as opposed to being grasped immediately as *gestalts*. In effect, Porterfield demonstrated that scansion—and, by extension, three-dimensional reality—is a brain-generated illusion made possible by saccades. "In viewing any large Body," Porterfield writes, "We are ready to imagine that we see at the same Time all its Parts equally distinct and clear; But this is a vulgar Error, and we are led into it from the quick and almost continual Motion of the Eye, whereby it is successively directed towards all the Parts of the Object in an Instant of Time."[30]

> Thus in viewing any Word, such as MEDICINE, if the Eye be directed to the first Letter M, and keep itself fixed thereon for observing it accurately, the other Letters will not appear clear or distinct . . . Hence it is that to view any Object and thence to receive the strongest and most lively Impressions, it is always necessary we turn our Eyes directly towards it, that its Picture may fall precisely upon this most delicate and sensible Part of the Organ, which is naturally in the Axis of the Eye.

Building on Porterfield's work, Hartley analyzed how complicated motor skills such as learning to walk or playing the harpsichord are made possible by their becoming automatized as looser, more sophisticated "decomplex"[31] ideas that synthesize multiple long and different trains of sensory associations into flexible new unities. Decomplex activities, Richard Allen explains, are "continuous and self-sustaining processes that integrate sensory, ideational, and muscular activity" (170). They are epitomized for Hartley by the speaking, hearing, and writing of language. Thus, the body is mindful—not just directed by mind as an independent spiritual substance but able through training, conditioning, and education, to become a physical extension of mind and even, in a sense, to develop a mind of its own (albeit, *Songs of Experience* shows how social conditioning redirects this learning process in favor of mindlessness). Allen sums it up: for Hartley, "the achievement of language marks the emergence of those processes, so richly developed in humans, that we call mind."[32] Compare Alfred North Whitehead: "If we like to assume the rise of language as a given fact, then it is not going too far to say that the souls of men are the gift from language to mankind. The account of the sixth day should be written, He gave them speech and they became souls."[33]

Accordingly, Hartley emphasizes,

> Children may learn to read Words not only in an elementary Way, viz. by learning the Letters and Syllables of which they are composed, but also by a summary one, viz. by associating the

> Sound of intire Words, with their Pictures, in the Eye; and must, in some Cases, be taught in the last Way, i.e. wheresoever the Sound of the Words deviates from that of its Elements; so both Children and Adults learn the Ideas belonging to whole Sentences many times in a summary Way, and not by adding together the Ideas of the several Words in the Sentence. (1.273–74)

This synchronous mental activity is a kind of conditioned reflex, but it is far from the rote memorization that Blake's School Boy deplores (in, variously, *Innocence* and *Experience* both). Hartley's model of learning resembles the kind of now-called "analytical" inductive inferences based on probable belief that Thomas Bayes was then investigating in order to uphold divine Design against the problem of induction Hume had raised in his explosive essay "Of Miracles" (1748).[34] In Hartley, these inferences are commonsense, despite that their calculation in Bayes' paper as edited by Price permutes mathematically in ways that would have appalled Blake. (Likely, it was the Bayesian formalization of Hartleyan mental growth that provoked Blake's later comment on Bishop Watson: "Hartley a Man of Judgment then Judgment was a Fool" [E 610]). Decomplex ideas, which enable learning via processes of repetition and self-correction, trial and error, are adaptable and constitute a manmade second nature. In the pietistic Hartley's overall scheme of redemption, where individual mental development is writ large as the progressive improvement of society, these ideas provide the physiological basis for human culture and art.[35] Blake's indebtedness to Hartley can be seen as the positive side of his parodies, in several Songs, of children's chapbooks by moralizers like Isaac Watts, Anna Laetitia Barbauld, and Hannah More.[36] No doubt, his interest in Hartley was reinforced by his proximity to the unprecedented experiment in childcare taking place at the time right around the corner from his Broad Street flat near Golden Square. As Stanley Gardner's research has shown, in this parish, the enlightened Governors of the Poor were beginning to establish provision of nursing for destitute infants. In 1782, they founded, in place of a workhouse, the King Street charity school—its haberdashery supplied by James Blake, the poet's father.[37]

Before continuing with Hartley, let us briefly track back to Hume's essay "Of Miracles" and see how it provides an unlikely context for the fractal visions of *Songs of Innocence*. Blake's acid Annotations to Watson indicate familiarity with Price's "On the Nature of Historical Evidence and Miracles," which draws heavily on the Deistical Bayes' probability theorem to refute Hume and atheism.[38] And yet it was Hume himself—Bayes' antagonist—who first framed the problem of miracles in Bayesian mathematical terms: "We must balance the opposite experiments, where they are opposite, and deduct the smaller number from the greater, in order to know the exact force of the superior evidence . . . the evidence, resulting from the testimony, results in a diminution, greater or less, in proportion as the fact is more or less unusual."[39] Hume plainly thinks the probability of miracles approaches zero. In "Of Miracles," he sarcastically defends "our most holy religion" against "dangerous friends or disguised enemies . . . who have undertaken to defend it by the principles of human reason" rather than "*Faith*"—namely, Royal Society divines and physico-theologians like Price and Priestley. He writes, "we may conclude, that the *Christian Religion* not only was at first attended with miracles, but even at this day cannot be believed by any reasonable person without one. Mere reason is insufficient to convince us of its veracity: And whoever is moved by *Faith* to assent to it, is conscious of a continued miracle in his own person" (*Enquiries* 130, 131; Hume's italics). Which is just what Blake with evangelical enthusiasm *affirms*: "Jesus could not do miracles where unbelief hinderd hence we must conclude that the man who holds miracles to be ceased puts it out of his own power to ever witness one" (E 616).[40] The paradoxical empiricism of Blake's position mirrors Hume's in reverse: to those who experience them, miracles are indeed certain, thus, their probable reality is infinite. The empiricist shrug with which Blake ends his discussion of miracles in the Annotations to Watson—"look over the events of your own life & if you do not find that you have both done such miracles & lived by such you do not see as I do" (E 617)—seems a recollection of Hume's take-it-or-leave-it following his annihilation of personal identity: "If any one upon serious and unprejudic'd reflexion, thinks he has a different notion of himself . . . [a]ll I can allow him is, that he may be in the right as well as I, and that we are essentially different in this particular" (*Treatise* 252). If Anselm's ontological proof of God as "that than which no greater can be conceived" has the effect of making the conceptual turn real before our very eyes—for, every time we try to out-imagine God's existence, he is already there—then Blake brings Anselm's deity further down to earth by psychologizing him, so bypassing Hume's objection in the *Dialogues concerning Natural Religion* that existence is not a predicate in favor of something very like present-day anthropism. To put it differently, Hume and Blake stand agreed that a miracle is "a violation of the laws of nature" (*Enquiries* 114) but draw opposite conclusions from it. Hume infers miracles can only be explained in subjective terms through "feeling" and "human nature"; Blake, that the implicit dualism underpinning the so-called "laws of nature" must be revised to include a non-Newtonian fourth dimension glimpsable through higher, self-annihilating Vision. As for Price's and Bayes' providential Design, Hume and Blake alike dismiss it as a tainted religious half-measure rooted in the failure to take epistemological doubt seriously.[41]

In the context of *Songs of Innocence* and Hartley, the convoluted eighteenth-century debate about miracles perhaps reduces to the simple Romantic point that children grow and learn through a "miraculous" combination of credulity, imagination, and practical (as opposed to deductive) reasoning. But the debate is worth pursuing because Blake's idea of miracles underwrites processes of imaginative projection that *Songs of Innocence* aims to instantiate and arouse. Against Hume's argument that ordinary reality and testified miraculous fact make up "two opposite experiences" (*Enquiries* 113), such that the testimony necessarily presents a full-blown epistemic barrier, Price had argued "a miracle is more properly an event different from experience than contrary to it" because "Testimony is truly no more than Sense at second-hand" (*Four Dissertations* 402, 416).[42] Blake embraces Price's point but then stretches "Sense" in Swedenborgian fashion to include "Spiritual Sensation" (E 703). Examined in its minute particulars, *every* event is "different from experience" because its very existence is necessarily new and unprecedented, in consequence of the underlying reality of temporal passage and becoming without which there could be no change. Thanks to passage, change, and "Times swiftness" (*M* 24:73; E 121), miracles occur all the time, although they mostly go unregarded.

After a century of relativity theory, it appears to be Blake, not Hume or Price, who has the better of this argument. Whereas Hume follows Berkeley in defining "laws of nature" subjectivistically as nature's observed constancy and regularity, Blake takes Hume's psychologizing empiricism a step further and self-reflexively recognizes nature's constancy, regularity, and ostensible design as no more than general macroscopic appearances arising from human habit, custom, and the ideology of mechanistic materialism itself, all alterable historical processes. In other words, man chooses what counts as laws of nature and what counts as those laws' miraculous contravention. On this view, Hume arrived at his atheistic stance because paradoxically he agrees with Blake's claim: "The manner of a miracle being performd is in modern times considerd as an arbitrary command of the agent upon the patient but this is an impossibility not a miracle" (E 616–17). The difference is that Hume lacks the revolutionary historicist awareness implied by Blake's qualifier, "in modern times" (compare *The Marriage*'s "that calld Body is a portion of Soul discernd by the five Senses. The chief inlets of Soul *in this age*" [*MHH* 4; E 34; my italics]).[43] Indeed, in *Songs of Innocence*, it is the outwardly most insignificant events that best serve as wonders, marvels, and miracles because the very act of detecting them expands human vision to a limit where imagination takes over: "Is it a greater miracle to feed five thousand men with five loaves than to overthrow all the armies of Europe with a small pamphlet. [L]ook over the events of your own life" (E 617). At this threshold, minute events *become* miracles simply by being apprehended as such against all odds, through the kind of bootstrapping self-consciousness Angela Esterhammer identifies as Blakean prophecy's "phenomenological performative" whose force derives from "an author's ability to 'create' reality through poetic or fictional utterance."[44] The epigraph to *Milton*, "Would to God that all the Lords people were Prophets. Numbers XI. Ch 29 v." (*M* 1; E 96), invites readers to envision prophecy's performative capability for themselves. The real miracle lies in people's inborn imaginative ability to apprehend one in the first place through fined-tuned reading of their natural surroundings: the sort of reading promoted by *Songs of Innocence*. The testimony of eyewitnesses and Deistic reasoning about probability are beside the point because the only believable miracles are the ones you see for yourself. "Therefore God becomes as we are, that we may be as he is" (*NNR* [b]; E 3).

As we might have expected, Newton's disciples Bentley, Clarke, and Whiston defended miracles on "Bayesian" grounds. Instead of being outright contraventions of natural law, miracles were simply, in the tradition of Augustine, "*something strange and difficult which exceeds the expectation of him who marvels at it.*"[45] Like Newton, these writers refrained from assigning mechanical causes, instead limiting themselves to observational effects. The paradox is that Blake doesn't reject this approach but rather carries its underlying phenomenology to a relativistic extreme. A miracle originates in the observer as an outward manifestation of indwelling divinity. Its perception therefore constitutes the passing of a Last Judgment.

And it is here Blake's position most closely resembles Hume's. They both affirm the power of subjective belief and feeling to rescue man from the paralysis of rationalist doubting and ratiocination. Hume figures his infinity limit negatively as Pyrrhonian skepticism—an abyss of "melancholy and delirium" from which he is famously saved, as we've seen, by the social pleasures of backgammon and dinner with friends (*Treatise* 269). Blake's limit is the inverse of this, a divine revelation that annihilates all conventional worldly enjoyment in a private and unconditional Last Judgment. For Blake, society is already so thoroughly infected with skepticism that Hume's escape back into it merely evades the crisis he nearly, by great effort, managed to confront. (Hume portrays the crisis in amusingly Shandean terms but it likely has autobiographical roots in his nervous breakdown a decade earlier.[46]) Hume stood on the brink of dialectical insight into both the self-contradictory nature of "the ratio of all we have already known" (*NNR* [b]; E 2; E 659) and, on the other hand, the authentically Contrary nature of philosophizing as a "Creat[ion]" of "System[s]" (*J* 10:20; E 153). But he pulled back. Hume would reply, undialectically, that the immoderateness of Blake's position renders the extremity of its truth no less improbable than Pyrrhonian skepticism's extremity of abstract self-doubt.

Hartleyan "Decomplex" Ideas in Reading

Hartley himself emphasizes how different his decomplex ideas are from reflection, defined by Locke as "*internal sense*," "the perception of the operations of our own mind within us."[47] This definition paved the way for Hume's epistemological doubt in the first place. Locke's "error," according to Hartley, was that "he called such Ideas as he could analyse up to Sensation, Ideas of Sensation; the rest Ideas of Reflection, using Reflection, as a Term of Art, denoting an unknown Quantity" (1.360). For Hartley as for Blake, Locke was a sensationalist. The mind, says Locke, royally reviews the testimony of the senses brought for an "audience" in its "presence-room" (1.149). Hartley objects: "But this is like supposing an Eye within the Eye to view the Pictures made by Objects upon the *Retina*" (1.379). If the mind possesses senses of its own with which to peruse the "testimony" of the bodily senses, then to avoid Locke's regress *ad infinitum* the mind must be considered as an embodied, organic extension of the nerves of the body itself.

In short, Hartley's system demonstrates how the mind can, by effort and practice, adapt and condition the body to its own higher purposes and ends, not by despotically forcing the body to act but by physically altering it:

> After the Actions, which are most perfectly voluntary, have been rendered so by one Set of Associations, they may, by another, be made to depend upon the most diminutive Sensations, Ideas, and Motions, such as the Mind scarce regards, or is conscious of; and which therefore it can scarce recollect the Moment after the Action is over. Hence it follows, that Association not only

> converts automatic Actions into voluntary, but voluntary ones into automatic. For these Actions . . . are rather to be ascribed to the Body than the Mind. (1.104)

Decomplex ideas are therefore a kind of lived psychological illusion, like the unitary objects effectuated by Porterfield's saccades, or indeed like personal identity according to Hume. In this way, Hartley appears as a key source for Blake's identification of Jesus with the faculty of Imagination and the mind's power to realize its dreams to create novelty and growth beyond mechanism. In the close-readings of Blake's Songs that follow, I argue that the residual traces of Hartleyan decomplex ideas—"the most diminutive Sensations, Ideas, and Motions, such as the Mind scarce regards, or is conscious of; and which therefore it can scarce recollect the Moment after the Action is over" (1.275)—are heard in the form of receding verbal echoes and seen in the form of the tiny human figures that disport themselves about the words of the volume's titles and texts. For, says Hartley, "the Words in passing over the Ear must raise up Trains of visible and other Ideas by Association" (1.275). These secondary ideas associated with language are "the Miniatures excited in [a Man's] nervous System by the Word" (1.284). That so many of the little human figures scattered around Blake's pages are piping, reading, and attending or listening indicates the essentially dynamic nature of his phenomenal "space of reading." Hartley points out that even words read in silence have a physiological effect upon the body—thus, he anticipates the discovery of subvocalization—and he portrays the interaction between text and reader as recursive: "In passing over Words with our Eye, in viewing Objects, in Thinking, and particularly in Writing and Speaking, faint Miniatures of the Sounds of Words pass over the Ear. I even suspect, that, in Speaking, these Miniatures are the associated Circumstances which excite the Action" (1.234). In *Songs of Innocence*, these diminishing vibrations of the image and its multiple associations imply the presence of an underlying ideality beyond all such imagery, a vanishing point of calm silence that constitutes revelation of an intense self-reflexivity not of this world—namely, as my later chapters will argue, the spatiotemporal fourth dimension of Jesus the Imagination.

However, unlike Hartley and Priestley—scientific reductionists, for all their utopian idealism—Blake does not locate mind's creative power somewhere in the physical brain in the certainty that better microscopes or closer analysis might someday render it observable. Yet neither does his position resemble that of Thomas Reid, one of Hartley's chief critics. Reid notes that what Newton merely infers in the Queries concerning the vibrations of aether around the optic nerve, Hartley purports to have demonstrated despite having no actual evidence for vibrations, vibratiuncles, or, indeed, aether; thus, "this system of vibrations [tends] to make all the operations of the mind mere mechanism."[48] And Coleridge later followed suit, notably in *Biographia Literaria,* chapter 6. Both men worried about the religious implications of what looked to them like a revolutionary materialist threat to the freedom of the will. What this sort of dualist objection misses—with respect to Blake, if not Hartley—is that Blake's adapted vibratiuncles are not particles vibrating within a given space. Rather, the vibratiuncles *are* the space; they constitute the space-time event of the vibration itself, which arises and passes without leaving behind anything more substantial than a new potentiality, namely, an attunement or disposition of the nerve to vibrate in accordance with similar vibratiuncles in the future. That is how learning occurs, and the miniaturized figures in Blake's *Songs* designs bring this process to light. It is our own perceptions that constitute the seemingly given and external space in which those perceptions are perceived to exist. "Vibratiuncle" is an idealization of a physical process. No two are identical, and there is not even a "same" vibratiuncle from one instant to the next. Implicit here is a pulsational universe. Consider Whitehead again: "In the process of self-creation which is an actual entity the genetic passage from phase to phase is not in physical time . . . the genetic process is not the temporal succession. . . . Each phase in the genetic process presupposes the entire quantum."[49] Already in *Songs of Innocence, Milton*'s cosmic space-time Moment of apocalypse and Judgment awaits unfolding.

Returning to "A Cradle Song," it becomes apparent that the relationship between the song and the unusually large picture at the end is strangely inverted. The picture is a solid externalization of the song's indefinite, in-between dimensionality based on associations of ideas. Stanza one's "dreams" are partaken by mother and infant alike. So, when the mother requests a dreamy angel of sleep to "Hover o'er my happy child" (l. 8; E 11), that angel implicitly includes herself gazing down at her infant (in stanza 3, the "smiles" invited to "Hover over my delight" are designated "Mothers smiles"). The purpose of this tertiary angel of sleep—a presence not literally inside the scene with the two humans but rather the immanent image of their shared intimacy—is to include the reader as a participant. This is seen in the last three stanzas, which transact a series of shifting identifications between mother and child that disclose a "Holy image" (l. 22) belonging to both and neither, a "Heavenly face that smiles . . . / . . . on thee on me on all" (ll. 28–29) from a displaced position that is *not* on high. The repeated "all" sweeps up the (already drowsily susceptible) reader into an apprehension of a universal deity made incarnate through human love and connectedness. To be sure, this is also an ideological seduction based on devout, perhaps Moravian belief, but it is a seduction in which the reader imaginatively cooperates—or else misses the whole experience.

Blake emphasizes the precariousness of his sleep angel by allusion to the Christ story, which continually threatens to turn the scene into a mere allegory of Nativity ("infant crown" = Crown of Thorns; "dovelike sighs" = Dove of Peace; "o'er thee thy mother weep" = Christ's battered body taken down from the cross; "All creation slept and smild" = nature's universal Peace in the moment of Christ's birth, per Milton's Nativity Ode). The song only tentatively "beguiles" its traditional religious overtones. Really, it could not exist without them since they supply the ancient materials it remakes and brings into the here and now. By constantly reassimilating orthodoxy's literalisms into the now of co-presencing, Blake's human form becomes divinely "decomplex." The poem's temporal instability is dramatized by the uncertainty of its grammar. Stanza 5 offers a vivid instance:

> Sleep sleep happy child.
> All creation slept and smil'd.
> Sleep sleep, happy sleep,
> While o'er thee thy mother weep. (E 12)

If the mother speaks the present-tense first line, what voice is it that narrates the past-tense second line? Our hesitation carries over into the third, which seems to lack a viable subject. Perhaps "sleep" has become a metonymy for "child," or perhaps the mother is now talking soporifically to herself as she, too, falls asleep. Either way, it seems "sleep" is no longer an imperative verb, as in line 1, but a noun welcomed with gratitude and relief. Thus, verb, subject, and object are confounded together. Yet they seem reseparated in line 4, which restores the mother to her watcher role—except that the third person here seems a continuation of line 2's descriptive narrative, even as the vocative case glances toward a future act. Indeed, by the stanza's end, it is retrospectively possible to hear every one of the four lines, including the first, as channeling some voice not the mother's own. The now of song dilates thickly to encompass past and future. The present proves to be, in William James's phrase, an extended, noninstantaneous "drop or bud of experience,"[50] an anticipation of *Milton*'s apocalyptic "Pulsation of the Artery" dense with all of history (*M* 29:3; E 127).

What the picture at the bottom of the second page shows, therefore, is the lapsing of the song's delicate spiritual "hover" back into a stance of traditional religious reverence. Evidently, the song and design portray "the Two Contrary States of the Human Soul" (*SI*; E 7) as sharing a single continuum. Each begins where the other leaves off. I noted earlier that the curtain forming the backdrop to the "Cradle Song" picture is literally pegged to the text plane hanging in the middle distance, which it covers over. So, one might imagine the song as continuing blissfully down to the bottom of the page, its labile self-presence invisible and inaudible from a three-dimensional perspective. What the picture mockingly shows in place of this imagining is the form of our own pious alienation from the song's unifying angel of sleep. Mother and child form a Nativity vignette (the babe's pillow resembles a halo, badge of spiritual aura), while viewers find themselves in the position of the shepherds peeping in voyeuristically from the "outward circumference" (*MHH* 4; E 34). Whereas the song induces participation in the human form divine, the picture conventionalizes the invisible Jesus as a domestic bourgeois relationship. The preponderance of woven objects in the room suggests a hardening of text into textile, feeling into ritual. In this light, the rattan cradle recalls the ark of woven bulrushes that contained the infant Moses, the giver of Law from which Jesus supposedly set us free. In a sociocultural sense, then, contemporary reader-viewers of Blake's Song are back where they started. They have passed through the poem's energetic Vortex of displaced and "self-annihilated" (Hartley 2.280–82) feelings and reemerged into the familiar world of Euclidean perspective, representational realism, and moral religion.[51] As Daniel Heller-Roazen points out, "There are many ways not to be someone." The "nonperson," "absentee," or "missing person"—a group that includes "lesser persons" and those officially or unofficially disappeared from public view, as well as the dead—is "not external to the category of person, but internal to it. This 'nonperson' names the depletion of the notion [person] to which it is bound."[52] At the opposite end of this Urizenic limit, Jesus—Blake's name for anonymous, nonself-centered human connection and caring, a name pointedly never uttered in *Songs of Innocence*—signifies a corresponding expansion of "nobody" beyond everyday public reality. Together, "A Cradle Song" and its design portray personhood as, not an essence, but a potentiality wobbling vulnerably between Contrary poles.

Even so, potentiality can be cultivated and taught. Blake's "Introduction" affirms a steady Hartleyan progression from inarticulate music to sung words to the "decomplex" activities of writing and reading. Blake acknowledges Hartley's point that language acquisition begins with hearing rather than reading: "Words heard, and their audible Ideas, have a prior Claim" over "the Pictures of Words in the Eye, and their Ideas" (1.235). In the end, the piper's "hollow reed" pen transmits his music by synchronously joining together hearing and seeing in reading, as in Hartley's scheme (*SI*; E 7). Thus, the poem portrays the "vanish[ing]" of song and vision into writing and reading. The tearful joy of the angel child who "wept to hear" the piper's singing is similarly carried over into language via the medium of water and ink. In other words, the angel's vanishing is a precondition for the emergence of a silent virtual audience beyond physical song. In *Observations on Man*, Hartley describes "the manner in which the first Rudiments are laid of that faculty of Imitation, which is so observable in young Children. They see the Actions of their own Hands, and hear themselves pronounce. Hence the Impressions made by themselves on their own Eyes and Ears become associated Circumstances, and consequently must, in due time, excite to the Repetition of the Actions" (1.107). Blake's poem brings out the self-reflexive nature of such repetition, and its difference from rote.

That said, unsettling aspects of "Introduction" have not escaped notice. The poem has long been seen to depict a fall from the immediacy of the initial vision ("On a cloud I saw a child") into the compromised "stain" of writing, whose "hollow" technology the piper elaborates in detail. Hence, perhaps, the tentativeness of the closing line, "Every child may joy to hear" (though more heavy-handed auxiliaries like "must," "shall," "will," or "should" do not sound inspiring).[53] This chapter argues the opposite. For Blake as for Hartley, intense, self-reflexive mediation—the mediation provided by illuminated printing, with its implicit invitation to physical movement and change—practically *is* vision. Blake gave his lead poem the bookish title of "Introduction" to suggest that *Songs of Innocence*, with its lavishly detailed designs, is an even richer material presence than audited song, notwithstanding the latter's sentimental charm and immediacy. The title advertises the anthropologist Alfred Lord's pioneering insight in 1960: "One cannot write song."[54] But, far from evincing Romantic nostalgia for oral culture, the poem attests Blake's Enlightened faith in the power of education. Unlike the balladry of Wordsworth, Coleridge, and Keats, the Songs of Innocence do not position themselves as belated adaptations of a bygone art. Scholars underestimate how much the ambitious and aspirational early Blake shared in the late Enlightenment's utopian ideology of progress through educational reform, an ideology very different from, and often at odds with, the working-class millenarianism with which his enthusiasm has been confused. (The confusion goes back to the millennialist-sounding rhetoric of Francis Bacon himself, whose sincerity and politics remain controversial.[55] If Blake deemed Bacon a "Machiavel" and denounced his *Essays Moral, Economical and Political* as "Good Advice for Satans Kingdom [E 620], arguably this was to distinguish the corrupt lord chancellor from his scientific project.) The Proverbs of Hell, "What is now proved was once, only imagin'd" (*MHH* 8:33; E 36), and "You never know what is enough unless you know what is more than enough" (*MHH* 9:46; E 37), express the same Enlightened Baconianism as Priestley when he argues in *History and Present State of Electricity*: "It may be said, that there is a *ne plus ultra* in every thing, and therefore in electricity. It is true: but what reason is there to think that we have arrived at it?"[56] Elsewhere, Priestley writes: "All things (and particularly whatever depends on science) have of late years been in a quicker progress toward perfection than ever.... [I]n spite of all the fetters we can lay upon the human mind . . . knowledge of all kinds will increase. The wisdom of one generation will be the folly of the next."[57] That last sentence recalls other Proverbs of Hell on the creative power of "excess," including, "If the fool would persist in his folly he would become wise" (*MHH* 7:18; E 36). Is Blake here suggesting that progress comes through the trial and error of scientific experimentation? Looking ahead to *Milton*, it seems clear Milton's journey back to the future of "1804" is an Enlightened rehabilitation project along the historicist path Priestley describes.

Blake's invention of etching in relief therefore suits the tradition of Bacon and the Royal Society much better than secretive dark arts like alchemy, with which it is sometimes linked. Relief etching combines the reproducibility of printing with the one-of-a-kindness of touching-up by hand, thus straddling the Platonic divide between presence and absence, aura and mechanical reproduction. The multisensory processes of association provoked by Blakean illuminated printing exemplify that organic symbiosis of mind and body which Hartley identifies with the decomplex motor skills of writing and reading.

In line with his sensational emphasis, Hartley argues that the priority of words heard over words seen "confirms . . . that in Writing one is often apt to mis-spell in Conformity with the Pronunciation, as in writing *hear* for *here*; for this may proceed from the audible Idea, which is the same in both Cases; cannot from the visible one [sic]" (1.235). Blake's "Introduction" deliberately rehearses this point. The repeated chear/hear rhyme instantiates the middle stanzas' idea that hearing the song is cheering, while hinting that the hearing is here, i.e., it occurs performatively through the text whose visibility generates echoes subtler than audible rhyme. The same holds for the more ominous (to some critics) "hollow reed"/"may joy to read" rhyme, with its apparent Platonic implication that reading lacks the presence of song. Regarded through the educational lens of Hartley, the reed/read homonym complicates the vowel sound of the chear/hear rhyme by extending it into the silence of reading, where the difference between the two words is first seen, then heard as a miniature idea in the mind's ear, as opposed to being sensed physically. One hears the difference in chear/hear while seeing the two words' anagrammatic similarity. One *sees* the difference in reed/read while *hearing* the two words' similarity. These recursive patterns of difference-in-similarity and similarity-in-difference can be extended still further, *à la* Derrida, though they soon lose immediate relevance to Blake. For example, the word, "read," is *self*-different when it is read—if you see, or rather hear, what I mean; and when correctly audited, its sound clinches the overall meaning of Blake's sentence. And notice how the spelling of "chear" calls attention to the "ear" in its rhyme—another visible instantiation of hearing the poem's cheer. Et cetera.

This is to say that Blake's "Introduction" employs Hartleyan associationism to depict writing and reading as vehicles of a generative power far greater than its creative manifestation in any one poem, line, or word. "This coalescence of simple ideas into complex ones may be illustrated, and farther confirmed," says Hartley, "by the similar Coalescence of Letters into Syllables and Words, in which Association is likewise a chief Instrument" (1.75). Richard Allen rightly stresses, "Hartley is not advocating here a brickbat plan of construction."[58] Rather, Hartley asserts, "the decomplex Idea belonging to any Sentence is not compounded merely of the complex Ideas belonging to the Words in it" (1.79), any more than the meaning of a word is made up of simple ideas of the letters that compose it. Far from being deconstructive, the word play of "Introduction" evokes Hartley's constructive speculations concerning language as "one Species of Algebra" (2.280). In the tradition of Baconian Royal Society projectors like John Wilkins, who sought to recover or reinvent "the language of Adam" the better to promote the Society's goal of "Physico-Mathematicall-Experimentall Learning,"[59] Hartley dreams of the "Happiness" afforded by "a truly philosophical Language" whose users would be "capable both of expressing their own Feelings, and of understanding those of others" in order to give "new Senses and Powers of Perception to each other" (1.320):

> Let *a, b, c, d,* &c. the several Letters of an Alphabet . . . represent respectively the several simple sensible Pleasures and Pains, to which a Child becomes subject upon its first Entrance into the World. Then will the various Combinations of these Let-

ters represent the various Combinations of Pleasures and Pains, formed by the Events and Incidents of human Life; and, if we suppose them to be also the Words of a Language, this Language will be an Emblem or Adumbration of our Passage through the present Life; the several Particulars in this being represented by analogous ones in that. (1.318–19)

That Blake drew lifelong inspiration from this passage is evident from its echoes in *Jerusalem*'s concluding vision of a redeemed language of "Visionary forms dramatic" able to communicate "Childhood, Manhood & Old Age," whose "every Word & Every Character / Was Human according to the Expansion or Contraction, the Translucence or / Opakeness of Nervous fibres" (*J* 98:28–37; E 257–58). Such a language is very different from Berkeley's static, God-given "language of nature" impressed upon the senses by habit and leading to appreciation of nature's underlying order and regularity. Berkeley's language of nature forms part of his argument for understanding divine Design. In contrast, Hartley's philosophical language is subjoined to a model of child development with the goal of mutual human understanding, as epitomized by the caring and compassion between mother or nurse and child. Hartley likens pronouns and particles to "the unknown Quantities in Algebra." For the child, they "answer, in some measure, to x, y, and z" (1.274). Blake takes the hint and portrays key words in *Songs of Innocence* like "duty," "angel," "weep," "pity," "joy," "bless," "call," "lamb," "night," "day," as elements of a Hartleyan "truly philosophical Language." The defamiliarized, ironical innocence-in-experience of these terms turns them into placeholders for "unknown Quantities" whose interrelations paint, in Hartley's phrase, "an Emblem or Adumbration of our Passage through the present Life," thereby enabling child and adult readers to impart "new Senses and Powers of Perception to each other" (1.319–20).

"Secondary Vocality" and Potential Association in *Innocence and Experience*

The notion of written language as a decomplex "Coalescence of Letters into Syllables and Words" (1.75) synthesized by supplemental associations that differ from one person to another and also, in virtue of their cumulative nature, from one reading to the next, may be regarded as an eighteenth-century anticipation of what Garrett Stewart calls "secondary vocality":

"writing that speaks through rather than to you," in "a force field of [words'] unique 'reverberation' as a resonance phonetic before [it becomes] referential or thematic."[60] The inscribed past of the text is made recurrently present through the wavering polyvalence of meaning that prevails from word to word as a result of what Stewart's guiding light, Giorgio Agamben, calls "the tension and the difference . . . between sound and sense, between the semiotic and the semantic sphere."[61] In *Songs of Innocence*, the shimmering aura of Blake's text planes, which are at the same time text *panes* the reader looks through to indefinite morning skies, evokes the backdrop of language itself as prior to any of its actualizations as meaning or message—as an inexhaustible semiotic generativity that suggests a realm of human presence beyond three-dimensional visualization. Hence, Blake's enigmatic remark to Crabb Robinson that he wrote only when his angels so commanded him, "and the moment I have written I see the words fly about the room in all directions."[62] As I'll argue in chapter 5, this is Blake's apprehension of Plato's *khora* or Receptacle—the prerational, spaceless "place of things" which provides, in Agamben's words, "an experience of language . . . at the limit of signification, . . . language's bare giving of itself" as "a pure quantum of signification, which, however, does not signify a thing or a concept, but only the giving itself, the pure 'taking place' of something."[63]

To take up again Hartley's point that, for children, pronouns and particles "answer, in some measure, to *x, y, and z*" (1.274), we notice that the title of "Nurse's Song" lacks a definite article. Hartley explains that children "are much at a loss for the true use of the Pronouns and Particles for some years, and . . . they often repeat the proper Name of the Person instead of the Pronoun" (1.274–75). Accordingly, Blake's title portrays the relationship with Nurse from a child's viewpoint as concrete and specific rather than typical or class based. Such concreteness is inherent to the process of association itself as Hartley conceives it:

> The Name of the visible Object, the Nurse, for Instance, is pronounced and repeated by the Attendants to the Child, more frequently when his Eye is fixed upon the Nurse, than when upon other Objects, and much more so than when upon any particular one. The Word *Nurse* is also sounded in an emphatical manner, when the Child's Eye is directed to the Nurse with Earnestness and Desire. The Association therefore of the Sound *Nurse*, with the Picture of the Nurse upon the *Retina*, will be far stronger than that with any other visible Impression . . . And thus, at

last, the Word will excite the visible Idea readily and certainly. The same Association of the Picture of the Nurse in the Eye with the Sound *Nurse* will, by degrees, overpower all the accidental Associations of this Picture with other Words, and be so firmly cemented at last, that the Picture will excite the audible Idea of the Word. (1.271; his italics)

In time, other sensory ideas will come to be associated with the picture and the sound of the word, supplementing and modifying the original, primarily visual association: "Thus an Idea, or nascent Perception, of the Nurse's Milk will rise up in that part of the Child's Brain which corresponds to the Nerves of Taste, upon hearing her Name. And hence the whole Idea belonging to the Word *Nurse* now begins to be complex, as consisting of a visible Idea, and an Idea of Taste" (1.272). The same process holds for "The Words denoting the sensible Qualities, whether Substantive or Adjective. . . . Thus the word white, being associated with the visible Appearances of Milk, Linen, Paper, gets a stable Power of exciting the Idea of what is common to all, and a variable one in respect of the Particularities, Circumstances, and Adjuncts" (1.273). Almost paradoxically, the very elusiveness of the associations conveyed in Blake's "Nurse's Song" find a source in Hartley's rich appreciation of the preverbal child's naïve appetite for activity and growth. If Hartley's model of the child's mind is simple and unsubtle, it is also dynamic, cumulative, and synthetic to the point of synesthesia.

Associations of whiteness indeed ramify, crisscross, deepen, and reverberate across different Songs of Innocence in much the way Hartley describes. Lambs and sheep, wool, the skin color of the Little Black Boy's young English friend, Tom the Sweep's hair "that curled like a lambs back" (E 10), God who "appeared like [a] father in white" in "The Little Boy Found" (E 11), the translucent-looking fairies in the designs to several copies of "Night," various shinings of light (white in color, according to Newton): all this imagery combines with conspicuously uncolored portions of the printed white pages themselves, which in some copies compel a self-conscious, almost modernist recognition of the processes of representation, not only that of the artist giving form to his materials, but also the viewer's perceptual formation of a three-dimensional image from hints on a flat surface. Conversely, the foliage and tendrils snaking their way into the stanza breaks in most of the designs invite reader-viewers to project themselves visually into formal spaces normally left blank by typographic convention. The spiritual origins of such self-projection are seen throughout the volume in child speakers' repeated identifications with the lambs and sheep they see (most obviously, in "The Lamb" and "Spring," ll. 19–22) and by their mothers' and guardians' unself-conscious identifications with their children ("A Cradle Song," ll. 21–28, "A Dream"). "Spring" illustrates this process by way of Hartley's account of how the child learning to walk is assisted by adults but motivated by innate "Desire": "If he be set upon his Legs, and his body carried forward by the Nurse, an imperfect attempt to walk follows of course. It is made more perfect gradually by his Improvement in the Rudiments, by the Nurse's moving his Legs alternately in the proper Manner, by his Desire of going up to Persons, Playthings, &c." (1.257). In Blake's design, the child held up by his nurse or mother recognizes the nearby flock of sheep as natural playmates.

But this describes only one minor subset of associations in the volume. Children in *Songs* are not only lambs but "tender plants" (E 31) and "flowers of London town" (E 13), even as they also appear *within* flowers (in the design of "Infant Joy"). And yet their maternal guardians are, as well, compared to flowers. For example, the Blossom-bosom shelters the small birds of "The Blossom"—one a mother, one a child—similarly to the design to "Infant Joy"; or the watchful Moon in "Night," "like a flower, / In . . . [a] bower" gazes upon "each bud and blossom, / And each sleeping bosom" (E 14), similarly to the mother of "A Cradle Song." Overall, the volume's imagery compares children both to themselves and to their adult guardians. Further, "The Blossom" intimates crosshatched differences-in-similarity between, on the one hand, its happy sparrow and sobbing robin, and on the other hand the smiling, moaning infant and smiling, tearful mother in "A Cradle Song." In turn, "A Cradle Song" and "Infant Joy" may be regarded as human translations of the mother robin's song in "A Blossom," renaturalized elsewhere as Tom the Sweep's birdlike cry of alienation, "weep weep weep weep" (E 10). Sometimes Blake's children act *in loco parentis*, as when the child narrator in "The Chimney Sweeper" comforts little Tom; at other times, the parent figure is a persona internal to the child, as when the prophetic Glow Worm lights the way home for the bewildered child speaker of "A Dream." Different Songs reconfigure one another's imagery kaleidoscopically on a variety of different scales in fractal fashion. The metaphors metamorphose.

The panexperientialism implicit in this rendering of every universal as a particular, and vice versa, perhaps owes something to Blake's quondam geometry tutor, Thomas Taylor, whose later *Elements of the True Arithmetic of Infinites* (1809) declares contra Natural Religion: "As one of the principal discoveries in this treatise is, *that in every infinite series of terms, whether integral or fractional, the last term multiplied by the number of terms is equal to the sum of the series*, I rejoice to find as the result of this discovery, that it affords a most splendid instance of the absurdity which may attend reasoning by induction from parts to whole, or from wholes to parts, when the wholes are themselves infinite."[64] To this Proclus-like affirmation of the One as underlying all its many emanations, Blake could add that the reality of fractals demonstrates the parts themselves to be wholes, and thus infinite. Even so, the formal, Contrary relationship of parental guardian and child persists in *Innocence* as an ontological structure able to keep the energetic ongoing self-presence of each different song from radiating out across the entire volume and undermining all the others by merging figure and ground perspectivistically. Such a collapse would indeed resemble the outcome Coleridge extrapolated from Hartley's theory of mind in *Biographia Literaria*: "absolute *delirium*," where "*any* part of *any* impression might recal [sic] *any* part of any *other*, without a cause present to determine *what* it should be."[65]

That said, the aging Trinitarian author of the *Biographia* was surely trimming when he portrayed Hartley as limiting the mind to entertaining passive "trains" of ideas like beads on a string. He knew better twenty years earlier when, like Blake, he regarded Hartley's theory as empowering a revolutionary reformation of society. The reason "*any* idea" does *not* "lead to any *other*" in *Songs of Innocence* is that so many of the ideas are not discrete and immediate but Contrary, cumulative, and "decomplex." Blake is at great pains to promote the reader's voluntary, self-conscious participation in the processes of association, consistent with Hartley's observation that when we read, "Even visible Trains do not appear as Objects of Consciousness and Memory, till we begin to attend to them, and watch the evanescent Perceptions of our Minds" (1.235). This is strikingly the case when a Song's imminent ending and the reader's corresponding urge to find satisfying closure demand some sort of synthesis, however partial, of the whole preceding sequence.

Consider, in this light, stanza 3 of "Nurse's Song" and its children's bedtime cry of protest:

> No no let us play, for it is yet day
> And we cannot go to sleep
> Besides in the sky, the little birds fly
> And the hills are all coverd with sheep. (E 15)

The last two lines convey, perhaps, a pantheistic image of hillsides wrapped in "Softest clothing, wooly bright" ("The Lamb"; E 8). The birds soaring above complement the peaceable lambs below in a suggestion of heaven on earth. Further, when "all the hills ecchoed" the children at play in the poem's last line, it is conceivably because the birds have served as vehicles to carry the sound through the air. On the other hand, since we learn elsewhere that the Lamb has "a tender voice, / Making all the vales rejoice" ("The Lamb"; E 8), those echoes may originate in the bleating hillside sheep as well as in the children (much as "The Lamb" speaker's identification with the Lamb, his addressee, allows the closing refrain to be heard as liturgical call-and-response between them). Sheep give milk, and so do nurses, Hartley points out; so maybe the surrounding sheep-covered (soft and breast-like?) hills in "Nurse's Song" image the watchfulness of Nurse, who in the design sits at the edge of the children's play even as visually she completes the circle they form. As viewed from those hills, the whole group might indeed resemble a small flock.

If these micro-responses to stanza 3 appear farfetched, they at least attest how the Song is not exclusively, or even primarily, Nurse's own. The voices and moods of Nurse and the children are projected throughout the poem into their surroundings and one another. Nurse's conventional mode of self-address in stanza 1 takes place in a generalized continuing present tense that evidently reflects her peaceful assimilation of the eternal now of children at play. The first line's "When" signifies whenever, as opposed to some definite point in time. The passive verbs suggest that her auditing of the children is a reverie, not voluntary, attended perception. In stanza 2, however, the latent chronological meaning of "when" suddenly activates. Nurse is startled to remember that time has been passing ("Then"), darkness is nigh, and responsibilities beckon, so she calls the children home. No longer a mind in solitary reflection, she is concretely situated in the now of temporal passage. Accordingly, when the children reply to her in the above-quoted stanza 3, their voice bursts into the poem without any narrative mediation or framing description, and this immediacy then extends to the Nurse's reply in stanza 4.

To be sure, suddenly changed speakers and unexpected shifts from narrative to dramatized speech are common enough devices Blake, like other Romantics, adopted from the (often highly fragmented) ballads in Percy's *Reliques* (1765). But the shift to narrative past tense in "Nurse's Song" in the last two lines is much stranger:

> The little ones leaped & shouted & laugh'd,
> And all the hills ecchoed.

The lines glance at an impersonal perspective in counterpoint to the opening stanza's mellow mood of evening calm suffused with overheard sounds of children at play. One recalls the mysteriously detached past-tense comment in stanza 5 of "A Cradle Song": "All creation slept and smil'd." In contrast to Nurse's initial, invitingly inward Hartleyan state of "self-annihilation," the poem's closing lines perform an optical pullback. The whole scene loses immediacy and takes on the feel of a small glass-globe diorama held in the hand. Far from arriving at an end, Blake's threshold between night and day lingers indefinitely in the manner of a Wordsworthian spot of time that endures not beyond time but across it and so remains liable, potentially, to nightfall and the death of vision. As in the similar fade-away at the end of "The Ecchoing Green," the "Nurse's Song" scenario will soon be "no more seen" (E 8). Readers thus bear witness to their own exiting of the poem whose world seemingly goes on without them, like the song in the pegged-up text of the "Cradle Song" design. Hartley stresses that "both the diminutive declining Sensations, which remain for a short Space after the Impressions of the Objects cease, and the Ideas, which are Copies of such Impressions, are far more distinct and vivid, in respect of visible and audible Impressions, than of any others" (1.57). Similarly, at the end of "Nurses Song" we watch our visible and audible sense perceptions of the poem comingle and interpenetrate to form a distinct idea in the mind. Building on Hartley's account of how children synthesize decomplex new ideas out of old associations, Blake renders self-conscious his readers' retrospective attempt to unify their passage through the poem into an aesthetic whole—what Coleridge would call a Symbol. At the same time, it is just this slightly directed effort to achieve closure and return our reading to the wider world that cuts us off from the spontaneous ongoing activity of the ideal scene itself.

Thereby, the reader witnesses, in Garrett Stewart's words, "the vanishing . . . of somatic voice into meaning, silent sound into sense, organic impulse into uttered thought" (42). Agamben's essay "The End of the Poem," which Stewart is following, emphasizes that this return to the negativity of text, which constitutes a text's perennial availability as possible speech, restores at the same time a real but unactualized potentiality for rereading. In dramatizing the event of its own passing, Innocent song's subvocalized closure repudiates the nihilistic absencing at the base of Derridean "deferral" and resurrects the possibility of rereading as a real presence. This, presumably, is what Simon Jarvis and others mean by "thinking through poetry" or "prosody as cognition."[66] In offering his relief-etched pages as force fields dense with their own reiterability, Blake complicates the mind/matter dichotomy behind the eighteenth-century doctrine that ideas must be "present to the mind" according to the Newtonian principle, no action at a distance.[67]

One model for the faintly elegiac closing stanza of "Nurse's Song" would appear to be Milton's *Lycidas*, another first-person song suddenly interrupted by voices from without, above all by the much-remarked sudden change to third-person past tense in the final stanza:

> And now the Sun had stretched out all the hills,
> And now was dropped into the western bay;
> At last he rose, and twitched his mantle blue:
> To morrow to fresh woods, and pastures new.[68]

For Romanticists, it can be difficult to avoid reading *Lycidas* in the light of Hegel and Wordsworth's *Prelude* as a pilgrim's progress or *bildung* of the unripe poet who seizes the time and after much striving emerges, a butterfly, as himself. "Nurse's Song" portrays the realization of potential in similarly organic fashion, as a metamorphosis or flowering of a hidden germ. But it also transfers this process directly onto the reader, who thereby gains an empowering self-reflexive perception of the psychological mechanisms by which mind synthesizes trains of associated poetic ideas into a complex unity, the better to move on tomorrow to pastorals new. Other valedictory tailpieces in *Songs of Innocence* include the passive participle in the closing couplet of "The Ecchoing Green," with its implied disappearance of the grammatical subject. Also, the last lines of "Holy Thursday" and "The Chimney Sweeper," each freighted with the force of a remembered moral precept applied in reductive summation of the richly detailed narrative that went before. In "A Dream," the lost speaker's final restoration to Innocence does nothing to resolve mother Emmet's grief-stricken search for her (notably plural) children, which continues subaudibly beyond the ending of the poem. The missing children are still sought, thus not quite lost. The mother recalls Rachel weeping inconsolably for her murdered babies in Matthew 2:18, but her despair is mixed with the reader's self-conscious "hope against hope" (Rom. 4:18). These closures—quasiclosures—stage the vanishing of the subvocalized secondary event of reading and the concomitant, faintly ominous return to the primary silence and fixity of text as a recovery of the Contrary possibility of rereading. The dialogical "ecchoings" of different vocal moods and tenses in *Songs*, the hearing of one voice inside another in poems such as "Infant Joy," "The Lamb," and "A Dream" (whose second stanza through the phrase, "All heart-broke," can refer to the wayward dreamer and wandering Emmet both), constitute a literary adaptation of the Newtonian "fractal lattice" of matter that Hartley had applied to the psychology of association of ideas.[69]

The multivalences of *Songs of Innocence*—phonemic, lexical, syntactic—present not ideal alternatives to what exists, but rather real potentialities heard kinesthetically in the mind's ear and envisioned in the displaced human miniatures depicted throughout the designs. To put it differently, the endings of these poems constitute an awakening of eros in Whitehead's sense of "the urge toward the realization of ideal perfection,"[70] that feeling of wholeness or togetherness in experience which aims not to restore order conceptually in the face of change but to intensify order by integrating change. Underpinning these subvocal phenotexts lies an analogy between Boscovich's graph and Hartley's nerve fibers, whose vibratory response to a stimulus gradually dies down like a musical string to produce low-energy representations without "ecchoing"—like the inert naturalistic picture at the end of "A Cradle Song." Several Songs of Experience portray the world of Experience as a flattening and hardening not only of affect but dimensionality. "The Clod & the Pebble" satirizes Deism's *concordia discors* as a just-so story of "fitting & fitted" (E 667), where some in the three-dimensional world are designated to receive two-dimensional imprints (Clods) and others to give them (Pebbles): a sadomasochistic match made in Heaven. Similarly, the difference between the near-identical first and last stanzas of "The Tyger" underscores the distance traveled from sublime open wonderment to Manichaean horror: "Did he who made the Lamb make thee?" (*SE*; E 25). The forbiddingly presumptive "Dare frame"—a rhetorical question—exposes the poem's own symmetry as imprecise, a collapsing of the ideal and the actual. At a self-reflexive level, he who made "The Lamb"—"William Blake, *sculpsit*," as he often signed engravings—indeed made "The Tyger." But the final question bypasses the human smith's sculpting process; it identifies the Tyger with the hard, unalterable perfection of the incised copperplate regarded as an instantaneous materialization of a two-dimensional blueprint of divine Design, similarly to Urizen's "Book / Of eternal brass, written in [his] solitude" (*BU* 4:32–33; E 72). The Tyger's intermediate materiality as an illuminated print available for rereading is denied, as the picture's noticeably flat, two-dimensional animal goes to show—visibly, a *paper* Tyger, in contrast to the tactile, fleshly imagery of "The Lamb."

A more striking example of "Tyger"-like Newtonian *concordia discors* is the inward-curling dangling participle of "Ah! Sun-Flower." The opening apostrophe never arrives at its main verb but merely peters out in a series of modifying subclauses, in parody of the Sun-Flower's spiral of natural growth and energy. The speaker's lapse into weary religious "aspir[ing]" (E 25) invites comparison with the Vortex-like structure just discernible on *The Marriage of Heaven and Hell* title page, with an inert material world at the top and a delocalized energetic realm below (figure C.1). Both appear to be modeled after Boscovich's spiral graph of continuously alternating spheres or shells of attractive and repulsive forces running from infinity on the left to the stable Newtonian world on the right (figure 2.2). For, throughout *Songs of Experience*, culture and psychology are reduced to contact physics; "The Clod & the Pebble" is emblematic in this regard. By contrast, the "fractal" imagery of *Songs of Innocence*—the poems' asymptotic oscillations between mother or nurse and child, child and lamb, child and older child *in loco parentis*—joins Boscovich's system to Hartley's psychology of ramifying associations so as to model the immateriality of matter. Of course, this runs counter to Boscovich's mathematic formalism. And yet, Boscovich did think his theory sufficiently physical to explain the geometric lattices in crystal molecules.[71]

The question then is how, if the perception of matter's immateriality is natural and innocent—as infants and mothers, lambs and birds all recognize—it somehow gave way to the prevailing belief of matter's impenetrable hardness and solidity. The cosmogony of *The Book of Urizen*, the focus of my next chapter, offers an explanation. *Songs of Experience* doesn't explain this fall so much as it offers demonstrations and instantiations of the fact. In "A Poison Tree," for example, the trope of prolepsis allows the speaker's hypocritical repression of his anger to spawn the very apple whose temptation destroys his enemy—a deadly inversion of the bootstrapping paradox explicated in "The Divine Image," where hopeful prayer in time of need finds fulfillment through the divine nature of man's actual feelings of "Mercy Pity Peace and Love" (*SI*; E 12), consolatory in themselves but also productive of charity toward others. Usually, the world of Experience appears as the end-product of a fall without prospect of further change. In "A Divine Image" (*SE*), the first stanza links "The Human" with wicked feelings toward others, but then the second stanza's constative "is" pre-empts education and improvement by linking those feelings with the inhuman Metallurgist behind the Tyger (though the title's indefinite article implies this fall may be merely contingent and local). *The Marriage of Heaven and*

Hell offers mythemes of "the ancient Poets" (*MHH* 11; E 38) and "the Giants who formed this world into its sensual existence and now seem to live in it in chains" (*MHH* 16; E 40), but they are not much more than cleverly compressed adaptations of Enlightened conspiracy theory by way of popular works like Volney's *Ruines des Empires* (1791).

Not until *The Book of Urizen* did Blake arrive at a full-blown myth of the collapse of Contrary relationship into mind-matter dualism. Thereby, his poet's concern to distinguish what things really are from how they are "calld" (*MHH* 3; E 34, "The Lamb"; E 8–9) moved beyond radical conspiratorial ideology into a deeply metaphysical attempt to reconceive scientific terms like "substance," "matter," "body," "distance," "causation," "perceive." In *Urizen*, Boscovich's theory would find additional uses.

CHAPTER THREE

The Book of Urizen as a Vortex of Perception

The Book of Urizen begins with the big bang of Eternity's implosion into "this abominable void / This soul-shudd'ring vacuum" (*BU* 3:4–5; E 70), upon a time when "Earth was not, nor globes of attraction" (3:36; E 71). That is, not Earth as *we* know it, especially not if "globes of attraction" refers sympathetically to eyes in Renaissance fashion. Three-dimensional Earth emerges as the outcome of what mankind's correspondingly contracted "little orbs / Of sight by degrees unfold" (13:29–32; E 77). The poem ends with humanity's whimpering, if hopeful, religiosity in the face of an enemy object-world: "They called it Egypt, & left it" (28:22; E 83). How did infinite Energy come to this?

Through the dialectical spiral of the Blakean Vortex, whose series of hypostases progressively filtrates the eternal War of Contraries into less titanic, finer nuanced, more mundane conflict. In Blake's Neoplatonic satire, the Intellect's "Dark revolving in silent activity" (*BU* 3:18; E 71)—Urizen's self-contemplation, per *Timaeus* 30c, under a régime of Lockean introspection—results in a Vortex. When the Vortex eventually everts into a solid globe, then the sensible world manifests as a materialization of the forms already present within the Intellect. In the end, mankind—the conditioned outcome of this fall: *we* are Blake's Eternals, and our Lockean introspection is happening now—no longer notice the consequent narrowing of perception, or the circular, *ex post facto* character of Urizen's appeals to the argument of divine Design to rationalize the whole process. Blake did not name the Vortex until his next work, *The Four Zoas*, but in *Urizen* it operates implicitly to organize a recursive series of subnarratives into an epigenesist account of Creation, thereby extending the eighteenth-century's controversial new biological theory into physics.¹ Hence, the poem's many analogies between wombs and planetary globes. If phylogeny repeats ontogeny, as the theory purported to show, then arguably biology repeats physics.

By means of the Vortex, *The Book of Urizen* exploits the temporality of narrative to vector the radially spreading, nested imagery of *Songs of Innocence* in a single, curved direction. The palimpsestic, evanescent "fractal lattice" of associations Blake adopted from Newton by way of Hartley in *Songs of Innocence*, transforms into a sequence of differently scaled subnarratives whose sum constitutes the Fall. From a mathematical perspective, the recursions of *Urizen*'s narrative generate what Whitehead calls "analogies of function," that is, a notion of "congruence" able to establish the metrical relationships that found any systematic geometry such as Euclid-Newton's at 22:33–40.² Politically, Hartley's utopian principle of "decomplex" ideas, which in *Innocence* demonstrates the child mind's bootstrapping power to learn by creatively synthesizing different perceptions into new unities, reduces under deepening repression in late 1794 to Humean habit and Burkean custom, the psychosocial equivalents of Newton's weak inverse-square law.

In chapter 2, I mentioned how the incomplete sentence that makes up Blake's "Ah! Sun-Flower" enacts, through its multiple dependent subclauses, an entropic winding-down of energy. This is the fall Steven Shaviro has identified as the "circular-regressive" pattern of Contrary relationship.³ *The Book of Urizen* describes a similar pattern through acts of naming that bring about creation through division, as in Genesis. The above-quoted reference to Egypt is last in a long series that begins when "Some said / 'It is Urizen'" (*BU* 3:5–6; E 70) and then continues through the naming of Death (6:9; E 74), Pity (19:1; E 78 and 24:25; E 82) and the Net of Religion (25:22; E 82). Angela Esterhammer observes that in *Urizen* "figures repeatedly take on literal existence."⁴ Blake implies this is due to self-ignorance and scapegoating. *Is* Urizen "It"? Should we credit what "Some" say? Impossible to know since the opening chapters are devoid of all perspective, as instanced by the narrative's punning first line: "*Lo*, a shadow of horror is *risen*" (*BU* 3:1; E 70; my italics). Even to entertain the question of what to call "Urizen" appears enough to set Blake's Fall in motion, reversing the immanency whereby, in *Songs of Innocence*, "We are called by [the] name" of one who "calls himself a Lamb" even as his proper name remains unsaid ("The Lamb"; E 8–9; and compare "Infant Joy").

These acts of naming generate, in turn, a set of subplots whose temporalized circular-regressive pattern delineates the shape of the poem's vortex of creation through division. First come Urizen's botched efforts to project a brave new world of independent material substance out of weltering bodily sensations. Then, Los's increasingly repressive attempts at fashioning a cohesive material body for Urizen, leading to Los's despairing collapse in a travesty of God's creation of man: "The Eternal Prophet was divided / Before the death-image of Urizen" (*BU* 15:1–2; E 78). Third, Los's subsequent desire for self-unity—a parody of Urizen's "[search] for a solid without fluctuation" (4:11; E 72)—drives him to repossess his divided female half (Pity) through rape, thus repeating his previous error in a darker new key: "Man begetting his likeness, / On his own divided image" (19:15–16; E 79). Fourth, Los, seized by Oedipal jealousy, chains his aspiring Promethean child Orc to a rock, so naturalizing the child's potentially revolutionary energies in a repetition of his earlier shackling of Urizen. But then, fifth, "The dead heard the voice of the child / And began to awake from sleep": "the odours of nature" sting the reinvigorated Urizen to "[Explore] his dens" (22:26–32; E 80) by means of science as, meanwhile, the offspring of Los and Enitharmon-Pity multiply into "an enormous race" (22:45; E 81) of self-divided Sons and Daughters of Urizen. All along, the defensive barrier erected by the Eternals against Urizen, Los, and their proliferating "death-images" gradually solidifies into the Net of Religion separating heaven and earth.

Through this process of recursive repetition, the primordial heroic Contraries weave themselves into the very fabric of earthly existence, generating finer and finer crosshatchings of likeness and difference such that the difference between them grows, itself, almost impossible to discern. One result of this secular flattening of myth is that Los and Urizen appear as increasingly realistic psychological characters. Los's attempt at mimesis, in imitation of God making man "in our image, after our likeness" (Gen. 1:26), generates swarms of free-floating liminal *mimemata* or simulacra, exactly as Plato feared. In consequence, the danger of becoming what one beholds becomes daily, microscopic, and banal. In the ensuing checkerboard world of minute oppositions, resistance and compliance, dissent and accommodation, appear reconciled in shades of gray. This is the zone Blake associates with Newton's physics, following Boscovich's *Theory of Natural Philosophy*, which identifies gravitation as a weak force and Newton's inverse-square law as a mere approximation. By its conclusion, then, *Urizen*'s narrative of spiraling fragmentation discloses the static unity of a *concordia discors*. Urizen's reconciling of the Contraries instates a moral order so deeply compromised as to set a new condition for genuinely human behavior: namely, an antinomian transvaluation of values only made possible by a crisis of conscience and the passing upon the individual of a non-relative, universal Last Judgment.

This implies the possibility of a redemptive countermovement to the poem's tale of outward defeat, much as its closing allusion to Exodus would suggest. Urizen's fall is not a *fait accompli*—otherwise, it could not have been told. To suppose it is over and done replaces the tumult of Blake's narrative as read forward in real time with a conventional epic-realistic or novelistic retelling of discrete past events. This is to grant Urizen's "assum'd power" (*BU* 2:1; E 70) the power to become a damaging Urizenic assumption of our own, by projecting back into the entirety of the narrative sequence the three-dimensional flattening that arises only at the end. To anticipate later chapters, let me frame my point in Neoplatonic terms. If the poem's recursively repeating subnarratives reflect the basic triadic structure of immanence, procession, and reversion whereby, says Proclus, "Every thing caused, abides in, proceeds from, and returns, or is converted to, its cause," then every effect remains in its cause, so the basic Plotinean intuition of reality's ultimate unity continues to be possible.[5]

In sum, readers of *Urizen* gradually discover that their thoughts are not directed outward from a punctual self into a stable, independent object-world, as in a realistic narrative. Rather, readers find themselves to be the final object of an evolving universe of increasingly fragmented and dualistic thoughts. Since the poem's overall sequence of representations undergoes a process of contraction in tandem with its narrowing of perspective, it isn't just Urizen but the narrative, too, that falls out of Eternity. Thus, there is no fixed, nonrelative perspective on events. However, this also means that the poem's ending marks a threshold between Urizen's discursive horizon and the ordinary world beyond his Book—a world designated by the familiar placename, "Egypt," just before that place is "left" (*BU* 28:22; E 83). Since this is a place left by the reader no less than by Urizen's Sons and Daughters, the poem's mythic allegory converges with its real-time earthly participants. In other words, the same synthetic power of decomplex associations that we saw actuated in *Songs of Innocence* by a range of different speakers persists, on the far side of *Urizen*, in the reader's ability to imagine an overarching unity for the poem's several subnarratives. By seeing Ur-Sin whole—that is, by retrospectively recognizing ourselves as the most recent participant-victims in his ancient history—readers can transcend the banality to which the War of Contraries has been reduced. The poem delivers us to a *latest* Judgment.

At the narrative level, bringing the moment to its crisis through an act of critical synthesis stands as an interpretive Contrary to the naïve realism of simply retrojecting the contingent present onto a story governed by absolute time and space, according to the principle of Milton's damaged Satan:

"We know no time when we were not as now" (5:859). To conceive the poem in its totality as a kind of Coleridgean Symbol that "partakes of the Reality which it renders intelligible"[6] is also to "leave" it. For, the irony is that this "Book of brass" is genitively "of" Urizen not only in the material sense of *belonging to* him. As a Symbol, it is also *made of* him, in laborious distillation of his spiritual indigence[7]—a literary communion wafer for religious skeptics. Whether Urizen ever finds his "solid without fluctuation" (*BU* 4:11; E 72) remains doubtful. Readers able to see through his claim of impenetrability—by embracing their own complicity in imagining him whole, namely, as a temporal product of subnarratives they themselves have traveled—thereby reconstitute Urizen as an authentic living Contrary, his Book no inert three-dimensional world bounded by an immoveable Horizon but rather an "immateriality of matter" in the physical, Priestleyan sense that arguably supplied the original prototype for the conservative later Coleridge's eucharistic "translucence of the Eternal through and in the Temporal."

The Book of Urizen's illustrations imply a similar idea of Urizen and his Book as being incomplete without supplementation from a living reader-viewer (not unlike Wordsworth's poems in *Lyrical Ballads* four years later, "Simon Lee" being the didactic exemplar). Critics have noticed the "striking . . . fusion of contrary dynamics"[8] in the poem's designs, certain of which appear indebted to Michelangelo's *non-finito* sculpture, the so-called Atlas (or Bound) Slave, particularly plates 8, 11, 13, possibly 26, and especially 9. In them, Urizen is portrayed as partially submerged in material mediums of water or stone or, in the case of 13, air. In plate 13 (figure 3.1), the stony cloud Urizen pushes against—or is it cloudy stone, as in Richard Wilbur's epigraph to this book?—mirrors the veil enwrapping his left leg, conceivably the sloughed skin of his sensitive Blakean body tunic. Like the figure in Blake's *Newton* print discussed in the introduction (figure I.1), Urizen appears to be shuffling off the mortal coil of his transparent body-soul, the better to discover and even become, himself, a stable "solid without fluctuation" (*BU* 4:11; E 72). In *Urizen* plate 9 (figure 3.2), the fissures in the rock suggest that the Atlas figure's occluded head and hands have already entered the juncture of these four converging lines, a vanishing point beyond perspective (thus the design anticipates Satan's zero-dimensional Pit of Hell at the bottom of Blake's Last Judgment painting, figure C.3). Urizen's head and hands not only seem to be inside a cave within the rock but inside some substratum of rock itself. Is he a hero striving to penetrate and explore this impossible realm of hypostasized matter, or is he pushing himself back out of it, hands and head last, to emerge as a whole human

Figure 3.1. William Blake, *The Book of Urizen* pl. 13, copy C, 1794. Yale Center for British Art, Paul Mellon Collection.

Figure 3.2. William Blake, *The Book of Urizen* pl. 9, copy C, 1794. Yale Center for British Art, Paul Mellon Collection.

form? Of course, Urizen wholly himself *is* Urizen self-destructively seeking after the transcendental purity of absolutes—so back inside he goes! If the aim of art is to endow objects with appearance, then Urizen's metaphysical cave is the essential form of anti-art: the void.[9] Similarly, plate 13 (figure 3.1) images matter and energy as opposing forces arrested at a threshold where something has got to give. As in plate 9 (figure 3.2), viewers face a moment of intense cognitive dissonance demanding judgment and decision. If Urizen were to push himself *through* 13's cloud-rock, wouldn't he just re-glove himself in another material bodysuit like the one still shrouding his leg? Likewise, if Urizen ever succeeded in fully cocooning himself in rock on plate 9, wouldn't that establish a new three-dimensional space able to contain him, such that his threshold position, half-in, half-out, merely repeated itself? Which interpretation will we import back into the actual world of our beliefs and commitments: vulgar Newtonianism's idealization of substantial matter as bedrock reality, or a more Contrary, less dualist materialism closer to the Neoplatonism of Newton's own thought? Behind these paradoxes lies the fractal, Hartley-Priestleyan idea of matter and spirit as interlocking like a honeycomb. The latter idea is, for Blake, the truth of the matter (so to speak), yet this truth cannot transcend its figure/ground relation to the materialist error which defines it.

Where did Blake get the idea of extending Hartley's fractal lattice of associations, based on Newton's lattice model of molecules, beyond the static nesting of analogous images into a dynamic parallelism of ongoing subnarratives at widely different scales? Again, the physics of Boscovich suggests an answer. Boscovich saw that the pattern of recursion that makes up his unified force law—a spiral leading from infinitely repulsive atom-points to weak gravitational attraction at the macroscopic level—applied not only to crystals. He also developed the further, astronomical possibility that on the other side of his force curve's infinite point (the vertical axis A' or A" in figure 3.3), there might be *another* force curve with corresponding spheres of resistance containing and delimiting another, different order of matter (AA' or A'A"), the two intervals being without communication. And he speculated that, within the boundaries of different infinity points, there might be whole universes of different sizes, some so small as to appear mere points within other of these universes. As the micro- and macroscopic converge in relativistic fashion, we seem to be approaching Blake's vision of worlds within grains of sand. All that separates the recursiveness of Boscovich's mathematical model from Blake's Infinite is Blake's refusal, following Priestley, to accept the *curva Boscovichiana* as imposing an absolute limit on

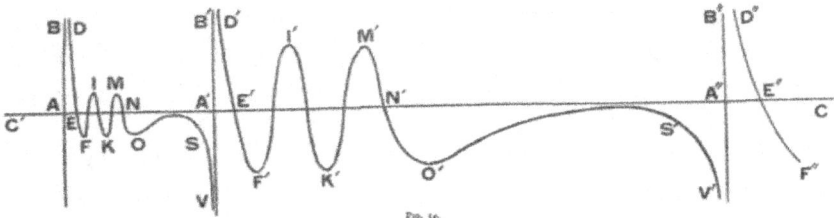

Figure 3.3. Roger Joseph Boscovich, *A Theory of Natural Philosophy* [*Theoria Philosophiae Naturalis*] (1758), figure 14, p. 138.

empirical reality, as in his later assertion: "There is no Limit of Expansion! There is no Limit of Translucence. / In the bosom of Man" (*J* 42:35–6; E 189). Of course, this difference is everything.

In *Disquisitions Relating to Matter and Spirit* (2nd ed. 1782), Priestley explicates what Boscovich graphed. Since geometrical principle forbids two points from coinciding in space, therefore in Boscovich's theory "they would only form *another center*, with different powers, those belonging to one center modifying those belonging to the other."[10] At the same time, Priestley is obviously trying to recruit Boscovich's suggestion of self-multiplying force points to support his own pantheistic doctrine of "the immateriality of matter." According to Priestley, Boscovich endorses the *"mutual penetrability of matter"*: "Provided . . . that any body have . . . sufficient momentum to overcome any powers of repulsion it may meet with, it will find no difficulty in making its way through any body whatsoever, for nothing will interfere, or penetrate one another, but *powers*, such as we know do in fact exist in the same place" (24–25). So, to glance ahead to *Milton*, a particle with extremely high velocity, like the spiritual Milton after he crashes out of Eternity into the aetherous "Sea of Time & Space" (*M* 15:39; E 110), would easily penetrate an earthly body like the Blake poet's left foot in Felpham. (Priestley concedes that a velocity significantly lower than light will cause particles of the target object to become "considerably agitated, and ignition might perhaps be the consequence" [25], which seems a fair description of the Blake poet's inflamed state after union with the Milton-Los-sun of plate 47 [copy C]; figure 4.12.)

As I argued in chapter 2, Priestley's illicit appropriation of Boscovich's force curve was made possible by the "nutshell" theory of matter's great tenuity and porosity, which underpinned both men's positions. Behind the nutshell theory lay, in turn, certain well-known speculations of Newton concerning the creation of matter. It appears that all his life Newton wanted to construe the apparent solidity and impenetrability of material substance as force effects arising at the time of Creation. In the best known of these speculations, Pierre Coste reported in his 1729 translation of Locke's *Essay* that Newton, in conversation with Locke, wondered if Creation might have come about through God's preventing the entrance of anything into certain portions of space.[11] By then applying forces of impenetrability or cohesion to those spaces, God could have generated hardness, one of the essential qualities of matter. The sensation of resistance dramatized in the design to *Urizen* plate 13, copy C (figure 3.1) can be seen to satirize Newton's speculation, since the thing Urizen appears to be pushing against and compressing may as well be cloud or water as rock (plate 11, copy C, makes the same point).

Space being uniform, according to Newton, God could have gone on to communicate impenetrability to successive portions of it to establish matter's mobility, another of its essential qualities. No doubt, a conceptual circularity is involved in Newton's suggestion that matter might be created by making space impenetrable to . . . matter. Moreover, as Jonathan Bennett and Peter Remnant go on to point out, since Newton-Locke's proposal presupposes the existence of an absolute space without anything in it, such that no other space can belong to or intrude upon it, therefore this space is *already* effectively impenetrable to every other portion of space.[12] It seems Newton's and Locke's imagined Creation is rather less dynamic than either man imagined. It fails to change anything. Notwithstanding, the comment of Locke's modern editor, Alexander Campbell Fraser, bears consideration: "Newton, it seems, suggested that 'creation of matter' means, God causing in sentient beings the sense-perception of resistance, in an otherwise pure space—a theory akin to Berkeleyanism in its recognition of the Supreme Power, and to Boscovich in its conception of the effect."[13]

Locke's allusion to Newton's speculation—his tantalizing hint of a "dim and seeming conception how matter might at first be made"—comes in a part of the *Essay* where he is trying to get around the materialists' argument that "creation out of nothing" is too inconceivable even for divine omnipotence to have done it, and therefore matter must be coeternal with God (2.311–22). Indeed, says Locke evasively, creation of matter is surely far *less* inconceivable than the creation of spirits. Compared to that, the human mind, despite its finitude, may well have some idea of creation from nothing. While that idea necessarily excludes pre-existing matter, it is not self-contradictory if it includes the power of the creator, who operates not mechanically as a natural cause producing natural effects but organically to make actual now what was potential before. Remarks Fraser, "This as little involves either a contradiction in terms, or contradiction to reason, as the transformation of chaos into cosmos."

When Leibniz read Locke's speculation, he shrewdly guessed Locke "had in mind the Platonists, who take matter as something fleeting and transitory, after the manner of the accidents, and had an altogether different idea of spirits and souls."[14] On this view, Leibniz suggests, matter is no substance at all but only a property or quality belonging to something else—for example, a spirit, perhaps even a resurrected "Spiritual Body" in Blake's sense ("To Tirzah"; E 30). Leibniz's position broadly resembles Blake's. In his debate with Newton's acolyte, Samuel Clarke, Leibniz claimed space is nothing but a set of relations among co-existing bodies. Remove the bodies, and "space" amounts to no more than the possibility of situating them someplace, much as Boscovich later reasoned.[15] As we shall see, Whitehead, whose modified relativity theory sheds light on Blake's conception of space-time as an evolving four-dimensional field of events, supposes similarly that events are ontologically prior to space-time while matter is just a contingent characteristic of events (in contrast to Einstein who, in Robert Palter's words, "supposes that matter [or the gravitational field] is ontologically prior to space-time, while events are simply intersections of world-lines of particles"[16]). Briefly to continue this excursus, Whitehead based his relativity principle on Maxwell's theory of electromagnetism, which rejected Newton's materialist "fallacy of simple location" and conceived of the universe as "a field of force—or, in other words, a field of incessant activity."[17] "[I]n a certain sense," Whitehead wrote, remembering Leibniz's monads, "everything is everywhere at all times. . . . Thus every spatiotemporal standpoint mirrors the world."[18]

In contrast to these Leibnizian implications, Newton in the early *De gravitatione et aequipondio fluidorum* had wondered if God, by "the sole action of thinking and willing" that certain regions of space be impenetrable, might create a "certain kind of being similar in every way to bodies."[19] Such a space would "equally operate upon our minds and in turn be operated upon, because it is nothing more than the product of the divine mind realized in a definite quantity of space." One commentator remarks: "In saying 'God can stimulate our perception by his own will,' Newton envisages Berkeleian matter as a possibility. But what he *actually* supposes is matter that mediates between God and our minds and is itself the immediate cause of our perceptions."[20] Newton concludes that the distinction between our ideas of thought and extension "will not be so great but that both may fit the same created substance, that is, but that a body may think, and a thinking being extend" (143), contra Descartes. Indeed, he says, "the analogy between the divine faculties and our own is greater than has formerly been perceived by Philosophers."

Locke himself proceeds, in the *Essay*'s next section following his "dim and seeming conception how matter might at first be made," to argue that our own experience of "all our voluntary motions" daily gives proof by analogy of something very like Newton's suggestion that the powers and properties of the material world are created by God through the continuous exercise of his will: "For example: my right hand writes, whilst my left hand is still: What causes rest in one, and motion in the other? Nothing but my will,—a thought of my mind; my thought only changing, the right hand rests, and the left hand moves. This is matter of fact, which cannot be denied: explain this and make it intelligible, and then the next step will be to understand creation" (2.322–23). Newton evidently regarded the analogy between Creation and human voluntary motion as implying that the physical universe might serve as God's sensorium. A less mechanistic inference would be that if there exists an analogy between the human mind in its body and God in space, it is because the universe is, itself, a Cosmic Animal.[21] So, when the poet of *Milton* rises to Locke's challenge by invoking his Muse, "Come into my hand / . . . descending down the Nerves of my right arm / From out the Portals of my Brain . . ." (*M* 2: 5–7; E 96), it emerges that his right hand moves not by an act of will but through the inspiring assistance of the Mundane Soul, an entire universe of spiritual agents within and without. Beginning with the Prophetic Books and *Urizen*, Blake's mythology opens the interface between divine mind and human bodily action and "makes it intelligible," as Locke all but admits his mechanistic view of the body cannot do.

Accordingly, Newton's speculation of how matter might first be made is systematically taken to task by the cosmogenesis in *The Book of Urizen*. Urizen's universe arises when an "all-repelling" force evacuates space to produce an "abominable void" and "soul-shudd'ring vacuum" (*BU* 3:3–5; E 70) within which material bodies are then fashioned, not instantaneously by divine fiat, but painfully by effort, trial, and error. The filling-up of space by bodies constrains the creator's own creativity in relational, Leibnizian fashion. Appalled at his bungled job of creation, Urizen tries to limit its effects by forming instruments of measurement, as in Blake's *Newton* print (figure I.1). Significantly, his final whereabouts remain a mystery. Has he become naturalized to the point of disappearing into his world pantheistically? Or has he absconded beyond it to a point of abstraction that is nowhere at all, becoming Old Nobodaddy, Blake's Euclidean-minded Ancient of Days? Everywhere or nowhere, the two alternatives reduce to the same thing: a god unavailable to finite, first-personal perception. Urizen's mind-body dualism, grounded in the materialism of Newton's absolute space and time, precludes any middle ground.

More specifically, *The Book of Urizen*'s narrative progression exposes how the illusion of a solid, three-dimensional object-world is generated by Urizen's constant arrest of the horizon at every instant to prevent movement and change.[22] The horizon comes into existence, we see, when Eternity's energy field of "all flexible senses" (*BU* 3:38; E 71) gradually inverts and "the eternal mind bounded beg[ins] to roll / Eddies of wrath ceaseless round & round" (10:19–20; E 75). Los's creation of a "Space undivided by existence" (13:46; E 77) then establishes the region within which the resulting spiral consolidates as "a round globe . . . / Trembling upon the Void" (14:58–59; E 77). Seen whole, this process forms a Blakean Vortex whose apex delivers that mode of perception which Whitehead terms "presentational immediacy," as contradistinguished from the primordial animal mode of "causal efficacy" residing at the rim.

Broadly speaking, Whitehead's distinction here is between the discerned object and the field of the discernible, the latter being a prior "vague" background sense of "the togetherness of things" that is normally ignored in rational perception"—"the lure for feeling," he calls it.[23] The contrast between the simplicity and indefiniteness of "causal efficacy," and the sharp clarity of "presentational immediacy" that derives from it, parallels that between topology and geometry. In Whitehead's doctrine of "symbolic reference," sense perception always involves two different kinds of awareness, one that discriminates objects or events qualitatively according to their individual peculiarities, and another that apprehends distant entities only as bare *relata* within the field of events. But these two opposing modes are joined together. The doctrine accords with Erasmus Darwin's cognitive theory, discussed in chapter 1, based on twin modes of analogy, one "unconscious" and "intuitive" and the other "rational." Peripheral awareness that one is, oneself, a bodily part of the immediate surround—another piece of furniture, as it were, in the very room whose perceptual reality exists in the mind and constitutes one's present self—operates to transmit physical reality into consciousness.[24] Thus, Whitehead argues, "feelings derived from causal efficacy ["the vector feeling-tone" of an actual experience that it inherits from others in its particular past] are 'precipitated' or 'projected' upon a region defined by presentational immediacy" (*Process and Reality* 184–86).

As Isabelle Stengers remarks, Whitehead was trying to explain the "vectoral" character of feeling, the fact that you feel this moment of experience to be your very own and yet derived from a world without.[25] Since the feeling *is* derived, solipsism is avoided. So, instead of "describing how subjective data pass into the appearance of an objective world," as Kant does, Whitehead declares that his chief epistemological concern—not unlike Newton, Priestley, and Blake, each in his own way—is to explain "how objective data pass into subjective satisfaction, and how order in the objective data provides intensity in the subjective satisfaction," creating out of the existing world "a 'superject' rather than a 'subject' " (*Process and Reality* 88). For example, the *Daughters of Albion* frontispiece (figure 1.1) offers a visual analogue of how, in perception, we remain dimly aware of the contours of our own face and head. The cavern's rocky wall portrays a peripherally visible cheekbone, the overhanging vegetation, an eyebrow. The presented image of a sun glaring down upon three arrested human figures forms part of a wider, indefinite region of sociocultural events (slavery, patriarchy) that includes our own persons as contemporary observers. Paradoxically, the frontispiece offers a "clear and distinct" perception of the reality of indefinite, largely unconscious perception in the mode of "causal efficacy."

To travel to the horizon, then, is to look over the edge of the known world, including the self, and glimpse the interrelated togetherness of things that underpins "the bound or outward circumference" (*MHH* 4; E 34) of rational perception. "The" horizon is no more absolute than its metaphysical analogues, "the" Fall and "the" Creation. Although the horizon is inescapable, since it is an aspect of the perceiver's finite location and limited viewpoint, we remain free to change it through movement in the world. Blake makes this point in a lovely pastoral passage in *Milton* asserting man is a nomad, the sky his tent, "And if he move his dwelling-place, his heavens also move" (*M* 29:12; E 127). The horizon necessarily implies a beyond, yet it cannot specify in advance just what it might be. As Stanley Bates emphasizes, wherever "our particular finitude-now," it cannot be separated from "the always present possibility of transcending that particular finitude-now."[26] The metaphor of the mind's horizon discloses not some fixed (Lockean or Kantian) limit to knowledge but the possibility of change within and through limits. It conveys "the closure required by individual finitude, while at the same time permitting the awareness that even the potentially infinite movement of the horizon will always leave something beyond it" (Bates 171). Thus, transcendence is always possible and never final.

To see through *Urizen*'s outer narrative of the Fall is, then, to recognize time's perpetual ending in an eternally returning now: Plato's "moving image of eternity" (*Timaeus* 37d) actually visible at the horizon seen across the face of spinning Earth orbiting in space, what Blake called "eternity's sun rise" (E 493, E 470) and the "Moment in each Day that Satan cannot find" (*M* 35:42; E 136), that constant utopian glimmer at the edge of the world as it unspools beneath the pilgrim's feet "in his bright journey of sixty years" (*M* 15:52; E 110). On the other hand, "that false appearance which appears to the reasoner, / As of a Globe rolling through voidness" (*M* 29:15–16; E 127), is a function of Newton's absolute space and time, which by restricting our limited but mobile vision to a series of finitude-nows of predetermined extent, like dates on a calendar, falsely implies the underlying existence of an impersonal Chronometer.

Accordingly, the narrative burden of *Milton* plate 29 is to argue that Beulah's mobile lower heaven offers the earthly pilgrim a home away from Home. Behind Beulah's moving space-time envelope lies Galileo's concept of inertial frames as set forth in his *Dialogue concerning the Two Chief World Systems* (1632). A ball dropped inside a steadily moving ship does not fall back relatively to the ship's forward motion but straight down. Similarly, inhabitants of the spinning, orbiting earth are not flung off into void space because we share the stability of its encompassing inertial frame. Clearly this, too, is a kind of relativity, and Einstein admired Galileo's theory as a precursor to his own. We might say that as Galileo's relativity is to Einstein's, Blake's Beulah is to Eternity. For, what *The Book of Urizen* and *Milton* contemplate is a perpendicular view on three-dimensional Euclidean perspective and its horizons: a four-dimensional viewpoint we can only imagine by analogy. (A few mathematicians claim to have perceived it for a moment or two, perhaps by means of Kurt Gödel's Platonic faculty of "mathematical intuition."[27] The many YouTube digital animations of topological deformation from three dimensions to four certainly make the task easier.) Blake as a visual artist well understood Euclid's method of building up perspective orthogonally. A point can be conceived as a line viewed on end, such that the line becomes a "point of points"; a line becomes a plane viewed from the side, that is, the plane is a "line of lines"; a plane or square becomes a solid figure (say, a cube) viewed from one of its sides, that is, the cube is a square of squares. And the cube of cubes? At its geometrical limit, the process of seeing past the horizon leads beyond sidelong viewing to a fully perpendicular perspective on the solid earthly world.

This perspective is, indeed, implied by the above-cited passage from *Milton* plate 29, where Blake claims that "every Space that a Man views around his dwelling-place . . . / . . . / . . . is his Universe; / And on its verge the Sun rises & sets. the Clouds bow / To meet the flat Earth & the Sea in such an orderd Space: / The starry heavens reach no further but here bend and set / On all sides" (*M* 29: 5–11; E 127). The passage echoes Robert Smith's explanation of the apparent concavity of the sky in *Compleat System of Opticks* (1738), summarized by the indefatigable Priestley in his *History and Present State of Discoveries Relating to Vision, Light, and Colours* (1772). "If the surface of the earth were perfectly plane," says Priestley, then "if we suppose a vast wall to be built at the extremity of the plane, beyond the point of visible distance, it will not appear straight, but circular, as if built upon the circumference of the horizon."[28] This is because the average person standing on level ground can see no more than about five miles on a clear day. "Extend this plane [the "vast wall"] upward," Priestley continues, "and it will appear as a dome overhead." In *Milton* Blake asserts, similarly, that to the local observer "the flat earth" (29:9; E 127) forms "one infinite plane" (15:32; E 109). The Beulah-like dome of heaven made visible by Smith-Priestley's supposedly material vertical wall may therefore be seen to constitute "Jerusalems wall" (*J* 77; E 231), the phenomenal outward circumference of Eternity conceived as an invisible four-dimensional sphere that wraps around the earthly globe at all points such that any spot on the globe can be connected to Eternity by a straight line, that is, a sightline. To anticipate the subject of my next chapter, Blake visualizes Eternity as a "sphere of spheres." This is the perspective needed to reimagine Albion's forbidding, Lockean-nominalistic "wall of words," the seemingly flat three-dimensional monolith in figure 1.2, as an indefinite space-time of reading.

One might still question the geometry of the cosmogenesis in *The Book of Urizen*. How does Eternity "give birth to" the three-dimensional material world, unless by the same illicit materialism as Priestley brought to Boscovich's mathematical theory? The answer is, through a process of eversion or turning inside out. *Urizen*'s title page (figure 3.4) implies as much. Urizen is shown to be an ideological "emanation" arising at the intersection of the three directional planes formed by his bookmaking: vertical (Mosaic tablets of Law, positioned in stony parody of angel wings), horizontal (the copperplate "Book of brass" [*BU* 4:44; E 72] he is busy inscribing with pen and burin, both hands at once: insane self-absorption,

Figure 3.4. William Blake, *The Book of Urizen* title page, copy C, 1794. Yale Center for British Art, Paul Mellon Collection.

the very opposite of divine dictation), and front-back (the axis formed by the brass book's printed copy outthrust before him). In other words, Urizen embodies the divine fourth dimension as it has been hideously misrecognized under Newton's Euclidean regime. His frantic scribbling is the discourse of materialism needed at every moment to prop up his authority, without which his illusion of control would collapse.

Yet the words of the book in the title page remain illegible in all copies. Relevant here is an observation Garrett Stewart makes of Western art generally: "the sharply painted replica of a page . . . wouldn't be a picture of its referent so much as its equivalent, a verbal document in its own right."[29] So it appears doubtful whether the squiggly, colored hieroglyphs in Urizen's open book resemble the writing we ourselves are about to read in *The Book of Urizen*. Blake here contradicts, or at least complicates, the *ut pictura poesis* tradition. Hence, he styled the Illuminated Books "prophecies": contrary to the tradition, the drawn lines and verse lines do not match up as (ideally stereoscopic) three-dimensional descriptions or illustrations of each other. Of course, *The Book of Urizen* is, itself, nonself-identical, not just because it treats poetry and painting as Contrary rather than complementary but because Blake's extension of the Contrary dynamic into the printmaking process renders each individual copy significantly different from the rest (though probably no one in Blake's day succeeded in comparing any two). In other words, the title, *"The" Book of Urizen*, names a three-dimensional abstraction from multiple books read in different times and places, similarly to "the" horizon. The title suggests, somewhat like *The Marriage and Heaven and Hell* plate 15, that designated "books" able to be "arranged in libraries" are a factitious institutional byproduct of imaginative activity taking place continually within the "infinite" four-dimensional interior of a printing house where referentiality does not exist (*MHH* 15; E 40).

My fuller answer to the preceding question concerning the geometry of Blake's cosmogenesis is reserved for chapter 4. Meantime, let us briefly consider the suggestive use Blake had already made of geophysical genesis as posited by Erasmus Darwin in *The Economy of Vegetation* (1791). There, Darwin argues that the earth and planets were created by ejection from exploding volcanoes in the sun and the moon created, in turn, by ejection from a volcano on earth. David Worrall has demonstrated in detail the influence of Darwin's poem upon Blake's *The French Revolution* and *America*, especially the following passage from *America* plate 5:

> Albions Angel stood beside the Stone of night, and saw
> The terror like a comet, or more like the planet red
> That once inclos'd the terrible wandering comets in its sphere.
> Then Mars thou wast our center, & the planets three flew round
> Thy crimson disk; so e'er the Sun was rent from thy red sphere.
> (*A* 5:15; E 53)

As Worrall points out, "we are given only the tyrant Angel's view" of the American rebellion as, first, an ominous comet, then as burning, warlike Mars.[30] From the Angel's fallen perspective, the Stone he guards (England-Albion's temple) was once the fiery universal center whose gravitational power held the comets and other planets in their orbits. Yet, Worrall brilliantly adds, "There is an implication that the fire was once an opening to Eternity," still dimly and distortedly recollected by the Angel (who is modeled, it seems, after the cognitively damaged devils of *Paradise Lost*). Indeed, if Mars once occupied the center where the Sun now stands, and if the Sun "was rent from" Mars, which now orbits around the Sun as the outermost of the so-called inner planets, then an inversion has taken place. The center has become the circumference.

Accordingly, when *The Four Zoas* relates, a few years later, how "All fell toward the Center in dire ruin, sinking down" (*FZ* 19:21; E 112), this isn't a mere three-dimensional collapse like the one William Herschel described in 1789. Herschel argued that gravitation's "clustering power" causes galaxies to compress such that, when "very aged, and drawing on towards a period of change or dissolution," they become Urizenically solid and "planetary."[31] Rather, Blake seems to imagine a four-dimensional sphere turning inside-out to form a globe of three dimensions surrounded by empty space (the topology of this process is explored in my next Chapter). Eternity's loss of energy to "expand inward" ejects it to the circumference where eventually it becomes viewed as traditional Heaven, while the center becomes occupied by Newtonianism's reified matter governed by mechanical laws of gravitation. In this way, Eternity's dynamically interlocking dialectic of Contraries degenerates into the stasis of dualism, as *The Marriage of Heaven and Hell* had previously argued.

This is much the same collapse as Worrall examines in Blake's *The French Revolution*. The king of France's "brows folded heavy, his forehead was in affliction, / Like the central fire" as he contemplates going to war, while "his "bosom / Expanded like starry heaven" (*FR* 5:79–82; E 289).

Noting that the imagery draws on Darwin's notion of "central fires" of lava at the earth's center as forming "a subterraneous sun" (*Economy of Vegetation* 1.139n), Worrall observes that Blake envisions these geophysics in cosmic human terms quite unlike anything in Darwin (407). Further still, Blake's dualistic imagery of a congested head surrounded by the "starry heaven" of its body suggests the Urizenic king is turned outside-in. The true center of the system would seem to lie in the energetic sun (mind) hidden within, which the King's head revolves around no less than the rest of his body, as shown for example in George Gamow's inverted universe-man in figure 4.2.

A question remains: Why was Blake so willing to adopt Smith-Priestley's thought experiment of a vertical wall rising at right angles to a flat earth, in the first place? Perhaps he was predisposed to it by Boscovich's graph of his mathematical force curve (figure 2.2), which shows a force of infinite repulsion rising perpendicularly to the weak force of Newtonian attraction spreading out on the right. If the graph's right side represents the three-dimensional world of intermingled Contraries and the weak conflict of Urizenic agreements-to-disagree that shore up liberalism's polite status quo, then the infinite vertical force corresponds to a higher-dimensional realm of Energy. And it is, precisely, man's limited distance vision and the conglobing effect of the human eye's roundness (also explained by Priestley) that converts Boscovich's vertical axis—a mathematical abstraction—into a sphere whose outer edge is perceptible. "Therefore God becomes as we are, that we may be as he is" (*NRR* [b]; E 3).

CHAPTER FOUR

A Brief Particular History of the Fourth Dimension of Space, with Special Reference to *Milton: A Poem*

Blake's belief that heaven lies all about us is generally put down to his critique of the "System" (*MHH* 11; E 38) of Natural Religion made up, very broadly speaking, of Locke's materialist psychology grafted onto Christian dualism. Or to his antinomian rejection of privileged sacred places and traditions. Or to his democratic Romantic celebration of all things vulnerable, small, and neglected. On the other hand, the mythmaking of the Prophetic Books is often regarded, especially by historicist critics, as Blake's allegorizing of these prior politico-religious concerns. This chapter seeks to flip perspective. Blake's radical convictions can be seen to flow from his antidualist cosmology based on an intuition, amounting even to a real perceptual glimpse, of curved non-Euclidean space and four-dimensional space-time. As we've already seen, for Blake social-revolutionary change is inseparable from the idea of physical movement.

My argument should occasion no surprise. The father of modern Blake studies, Northrop Frye, regarded God as "a center who is everywhere," partly owing to his readings in Mayahana Buddhism and the *Avatamsaka Sutra* but also because he was greatly influenced by Alfred North Whitehead's relativistic idea of the beyond as "interpenetrating" the spatiotemporal here and now, the two fusing together like Contraries in an endlessly recapitulated "dialectic of transcendence and immanence" (Frye's phrase to describe the necessary provisionality of the "final synthesis" with which *Fearful Symmetry* concludes).[1] Elsewhere Frye fantasized, "If I ever get a big enough office, I shall have the hundred plates of my *Jerusalem* reproduction framed and hung around the walls, so that the frontispiece will have the second plate on one side and the last plate on the other."[2] But what Frye really wanted, I surmise, was to position each plate adjacent to *all* of the others—an ordering that would have required a wider, more four-dimensional office than any at the University of Toronto.[3]

More particularly, my argument builds on Donald Ault's groundbreaking *Visionary Physics* (1974). Ault's achievement was to demonstrate the "coherence"—a Whiteheadian term[4]—of Blake's response to Newton, which he saw as developing out of inconsistencies in Newton chiefly concerning limits, infinitesimals, and the doctrine of fluxions. Ault evidently thought Blake applied to Newton an Einsteinian critique focused on the misfit between Newton's mathematics and his experimental empiricism much as Whitehead had done, notably in *Science and the Modern World*.[5] Accordingly, Ault pursued "the *structural* relationships that obtained between the worlds of Blake and Newton."[6] As for topology, "'spatial' relationships in Blake's poetry are," he says, "a kind of symbolic shorthand and are not themselves spatial in a Euclidian sense" (34). Blake's Eternity thus serves Ault as a negative critical concept or limit. "Blake views the world that Newton's system characterizes [as] a kind of topological mapping of the structure of Eternity onto three-dimensional space" (33)—while Eternity itself, as Ault sees it, is nonmetric and nondimensional. The trouble with this is that Eternity thereby seems so noumenal—so Kantian—one struggles to see how it could supply the basis for a "coherent" and "structural" response to specific physical inconsistencies in Newton. Topology describes the simpler, indefinite, more abstract structures underlying metric spaces and their quantitative measurements based on distance. But if Blake's Eternity is nonabstract and without topological properties, as Ault seems to imply—if "'spatial' relationships" are purely "symbolic" for Blake—then how would Newton's Euclidean "mapping" pose a threat to it? How would such mapping even be possible? And how could Newton's physics serve Blake as an inspiration, as plainly it does on Ault's own showing—to say nothing of the enormous respect manifest in Blake's color print, *Newton* (figure I.1)? These are questions I broached in my introduction. This chapter aims to take a closer look at the topology of Blake's myth—its "rubber-sheet" deformations and reconfigurations of space—and the accompanying geometrical methods he used to visualize Eternity as a fourth dimension.

Blake's Kleinian Four-Dimensional Space and Some English Literary Antecedents

First, how is it that Blake seems to depict four-dimensional objects in advance of their explicit theorization by twentieth-century science? As is well known, Whitehead, Henri Bergson, William James, and Maurice Merleau-Ponty themselves all struggled alongside Einstein and Bohr to understand quantum theory's scandalous resistance to the kinds of mental picturing that had long been an accepted part of experimental science. Bohr, for example, spoke notoriously of "the ingenious formalism of quantum mechanics, which abandons pictorial representation and aims directly at a statistical account of quantum processes."[7] In his own day, Blake plainly abhorred the new science of statistics. He rejected Bayes and Hume's probabilistic approach to traditional testimony about miracles, arguing instead that miracles are always necessarily first-personal and concrete (see chapter 2). He vilified Reynolds for confounding Platonic Forms with "General Forms," that is, averages, in the *Discourses on Art* (E 648–49). The answer to my opening question is that Blake—like Bohr, if not Einstein—understood pictorial representation as, itself, an empirical formalism or generalization. As in Blake, the twentieth-century scientists' efforts led inward to phenomenology and the conclusion that a "Democritean unconscious," operating within a biologically determined "zone of the middle dimensions" sandwiched between the microscopically small and the astronomically vast, drives human cognition ineluctably into one or another form of atomism.[8] Bergson, for one, argued that Cartesianism had abstracted its classic dualisms out of the more plastic, more permeable perception of space-time delivered by really intense introspection—not Lockean ratiocination or Cartesian meditation but a process akin to panexperiential Blakean "self-annihilation" (*M* 14:22, 41:2; E 108, 142) and that awareness of elemental "intercorporeity" that Bergson's successor, Merleau-Ponty, enigmatically denominated as "the flesh": "the return of the visible upon itself, a carnal adherence of the sentient to the sensed and of the sensed to the sentient," "a coiling-over of the visible upon the seeing body, of the tangible upon the touching body."[9]

So, Blake anticipated twentieth-century relativity because he was aware of the many ways in which time and process had already long been warping and expanding the imagining of *three*-dimensional space. He was alert to the malleability and constructedness of dimension, not only as an artist trained to create three-dimensional perspective from a flat surface, but as an engraver

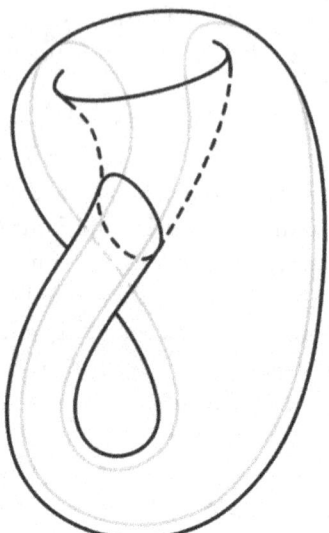

Figure 4.1. Klein bottle. Wikipedia. The Klein bottle portrays a geometrical version of Merleau-Ponty's phenomenology of "double touch," where the left hand in touching the right one gives rise to "a kind of reflection" with the touching hand experienced as subject and the touched hand as object. *Phenomenology of Perception*, trans. C. Smith (New York: Routledge, 1962), 107. Such nonlinear loops can be seen as lower-dimensional forms of Blake's spatiotemporal Vortex.

skilled in fashioning copperplate into a surface suitably in-between two and three dimensions to transfer his carved perspective scheme onto a printed page, where it could become an illusion. Moreover, geometrical awareness of a spatial fourth dimension was pre-Newtonian and readily available to Blake through Dante, Milton, and others. Like Dante, Blake's topological refigurings of Euclidean space imply an inside-out universe. As C. S. Lewis explained, gazing into the sky, one isn't necessarily looking out, as "a modern" tends to suppose. "If you accepted the Medieval Model, you would feel like one looking *in*. . . . The universe is thus, when our minds are sufficiently freed from the senses, turned inside out."[10] In Blake's Romantic revisiting of this model via Dante, the three-dimensional object world is contained within four-dimensional Man, the giant Albion, whose unfallen form is that of a torus or donut (figure 4.2).[11]

Hence, *Milton*'s extensive referencing of Bowlahoola and Allamanda as the stomach and bowels and alimentary canal of "every Generated Body

A Brief Particular History of the Fourth Dimension of Space / 109

Figure 4.2. "Inside-out universe." George Gamow, *One, Two, Three . . . Infinity* (London: Macmillan, 1962), figure 20. "Inside-out universe. This surrealistic drawing represents a man walking on the surface of the Earth and looking up at the stars. The picture is transformed topologically . . . Thus the Earth, stars, and sun are crowded in a comparatively narrow channel running through the body of the man, and surrounded by his internal organs" (Gamow, 59).

in its inward form" (*M* 26:31; E 123). Blake imagines the space of the gastrula, by which nutrients pass into and out of the human body, as an organic surround that nevertheless appears to that body's eyes as an *outer* space.[12] So, when *Jerusalem*'s Daughters of Albion sadistically bind the "Infant" Albion's human form divine to nature in a kind of crucifixion by weaving it "a Cradle of the grass that withereth away"—namely, a mortal body ("All flesh is grass": Is. 40:6–7)—they spin their "Cord of affection" from guts pulled out of their Samson-like prisoner, as shown in the illustrations to plates 25 and 59 (see figure 4.3; *J* 56:3–23; E 206). The passage bears comparison with the materialization of text into textile and of spontaneous song into a dense body of knowledge and foreboding, in "A Cradle Song," examined in chapter 2 (text, from L. *texere*: to weave).

110 / A Bastard Kind of Reasoning

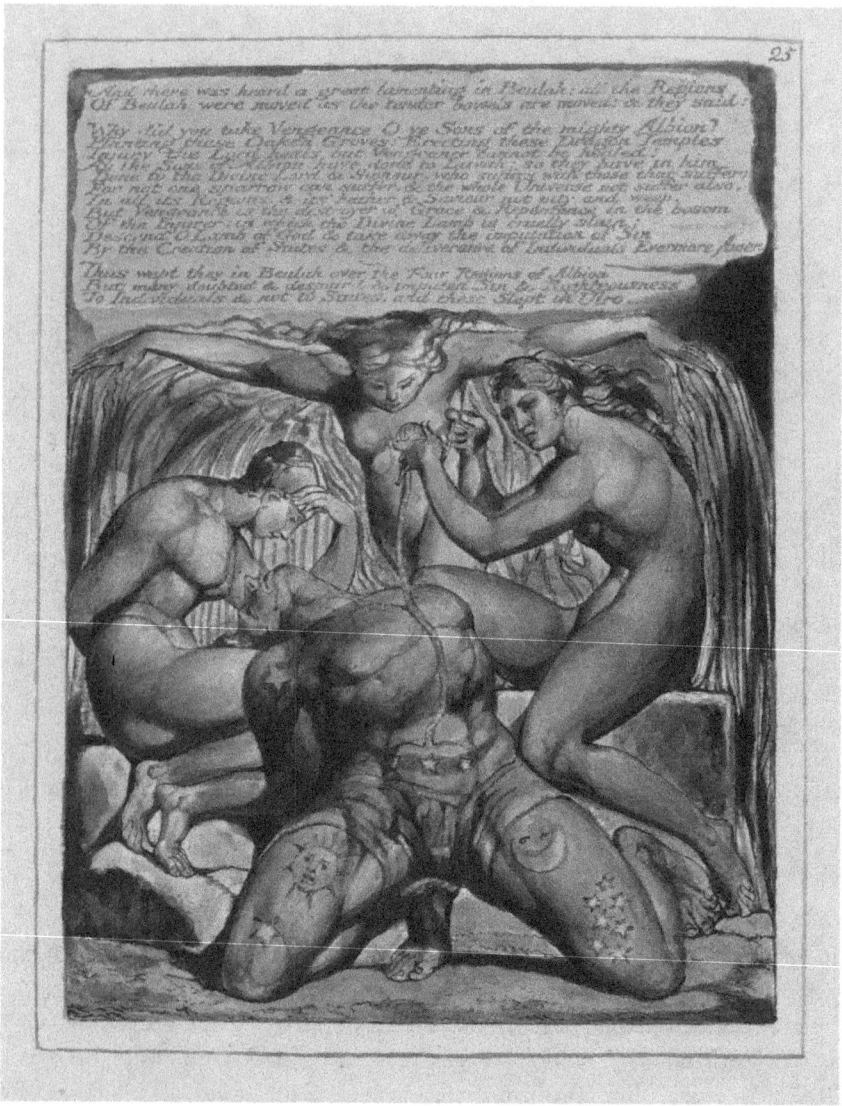

Figure 4.3. William Blake, *Jerusalem* pl. 25, copy E, c. 1821. Yale Center for British Art, Paul Mellon Collection.

The Daughter with outstretched arms at the top of figure 4.3 is evidently extending this human physiology to form an encompassing sphere, like the protective but confining earthly Spaces of Beulah which constitute man's "middle zone of perception."

Before Gauss, Reimann, and Einstein, *Milton* investigates in detail the topology of the curved space portrayed in the *Daughters of Albion* frontispiece, as seen in chapter 1. Blake explains that the outside world shifts according to the earthly nomad's traveling:

> The Sky is an immortal tent built by the Sons of Los
> And every Space that a Man views around his dwelling-place:
> Standing on his own roof, or in his garden on a mount
> Of twenty-five cubits in height, such space is his Universe;
> And on its verge the Sun rises & sets. the Clouds bow
> To meet the flat Earth & the Sea in such an orderd Space:
> The Starry heavens reach no further but here bend and set
> On all sides & the two Poles turn on their valves of gold:
> And if he move his dwelling-place, his heavens also move.
> Where'er he goes & all his neighborhood bewail his loss:
> Such are the Spaces called Earth & such its dimension.
> (*M* 29:4–14; E 127)

Like a modern physicist, Blake employs concepts of "neighborhood" and "locality" to make movement and, thence, time a "dimension" of "the Spaces called Earth." The passage offers a visionary perspective on the secluded, often cozily domestic Romantic "spot" celebrated in the conversation poems of Wordsworth and Coleridge.[13] In *Milton*, the closing pastoral scene at Felpham dawns into view not as a reassuringly concrete spot in enduring nature but as a human-scaled perspective on the world that has concresced, provisionally, out of an indefinite cosmic background of ongoing events. The heterogeneity implied by the plural "Spaces" contradicts all three main features of Newton's absolute Euclidean-Cartesian space: homogeneity, isotropy, and infinite extension. In Newton's system, these features render space a pre-existing container available to be occupied by inert matter. Blake suggests Earth and "the Starry heavens" share a more intimate inter-involvement, as for example in *Daughters of Albion* when Oothoon gazes at "the morning sun" in "the moment of desire" and enjoys "happy copulation" through her (similarly spherical) "eyes" (*VDA* 6:23–7:3; E 50), so realizing the Kleinian union of subject and object that underlies that poem's frontispiece.

Where did Blake get his idea of four-dimensional space? Well, the requisite analogy had been around at least since Plato compared ignorant Cave man's ideas of three-dimensional existence to flat shadow projections, while arguing that knowledge leads to apprehension of a higher realm of Forms. By analogy with Boscovich's nested shells of force, you might find it easiest to visualize four-dimensional space as a ball whose center has been cut out and replaced with another ball. As the radius of the inner ball goes to zero, the radius of the outer one goes to infinity (figure 4.4). Blake's Eno seems to apply this method when she "took an atom of space & opend its center / Into Infinitude" (*FZ* 9:12–13; E 305, also *J* 48:30–39; E 197). Similarly, *Jerusalem*'s address "To the Christians" asks readers to "wind" his "golden string" of narrative "into a ball," thereby turning allegory's linear time into a Symbol of pure duration, Eternity (*J* 77; E 231). In like analogical fashion, English Christians had for centuries exploited the Son-sun pun to portray Christ as "the sun of suns." A series of paradoxes arose based on the supposed existence of a higher dimension outwardly at cross-purposes with man's own while inwardly subsuming it into a larger design, as in Pope's callow Deistic summary:

> All nature is but art unknown to thee;
> All chance direction, which thou canst not see;
> All discord, harmony not understood;
> All partial evil, universal good.[14]

Any tension between higher and lower dimensions was put down to man's self-enclosing pride. As C. S. Lewis recognized, this tension corresponds to

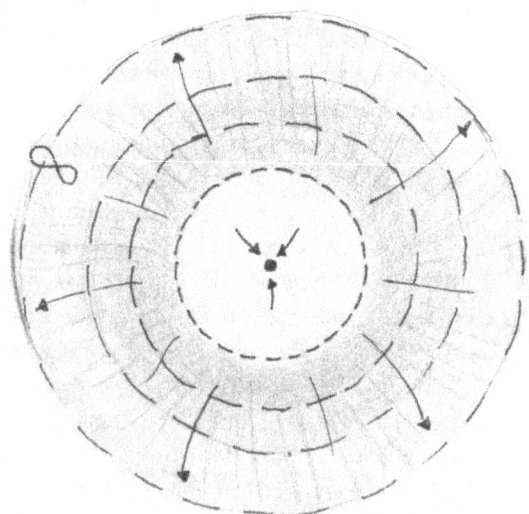

Figure 4.4. Center of a 3-sphere. Illustration, Rob DuToit.

that between the timeless Symbol and the temporality of allegory: "for the symbolist it is we who are the allegory. We are the 'frozen personifications'; the heavens above us are the 'shadowy abstractions'; the world which we mistake for reality is the flat outline of that which elsewhere veritably is in all the round of its unimaginable dimensions."[15] Blake's receptiveness toward this dual perspective in real life is seen in his letter to Butts just after his removal to Felpham in search of patronage and inspiration: "Work will go on here with God speed . . . going out at my gate the first morning after my arrival . . . the Plowboy said to the Plowman. 'Father The Gate is Open'" (E 711).

Consider, in this light, one of Blake's illustrations to Milton's "On the Morning of Christ's Nativity." Sitting in the snow, Milton's shepherds hear the harmony of united heaven and earth, and then suddenly they witness the source of this music in the angelic choir:

> At last surrounds their sight
> A globe of circular light,
> That with long beams the shame-faced Night arrayed;
> The helmèd Cherubim
> And sworded Seraphim,
> Are seen in glittering ranks with wings displayed.[16]

The light beams raying far into the night appear as emanations from a globe whose heavenly light is, itself, "circular." How, then, does this globe "surround" the shepherds' sight? Thomas Banchoff's introduction to Edwin Abbott's classic science-fiction romance, *Flatland* (1884), suggests an answer:

> Just as a sphere penetrating Flatland is viewed . . . as a circle growing and then shrinking in time, so also if we were visited by a hypersphere from a space of four dimensions, we might see a sphere growing and then shrinking in time. The ability to treat such a sequence of impressions [allegory, in Lewis's terms] as the gradual revelation of an entity from a higher dimension [i.e., a Symbol] is the first exercise for anyone who wishes to accept the challenge of Flatland.[17]

In her study of the ode's imagery, Mother M. Christopher Pecheux invokes several Augustinian paradoxes comparing the visible sun with its invisible creator.[18] For example, in arguing that the day of Christ's birth discloses the mystery of his light, Augustine asserts: "Among time-bound days He has His day in time; but he himself is the Eternal Day of the Eternal Day."[19] As a higher-dimensional being, Christ is not just a sphere of light; he is the sphere of spheres (and king of kings), analogously to a four-dimensional sphere whose surface is composed at all points of three-dimensional spheres—in mathematics, a "3-sphere." In Pecheux's explication, the Miltonic nativity occurs on an earthly timeline that it divides in half; thus, it stands at the center of the eternal now that surrounds it, as in figure 4.4. The nativity is a typical Renaissance meditative "composition of place," a locus surrounded by shepherds and, by extension, all mutable earthly history. Yet, as a moment of eternity in time, it also surrounds the shepherds and history, in its turn.

The outwardly severe geometry of Milton's "globe" derives from L. *globus*: "troop" or "phalanx." Thus, the pun is four-dimensional: a globe within a globe. Editors commonly trace its source to Giles Fletcher the Younger's *Christ's Triumph after Death* (1610):

> So long he wandred in our lower sphere,
> That heav'n began his cloudy starres despise,
> Half envious, to see on earth appeare
> A greater light, then flam'd in his own skies:
> At length it [heav'n] burst for spight, and out thear flies
> A globe of winged Angels, swift as thought,
> That, on their spotted feathers, lively caught
> The sparkling Earth, and to their azure fields it brought.[20]

Earth's "sphere" is seized and carried off by a higher "globe." "Swift as thought"—a Lucretian phrase—signals that while this is a physical event and takes place in time, it appears instantaneous to earthly imagining.[21] Similarly, Blake's illustration to Milton's ode strives to expand our powers of visualization (figure 4.5). His choir appears inside a ball shown in cross-section whose outer shell is, itself, composed of a troop of flying, singing angels. The convergence of their lumpish-looking heads at the top implies the flying angels are rising longitudinally not only along the edges of the cutaway but also across the ball's front and back. Perhaps they also descend back down it into and through one another across the spherical surface in that "redoubling" of the flesh which Milton's contemporary, Henry More, termed an "inspissation." Our ability to see *into* the densely angelical yet translucent airy sphere indicates the kind of transformation described by

A Brief Particular History of the Fourth Dimension of Space / 115

Figure 4.5. William Blake, *The Annunciation to the Shepherds*, Illustrations to Milton's "On the Morning of Christ's Nativity," The Butts Set, c. 1815. Huntington Library and Museum.

the mathematician Charles Howard Hinton in an 1880 article that laid the basis for Abbott's *Flatland*: "A being in three dimensions, looking down on a square [or circle], sees each part of it extended before him, and can touch each part without having to pass through the surrounding parts. . . . So a being in four dimensions could look at and touch every point of a solid figure," such as Milton's globe, and "see the insides" of lower-dimensional beings.[22] Blake's picture offers a sidelong glimpse of what it might be like to see as angels do. When *Europe* echoes Milton's ode to describe the birth of French Revolutionary Orc—"What time the secret child, / Descended thro' the orient gates of the eternal day" (*E* 3:2–3; E 61)—the metaphor similarly lifts the eyes up from this emanated earthly incarnation to its higher-dimensional origin in eternity.

Conceivably, Fletcher's earthly "sphere" within a "globe" is indebted to the Shakespearean conceit that "All the world's a stage," that is, a Globe Theatre. There is a similarly transcendental "ball" at the end of Marvell's "To His Coy Mistress":

> Let us roll all our strength, and all
> Our sweetness, up into one ball:
> And tear our pleasures with rough strife,
> Thorough the iron grates of life.
> Thus, though we cannot make our sun
> Stand still, yet we will make him run.[23]

The stanza effects a heroic reversal: instead of running from the sun and its associates, time, death, and fear, "*We* will make *him* run." This miracle cannot rival Joshua's sun stopper at the gates of Jericho, but it doesn't need to because the "ball" of compacted passion arises *within* the diurnal cycle, relegating the sun to the circumference as a natural fact of no great significance.[24] In this way, the closing couplet's shared "our sun" emerges as a spiritual antitype to "the" sun (compare the fallacy of "the" horizon, discussed in the preceding chapter). In the end, then, Marvell's libertine speaker does not escape from chronological time, though that was his initial, foolish *carpe diem* hope. Instead, he travels *into* time by opening it up as indefinite duration. Rather than overthrowing death's cage-like "iron grates"—an evident impossibility—he and his lover will pass "thorough" them like spirits (the better-known folio gives "gates," but they are probably still slatted). This transcendence is already implicit in a likely influence on Marvell, Edward Benlowes's *Theophilia* (1652), a "Heroick Poem" of "Spiritual Warfare," which uses the phrases "Times Ball" and "Universall Ball" to express a totality of perspective beyond ordinary time and space. Drawing on the familiar idea of God as an infinite sphere with center everywhere and circumference nowhere, Benlowe addresses God's omnipresence in much the same terms as Milton's ode:

> Were HE *Material*, then HE *local* were;
> All *Matter* be'ing in *Place*; So, there
> Th'INCIRCUMSCRIPTIBLE would *circumscrib'd* appear.
> HE's so *diffusive*, that HE's All in All!
> All *in* the Universall *Ball*!
> All *out* of it! The only WAS, the IS, the SHALL.[25]

Blake seems to have discerned in such late-Renaissance conceits the intuition of a fourth dimension able to impart depth, space, and freedom of motion to the rigid Euclidean geometry of the New Science. In the Rosenwald drawing of the Last Judgment, he locates the fourth dimension in Christ's head, the center of a rainbow-like series of concentric bands segmented into curved trapezoids by beams of light radiating from Christ's halo (figure 4.6). Several of these trapezoids contain human forms standing or flying in a direction parallel to the light rays and yet at right angles to Christ and the landscape with sun portrayed continuously across one of the middle bands. We seem to see Jesus the Imagination's inward light actively in process of forming "new Expanses, . . . / Creating Space, Creating Time" orthogonally out of Euclidean space (*J* 98:30–31; E 258). It is no accident that the geometry of Blake's image resembles the diagram of a four-dimensional, emanative "sphere within a sphere" in figure 4.4. And notice the sun's relegation to one of the outer bands, similar to Marvell's poem.

Other, less exalted techniques of four-dimensional visualization lay close to hand. Blake would have known William Hogarth's handbook, *The Analysis of Beauty* (1753), which advises artists to

Figure 4.6. William Blake, *The Last Judgment*, c. 1810, top portion. National Gallery of Art, Rosenwald Collection.

> let every object under our consideration, be imagined to have its inward contents scoop'd out so nicely, as to have nothing of it left but a thin shell, exactly corresponding both in its inner and outer surface, to the shape of the object itself . . . the oftner we think of objects in this shell-like manner, we shall facilitate and strengthen our conception of any particular of the surface of an object we are viewing . . . because the imagination will naturally enter into the vacant space within this shell, and there at once, as from a center, view the whole form within, . . . and make us masters of the meaning of every view of the object, as we walk round it, and view it from without. Thus the most perfect idea we can possibly acquire of a sphere, is by conceiving an infinite number of straight rays of equal lengths, issuing from the center, as from the eye, spreading every way alike; and circumscribed or wound about at their other extremities with close connected circular threads, or lines, forming a true spherical shell.[26]

Hogarth's artist constructs the sphere's shape progressively from within rather than representing it as a unitary external object. One recalls Abbott's point that a four-dimensional perceiver would be able to penetrate and see into the interiors of three-dimensional objects without tearing them. The suggestion of a higher-dimensional vision able to see round the back of solid figures is stronger still when Hogarth asserts that "by these means we obtain the true and full idea of what is call'd the *out-lines* of a figure . . . ; for, in the example of the sphere given above, every one of the imaginary circular threads has a right to be consider'd as an out-line of the sphere, as well as those which divide the half, that is seen, from that which is not seen" (9; his italics). Blake was no admirer of Hogarth's art of graphic realism. Nevertheless, his lifelong emphasis on perceived "Outline of Identity" (J 18:3; E 162) as embodied in "the bounding line" (*VLJ*; E 550) attests his appreciation of the enhanced dimensionality afforded by such practical methods of imaginative projection.

Hogarth's suggestion that a sphere's center be envisioned as "the eye" employs a model common in eighteenth-century popular treatises on astronomy, cartography, and projective geometry. As Hannes Ole Matthiessen points out, contemporary introductions to astronomy frequently asked readers to place themselves in the center of a transparent globe and then mark on the inside of the globe's surface all points of intersection between the fixed stars and their eye.[27] The globe thus becomes a representation of all the apparent positions of the fixed stars, including their angular distances from one another and degree of elevation from the celestial equator. The model works because the eye cannot gauge distances between itself and the stars.

Further, Matthiessen demonstrates this was the same model Thomas Reid deployed to arrive at his non-Euclidean "geometry of visibles" in chapter 6 of the *Inquiry into the Human Mind on the Principles of Common Sense* (1764). Reid argues there that the "visible figures" of objects—their perspectival appearances—are curved due to the eye's spherical shape (granted, "the" eye is a geometric idealization; the binocular phenomenon of stereopsis was not explained until 1838[28]). Starting from Berkeley's contention that depth or distance is an "intellectual construction," not an immediate object of sight—a position Blake evidently shared—Reid developed a simple, and metric, science of the perspectival figures of objects that refuted Berkeley's way-of-ideas claim, in *Essay towards a New Theory of Vision*, that because "visibles" exist within the mind and "tangibles" outside it, "visible extension and figures are not the object of geometry."[29] Based on the eye's spherical shape, Reid's realist phenomenology of the slightly curved "sphere of vision" shows, contra Berkeley and Hume both, that visual and tactual space (i.e., an object's mind-independent "material impression" on the retina) disclose two different geometries. The geometry of objects seen from the center of the eye's sphere of vision describes a different space from Euclidean geometry. Lie down in the middle of a large room, says Reid, and you'll notice all four corners of the ceiling form angles apparently greater than ninety degrees, as though projected onto a sphere. To be sure, Reid never went so far as to claim that we can actually *see* the curvature of visible objects. Blake, however, aims to raise this underlying condition of vision into consciousness, as for example in the implicitly spherical design of the *Daughters of Albion* frontispiece (figure 1.1).

Dante, *Milton*, and the Gluing Together of Two Spheres

Then there is Renaissance illusionist painting on the interior walls of temple domes. Most of these pictures invite analogy: as a dome's painted surface of saints and angels stands in relation to the real three-dimensional space occupied by the viewer, so that space stands in relation to heaven represented at the dome's pole. One of the first and most renowned of these domes was made by Coppo di Marcovaldo for the Baptistery of Florence. Which brings us to Dante, who saw the baptistery as its mosaics approached completion before he left Florence in 1301—Dante, arguably *the* major influence on the cosmology of the later Blake (whereas, Milton's cosmography in *Paradise Lost* lies halfway between Ptolemaic/Aristotelian and

Copernican/Galilean). The parallel in *Milton* between Dante's guidance by Virgil and Blake's by Milton is almost too obvious to mention.[30] Above all, it is the perfect unprecedentedness of the Dante pilgrim's journey through supernatural realms, his Christ-like singularity as a real historical event with everlasting repercussions, that Blake absorbed: "This thing / Was never known," exclaims Los, "that one of the holy dead [Milton] should willing return (*M* 23:57–58; E 113). *Milton* fuses the world-historical uniqueness of the French Revolution, whose violence has triggered the hero's return, with a traditional emphasis on the uniqueness of Christ in whom he must submerge his still-strident selfhood.[31] Coppo's dome is generally accepted as a key source for the *Commedia*'s eye-opening conception of a four-dimensional heaven. Long before Newton had convinced everyone to map the infinite space of the cosmos onto flat Euclidean geometry, Dante, drawing on the teachings of his mentor Brunetto Latini, anticipated Einstein's work in 1917 on the curvature of space by portraying an explicitly non-Euclidean universe.[32] As the mathematician Mark Peterson recognized in a 1979 article in *The American Journal of Physics, Paradiso,* cantos 27–28, solved the nagging problem of Aristotle's finite universe—what lay on the other side of it?—by "gluing" the standard Aristotelian hemisphere containing the nine concentric spheres of the planets and fixed stars with Earth and Satan at the center, to that of the Empyrean containing the corresponding spheres of angelic orders with God at the center.[33] When Dante's pilgrim stands on the Primum Mobile, which as the last sphere of the physical universe moved directly by God forms the equator where the physical and heavenly spheres meet, he sees the nine spheres of angels narrowing to a distant point of light at the pole, God:

> One circle, light and love, enclasping it.
> As this doth clasp the others, and to Him,
> Who draws the bound, its limit only known.[34]

Thus, Henry Cary in the 1819 translation Blake evidently came to prefer. "Enclose[d]" and "enclosing," or "surrounding" and "surrounded by," would be more accurate renderings of Dante's *d'un cerchio lui comprende, / sì como questo lui altri*. Blake's pencil drawing of the Last Judgment (figure 4.6) accordingly shows no fewer than nine distinct bands of space surrounding Christ's head, counting left from his halo.

One standard method that physicists use to visualize a sphere "enclosing and enclosed by" another is by means of a "locally glued area" whose relatively small surface is mathematically smooth and close to Euclidean (figure

4.7). Gustave Doré's 1868 illustration to *Paradiso* canto 31 depicts Dante's rocky Primum Mobile in between the physical and heavenly hemispheres as just such an area (figure 4.8). I've been suggesting the intermediate

Figure 4.7. "Gluing" two spheres together. "The white area near the black dot is a 'locally glued area.'" This area forms a pair of Blakean Vortexes extending into each sphere with their shared tip—the black dot—in 3-d space. Stephen Leon Lipscombe, *Art Meets Mathematics in the Fourth Dimension*, 2nd ed. (New York: Springer, 2014), fig. 2.7.

Figure 4.8. Gustave Doré, *Canto 31: The Saintly Throng in the Form of a Rose*, Illustrations to Dante's Divine Comedy, 1868. Commons Wikimedia.

peninsula of Blake's *Daughters of Albion* frontispiece portrays another such (figure 1.1). Blake's Dante illustration, *The Deity from Whom Proceed the Nine Spheres*, gives a side view of the Empyrean as rising at right angles to the earth's surface (figure 4.9). Blake uses the method of sequential "slicing"

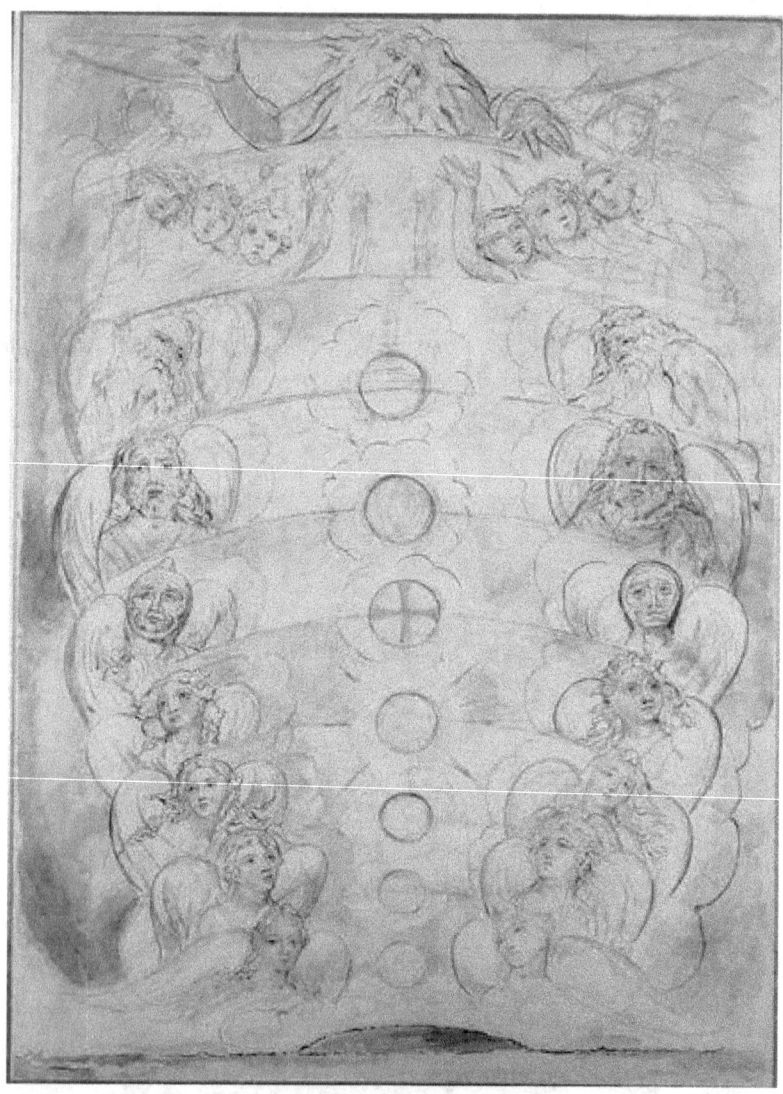

Figure 4.9. William Blake, *The Vision of the Deity from Whom Proceed the Nine Spheres*, Illustrations to Dante's Divine Comedy, 1824–27. Ashmolean Museum, Oxford University.

to construct Dante's four-dimensional sphere as a loaf. The row of badges running like buttons up the design's middle reminds us that each slice is really an orb, each line separating the slices, really an orbit. The loaf's God is shown as greater than the sum of his geometrical parts because he embodies an extra coordinate in addition to the triad of "Length: Bredth & Highth," which Blake associates with Satan-Urizen and Newtonian space (*M* 32:18; E 132). In *Paradiso*, canto 28:41, Dante designates this fourth coordinate as "swiftness."[35] There, Dante is at pains to depict the pilgrim's confusion over his vision's metaphysical significance. Why do larger spheres rotate with greater speed among the planets and fixed stars, whereas the Empyreal angels rotate faster the *smaller* they are? Beatrice explains that, even though the angelic spheres diminish in size and outward greatness as they approach their limit, God, they also gain in speed—namely, because the circumference of the overall cosmic sphere, which increased in size as it went from Satan to the Primum Mobile, is now decreasing as it reaches its point of completion at the opposite end (figure 4.10).

Says Peterson, "[Dante's] fourth dimension is speed of revolution" (1033). Indeed, Beatrice calls the Primum Mobile "The vase, wherein time's roots are plunged," whose "leaves" are "elsewhere" (27:118–19). "Swiftness," then, is the medium anchoring the Primum Mobile to its twin exterior points, Satan and God, much as Blake recognizes when he says, "Time is the mercy of Eternity," without which redemptive pilgrimage through space

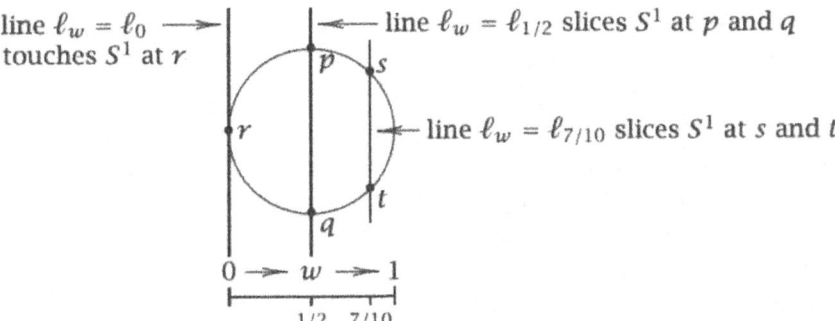

Figure 4.10. "Slicing" a sphere. "As the values of w increase, the line ℓ_w moves across S^1." Stephen Leon Lipscombe, *Art Meets Mathematics in the Fourth Dimension*, 2nd ed. (New York: Springer, 2014), fig. 2.8.

would be impossible: "without Times swiftness / Which is the swiftest of all things: all were eternal torment" (*M* 24:73–4; E 121).[36] So, in addition to the fixed up/down, forward/back, right/left coordinates that define the three-dimensional world, Dante adds movement both inward (expanding) and outward (contracting/conglobing).

Milton undertook his study of Italian in the fall of 1629, the year of the Nativity Ode and its "globe of circular light."[37] Blake probably did not read Dante intensively (likely in Boyd's translation) until he wrote *Milton* in 1804–11. So great was Dante's impact on him, one wonders if the core error Blake wanted to redeem in his poem wasn't Milton's moralism so much as *Paradise Lost*'s confusing tendency to conflate Dante's four-dimensional, pre-Newtonian universe of concrete place with a Cartesian-Newtonian analytic geometry of space.[38] Geometrically, Blake's elaborate mythology derives its coherence less from redemptively ironical parodies of Newton, as Donald Ault's *Visionary Physics* suggests, than from the positive cosmic structure of the *Paradiso*. For, in the *Paradiso*, the Old World Aristotle universe with a Ptolemaic Earth at its center is not rejected but *subsumed* by the Christian vision of a New World with an all-encompassing God at the center. It was Jacob Bronowski who first saw, in 1944, that in Blake's cosmos, "the centre meets in the circumference." Bronowski cites the following passage:

> The Vegetative Universe, opens like a flower from the Earths center:
> In which is Eternity. It expands in Stars to the Mundane Shell
> And there it meets Eternity again, both within and without.
> (*J* 13:34–36; E 157)

Then he tersely observes that this construct "cannot be put into a space of only three dimensions. And Blake knew this."[39] Blake's "both within and without" corresponds to Dante's "surrounds and is surrounded by."

I will return to the crucial issue of time and movement raised a moment ago. Meanwhile, let us pursue the consequences of the spatial technique of "locally gluing" the earthly sphere to a four-dimensional heavenly one. This move places God at the vertical directly above any earthly point, as in Blake's *The Deity from Whom Proceed the Nine Spheres*. Since Earth's surface is surrounded by heaven, one can touch the Empyrean anywhere. A recent commentator says that in Dante "the Empyrean is not just perched on top of the mundane cosmos in one specific place, as it necessarily appears in any two or three-dimensional rendition. Rather, the edges of the Empyrean are everywhere around us."[40] Accordingly, Blake portrays *Milton*'s hero as "descending perpendicular" from "the zenith" to the Blake poet in Felpham below (*M* 15:47–48; E 110). In the diagram of "Miltons Track" (figure 4.11), Milton passes through the tangent where the four worlds of Blake's myth

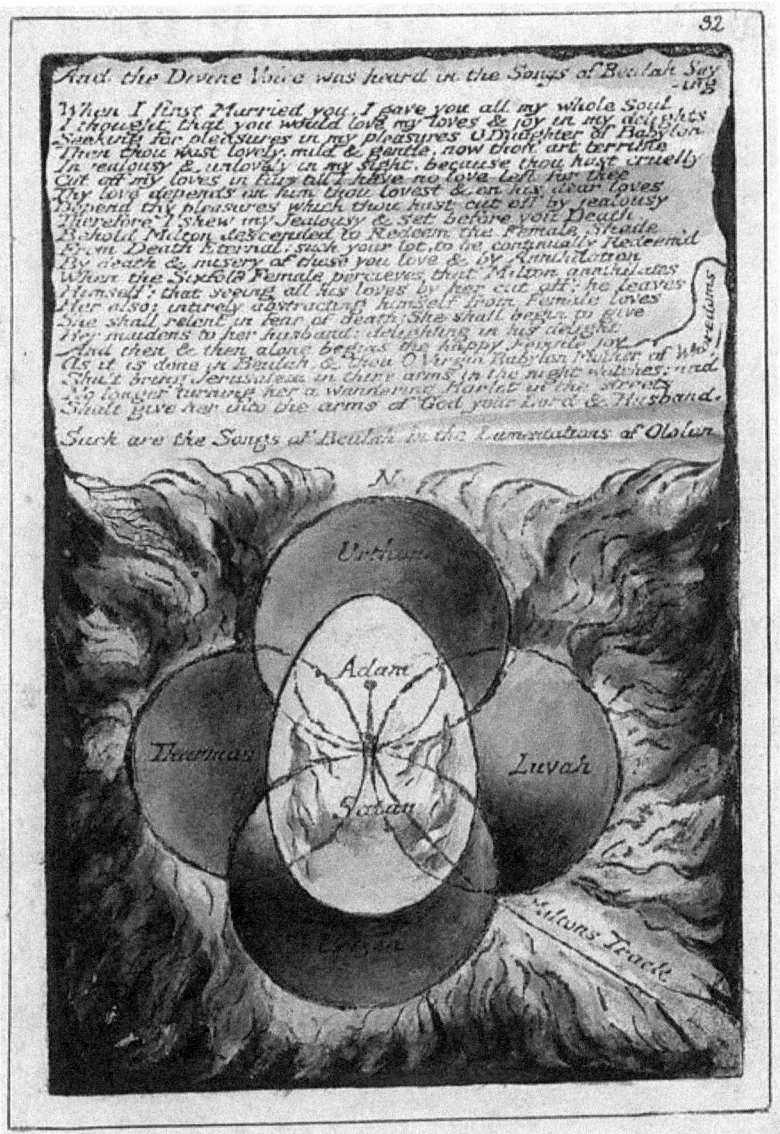

Figure 4.11. William Blake, *Milton*, pl. 32, copy C, c. 1804–11. New York Public Library.

(the "Four Universes round the Mundane Egg" [*M* 19:15; E 112]) touch: the "Moment in each Day that Satan cannot find" (*M* 35:42, E 136) where earthly vision opens onto the cosmos. Then he drops straight down through the now apocalyptically cracked Mundane Egg into Adam-Blake's sphere of perception. Blake draws a Venn diagram of how Eternity panentheistically includes the Egg of earthly existence while transcending it at every point.

Likewise, the nondescript, bulgy green globe in the design to *Milton* plate 47, where Milton-Los makes contact with the poet, implies that *any* earthly place can serve as Dante's Mount Purgatory, perhaps even one of the hills outside the Sussex village of Felpham (figure 4.12).[41] The design

Figure 4.12. William Blake, *Milton*, pl. 47, copy C, c. 1804–11. New York Public Library.

recalls the undistinguished-looking bump of Mount Purgatory in Blake's Dante illustration. Indeed, once we imagine the universe as a single spherical surface, any point on that surface may be regarded as its center. From that point, all lines, including lines of sight, must be drawn as meridians on a globe converging to a single point on the sphere's opposite side. This helps to explain why, after Blake's Milton defeats Satan and becomes one with Jesus, he reascends not to the Calvinist heaven he left behind but rather disappears into an inward Eternity more closely corresponding to Satan's opposite. In *Paradise Lost*, Milton's Satan wanders on the solid surface of the sphere of the Primum Mobile ("the bare outside of this World") before discovering at its zenith a Jacob's Ladder spiral stair connecting, that is, "gluing" Chaos to the Empyrean (3:74, 3:501–25). In embracing the Miltonic Satan, Blake's Milton overthrows the transcendence upward and out not only of *Paradise Lost* and Dante but their ultimate source, *Phaedrus* 24c, where Plato relates that the gods sometimes drive their chariots up through the opening at the apex of the sky's dome and then take up stations on the outer, convex surface of the cosmos, "the place above the sky." That is just the place Blake's Eternity revalues and inverts: for "beyond the skies: / There Chaos dwells & ancient Night" (*M* 20:32–33; E 114).

Considering all this, the charming anecdote recounted in Gilchrist's *Life of Blake*, in a chapter titled "Mad or Not Mad," quite exceeds the biographer's polite Victorian backpedaling:

> Some persons of a scientific turn were once discoursing pompously and, to him, distastefully, about the incredible distance of the planets, the length of time light takes to travel to the earth, &c., when he burst out, "Tis false! I was walking down a lane the other day, and at the end of it I touched the sky with my stick;" perhaps with a little covert sophistry, meaning that he thrust his stick out into space, and that, had he stood upon the remotest star, he could do no more; the blue sky itself being but the limit of our bodily perceptions of the Infinite which encompasses us.[42]

Rather, Blake's stick points resolutely forward to his strong, detailed intuition of non-Euclidean geometry. In the fourth century BCE, the philosopher Archytas had already taken Aristotle's closed universe to task: "If I found myself in the furthest sky, that of the fixed stars, would I be able to stretch my hand, or a rod, out beyond it—or not? That I should not be able to is absurd; but if I am able to, then an outside exists, be it of matter, or space."⁴³ The answer to that question, Blake like Dante found, is physically curved space—not a space embedded within another space external to it, but one whose intrinsic fabric is curved in the sense that the network of distances between its various points is not flat. In a *Jerusalem* passage that stands in dark counterpoint to the one Bronowski mentions, Blake associates Archytas's contradictory "outside" with Satan's temptations toward otherworldly transcendence:

> There is an Outside spread Without, & an Outside spread Within
> Beyond the Outline of Identity both ways, which meet in One:
> An orbed Void of doubt, despair, hunger, & thirst & sorrow.
> (*J* 18:1–4; E 162)

These twin outsides form the obverse of the "two Eternitys" that "meet together" at *Milton* 13:10–11. One of the outsides projects "Without" a three-dimensional world rooted in the solid Newtonian globe of Earth. This looks like an adaptation of Berkeley's idealist account of the chronic perceptual habit of "outness,"⁴⁴ which Whitehead saw to imply a critique of the Newtonian "fallacy of simple location." The other outside projects "Within" an anxiety-inducing loss of "fleshly" spatiotemporal location arising from the imposition of a coordinate system of abstract points, without which the externalized globe would lack a void in which to roll. A related passage tells how the lovely Daughters of Albion

> shall fold & unfold
> According to their will the outside surface of the Earth
> An outside shadowy Surface superadded to the real Surface;
> Which is unchangeable for ever & ever Amen: so be it!
> (*J* 83:45–48; E 242)

Since Earth's "real Surface" is three-dimensional and spherical—Earth enwrapped by Eternity forms a 3-sphere—therefore the "outside shadowy Surface" unfolded by Newtonianism is formless, as shown by Satan's ultimate collapse in *Milton*'s conclusion.

Yet, nature can be beautiful. *Milton*'s invocation identifies the poet's very muses as seductive Daughters of Beulah whose pastoral comforts entice the pilgrim to mistake "all this Vegetable World" (*M* 21:12; E 115) for heaven and so abandon his epic journey. This danger follows from the spatialization of time. Not even Los's redemptive realm of Art described on *Milton*

A Brief Particular History of the Fourth Dimension of Space / 129

plates 28–29 avoids escapist, Beulah-like overtones of religious consolation and aesthetic ideology. For, in giving form to the segments of chronological time, Los reifies their qualitative character of movement, passage, change, freshness, novelty, perishing. This spatializing tendency appears at the very base of Blake's *The Sun at His Eastern Gate* (figure 4.13), a diorama-like illustration to Milton's "l'Allegro" that recalls *Milton*'s fiery Los descending

Figure 4.13. William Blake, *The Sun at His Eastern Gate*, Illustrations to Milton's "l'Allegro" and "Il Penseroso," c. 1816–20. Morgan Library and Museum.

upon the poet at daybreak in the plate 47 design (figure 4.12). The Sun is just stepping forth from an enormous fructifying globe onto what looks like a plinth or ledge of dawn sky folded back at right angles to the lightening rural scene below. One is reminded of the baldachins to the High Gothic cathedral statues that Blake studied during his apprenticeship to Basire. As Erwin Panofsky observed, the baldachin "delimits and assigns [the statue] to a particular chunk of empty space . . . making [its] field of activity into a veritable stage," in contrast to the homogeneity of rational perspectival space.[45] Blake's rectilinear shelf makes plain that his outsize, all-surrounding Sun visits perpendicularly from beyond the third dimension. The back-to-front clockwise swirl of angels locates the Sun at the tip of a Vortex (or spindle torus) emerging from Eternity—namely, now, a repetition of the Creation–Fall at the heart of Blake's myth, when light creates from darkness a world of visible three-dimensional objects. This is the Moment of moments that *Milton* portrays as descending through the Vortex's other, open end from indefinite, measureless duration beyond all the chronological bits of time (also beyond Los's Golgonooza, whose gleaming gold and silver palaces are finally just imitations of Creation's light). In short, the "l'Allegro" design portrays sunrise as the "gluing" together of heaven and earth in a new Glad Day. Since light rays shoot across the plinth at bottom from *behind* the flat edges of the sun, manifestly, that disc, too, is just a painted outward proxy of Eternity.

Elsewhere as well, Blake uses noticeably flat, childlike diorama drawings to emphasize his designs' three-dimensional theatricality and constructedness, in contrast to the apocalyptic revelations described in the poems. Examples include the endearingly primitive *Milton* plate 40 ("Blake's Cottage at Felpham"; figure 4.14); the angular urban scene of "The Chimney Sweeper" (*SE*); the profile of "The Tyger" (whose tuck-up looks distinctly like a parody of Hogarth's "line of beauty" or "grace"); the x-y-z Cartesian directional planes on *The Book of Urizen* title page formed by Urizen's plinth-like open books and upright Tablet-tombstones (figure 3.4); and the stagy step or platform at the bottom of "A Cradle Song" (figure 2.3) and its companion "Infant Sorrow," which links the unusually dense realism of both pictures with a claustral, externalizing Euclidean perspective (and compare the S-curve of "Sorrow"'s bending female with that of the Tyger).

These blatantly rectilinear designs turn Blake's early, Flaxman-influenced style of "Romantic-neoclassical" linear abstraction against itself, exaggerating its formal reductions of space to generate dissatisfaction with the constraints of three-dimensional representation. The mathematician George Yuri Rain-

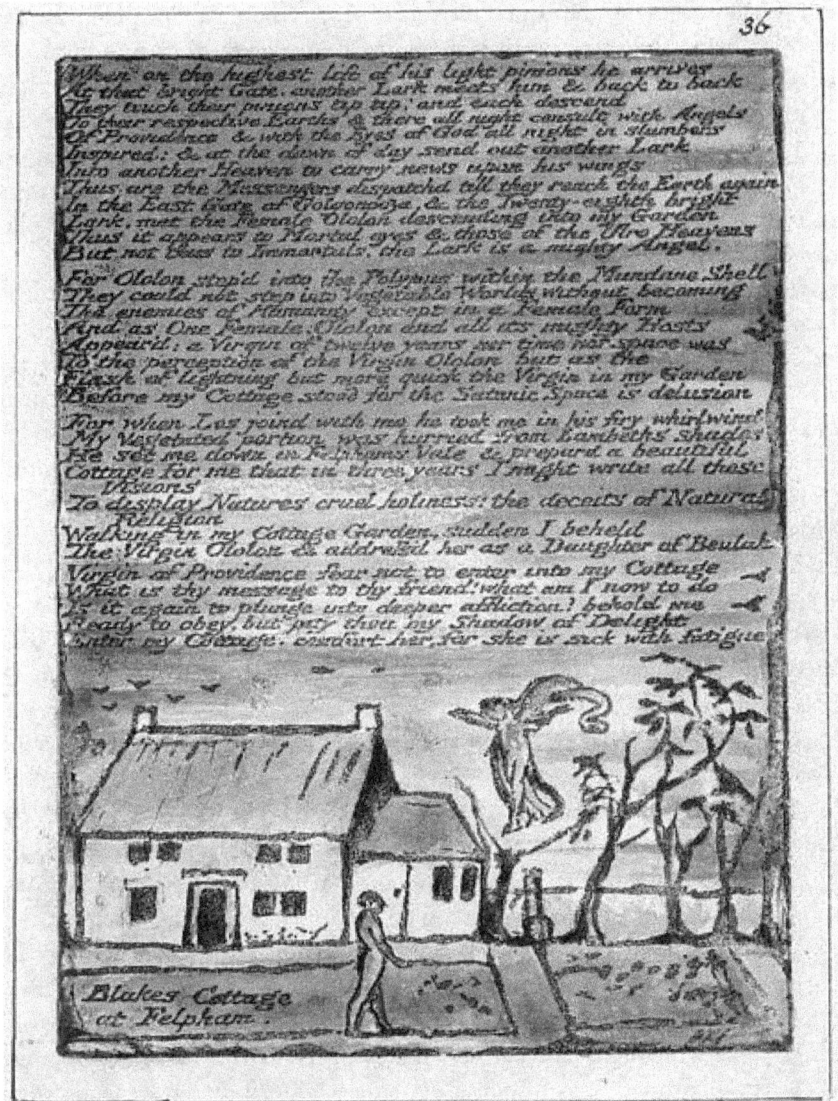

Figure 4.14. William Blake, *Milton* pl. 40, copy C, c. 1804–11. New York Public Library.

ich once told his students, as he prepared the room to teach Einsteinian Relativity: "We cannot imagine (or 'diagram') this space, for Euclid is in the blackboard."[46] Blake's flat-looking, right-angled illustrations emphasize that Euclid is also in the copperplate and paper. These parodies of three-dimensional representation anticipate the early twentieth-century critique of classical Cartesian space that accompanied cubism and the emergence of relativity. Milič Čapek remarks that in Cartesian space "no matter how minute a spatial interval may be, it must always be an *interval* separating two points, each of which is *external* to the other."[47] Space as an extensive continuum is infinitely divisible because it can accommodate an infinite density of constituent point-elements that have no extension themselves, no inside region or properties of their own, and are therefore unbounded. Hence, the epiphany of Blake's Satan-Urizen in *Milton*'s finale proves him to be empty externality incarnate, "an outside which is fallacious" (*M* 37:9; E 137) and whose location is "In the Newtonian Voids between the Substances of Creation" (37:46; E 138). In the Cartesian system, the emptiness of space is what enables finite objects to appear, while space itself eludes perception; Newton's notion that space's primacy might make it a divine sensorium followed naturally. In other words, Cartesian space appears boundless because it embodies the very condition of boundedness. Satan-Urizen's installment as a Space "Limited / To those without but Infinite to those within" (10:8–9; E 104) illustrates this paradox, as does the *Daughters of Albion* frontispiece (figure 1.1). Once Milton stands forth from his spectrous Shadow in *Milton*'s conclusion, thereby depriving it of spiritual substance, the delusory Urizenic Space immediately "fall[s] apart" to expose Satan's boundless emptiness as an abstractive self-contradiction, a State of "Eternal Death": "trembling round his Body, he incircled it / . . . / Howling in his Spectre round his Body hungring to devour / But fearing for the pain" (39:16–26; E 140).

By contrast with Satan, the concreteness of *Milton*'s conclusion demonstrates the truth of Hermann Minkowski's claim: "Nobody has ever noticed a place, except at a time, or a time except at a place."[48] If the spheres of Earth and Eternity "surround and are surrounded by" each other, then introducing time, motion, and change into this spatial geometry necessarily activates a certain unequal reciprocity between them. The ancient contraries, being and becoming, enter upon a dialectic whose asymmetry ensures they will continue to produce real difference or, in Whitehead's terms, freshness and novelty. "God becomes as we are, that we may be as he is" (*NNR* [b]; E 3). If Blake's God is being, and Jesus becoming, then the first "as" in Blake's sentence above designates a descent into self-sacrifice that is already occurring continually, while the second looks toward a redeemed condition that remains no more than prospective.

So, in addition to the up/down, forward/back, right/left coordinates of three-dimensional reality, Blake like Dante adds the dimension of divinely expanding inward/Urizenically contracting or "conglobing" outward. Blake's Creation-Fall occurs when Albion's eternal human torus becomes "unglued." The four-dimensional sphere at the center ceases to "enclose and be enclosed by" its virtual three-dimensional surface. In a reversal of the inside-out configuration noted earlier by C. S. Lewis, the material outer ring of the donut's surface contracts into a solid ball, expelling Heaven to the periphery as inert, empty space. This is the perspective parodied in the *Daughters of Albion* frontispiece, with its implication that the three characters are really estranged bits of the reader-viewer who contains them.

As we've already seen, though, this fall can be reversed by gluing two spheres together to create an area where they interpenetrate. Blake identifies the form of this gluing as a Vortex:

> The nature of infinity is this: That every thing has its
> Own Vortex; and when once a traveller thro Eternity.
> Has passd that Vortex, he percieves it roll backward behind
> His path, into a globe itself infolding; like a sun:
> Or like a moon, or like a universe of starry majesty,
> While he keeps onwards in his wondrous journey on the earth
> As the eye of man views both the east & west encompassing
> Its vortex; and the north & south, with all their starry host;
> Also the rising sun & setting moon he views surrounding
> His corn-fields and his valleys of five hundred acres square.
> Thus is the earth one infinite plane, and not as apparent
> To the weak traveller confin'd beneath the moony shade.
> Thus is the heaven a vortex passd already, and the earth
> A vortex not yet pass'd by the traveller thro' Eternity.
> (*M* 15:21–35; E 109)

Since Eternity is "glued" rather than existing independently, journeying to the tip of the Vortex will always lead back to the fallen three-dimensional world. The concept is topologically precise. As the distance from the axis of revolution decreases, a "ring" torus deforms first into a "horn" and then into a "spindle" torus, until finally it collapses in on itself to form a sphere: the point where, for Blake, Newton's weak law of attraction takes over. Albion, "the Immortal [who] expanded / Or contracted his all flexible senses" at will (*BU* 3:37–38; E 71), inverts into a horde of isolated, puny earthlings ruled by an external God. Satan-Urizen creates a space that "shrinks the Organs / Of Life till they become Finite & Itself seems infinite" (*M* 10:6–7; E 104).[49] By contrast, if "the eye of man" in the above passage "views both the east & west . . . / . . . and the north & south . . . / Also the rising sun & setting moon" all simultaneously, so that he sees in every direction at once, then his perspective can only be four-dimensional.

But the Vortex also defines the process of "infolding . . . into a globe" as a temporal interval. It takes place "within a Moment, a Pulsation of the Artery" (*M* 29:1–3; E 127), which then releases the space-time traveler to continue into the "infinity" of the next Vortex-Moment. As opposed to mathematizing change and motion by means of nonphysical instants such as Newton's fluxions, Blake introduces a process of extensive, "thick" becoming. For, as A. A. Robb remarked with respect to Einstein's relativity of simultaneity, "an instant cannot be in two places at once."[50] Unlike Dante's purely spatial fourth dimension, *Milton*'s Vortex-Moment establishes a relativistic space-time where "now" is always "here," and "there" is always "then." For Blake, therefore, "space" merely designates the class of simultaneous events, while "distance" names the one-dimensional series of coexisting points. In his mature myth, connections between events are necessarily successive and time-like rather than instantaneous geometric cuts in space like the peninsular promontory in the *Daughters of Albion* frontispiece. The spatial distance between events is redefined as the causal independence of different world-lines—albeit these are not geometric "lines" but series of events, some concretely realized like "Blake's Cottage at Felpham" (*M* pl. 40), some seeking realization like Milton and Ololon, and others like Satan doomed to irreality by the structural logic of Blake's myth.

Northrop Frye's observation that *Milton*'s Bard's Song offers "not the story of a sequence of events, but rather a series of lifting backdrops" is exactly right, provided we recognize that the Song's events are not dimensionless instants but possess "contemporaneity" or what Čapek calls "transversal extent or *width*" (161, his italics; and 220).[51] They exist on the world line of the vast series of causally independent events that make up "the contemporary world" (even if, viewed in conventional three-dimensional perspective, Blake's "lifting backdrops" mostly resemble stage mayhem in the Marx Brothers' *Night at the Opera*). In other words, the Bard's Song's different scenes and settings form a set of successive "nows-elsewhere," and only the Blake poet's ultimate frame of reference, "Felpham," can decide which ones get to be considered as simultaneous with the here-now. As *Milton*'s narrative moves beyond the Song, each new subnarrative frame of reference progressively narrows the indefinite cosmic timeline of nows-elsewhere to the privileged earthly here-now of the Blake poet one spring morning in "1804," the date carved on the title page, outside the "beautiful / Cottage" which was provided for him, he is able in the end to disclose, "that in three years I might write all these Visions" (*M* 36:23–24; E 137).

Conversely, what Blake's Milton needs to do is break through the Euclidean barrier of Newton's absolute time and space, which erects a single selective class of there-nows as simultaneous with the here-now, namely, those that form the "middle zone" of ordinary human perception. From an Einsteinian perspective, Newton illicitly merged Minkowski's successive, instantaneous cuts of space together to form a three-dimensional universe. In this way, he collapsed the multiplicity of virtual nows-elsewhere into concrete there-nows, so dividing the here-now from the possible world-lines of its past and future. The result is a purely spatial present capable only of mechanical change. Accordingly, Blake links the predestinarianism of the dead historical Milton "pondring the intricate mazes of Providence" (*M* 2:17; E 96) with Laplace's Demon of determinism. So, *Milton* tells what happens when, thanks to the supervening Blake poet, Milton comes to *know* he was "of the Devil's party without knowing it" (*MHH* 6; E 35). Such self-identity beyond all temporal displacement constitutes an instantaneity, "a State . . . / Called Eternal Annihilation" that points toward the paradoxically displaced and perpetually perishing nature of "the Human Existence itself" (*M* 32:26–32; E 132). Blake's redeemed Milton can therefore say with *Jerusalem*'s living Jesus: "I Die & pass the limits of possibility, as it appears / To Individual perception" (*J* 62:19–20; E 213).

The Blakean Milton's task, then, is to open his arrested present to the causal array of independent nows-elsewhere which Newtonianism had dismissed as mere indeterminate flux (the oozy world of Blake's Vala and his *Newton* color print; figure I.1) and replaced with a measurable, hard, and solid here-now (the reified world of Ulro, as represented by the diagram Newton is making in the print). Ololon, the human potential Milton had denied in his wives and daughters, and thence in himself (the patriarchal poet's "female side"), is one such now-elsewhere; David Erdman calls her "history-as-it-might-have-been."[52] Milton trapped in his own predestinarian Heaven is another. Gradually, these two world-lines converge to achieve causal connection in the mind of the inspired Blake poet. Far from being separate bodies moving through the container of Newtonian space, Milton and Ololon appear as evolving portions of a cosmic dialectic in which classical material substance has been resorbed into curved space-time as a local deformation, namely, a Vortex. Each figure gains progressively more definite human form until, married in one flesh, they acquire together the character of an event here and now in the garden outside "Blake's Cottage at Felpham." *Milton*'s Vortex is the poet's personal light cone.

Look again at the illustration so captioned (figure 4.14), a mild pastoral counterpart to the schematics of "Miltons Track" (figure 4.11). Blake's primitive little diorama built out of right angles invites us to try to plot Ololon's position against the x-y-z coordinate axes formed by the garden paths and cottage walls. Yet, she herself glides balletically along an invisible tightrope running obliquely from the text's bottom right corner in the middle distance straight toward the eye of comically unsuspecting, plumply waistcoated "Blake." As throughout the Illuminated Books, the text plane hanging against the dawn sky backdrop is a real material presence, within the picture though not *of* it (like the text pegged against the curtain in "A Cradle Song"'s design; figure 2.3). It seems Ololon has materialized out of a vertex—possibly a vortex—at the page margin just out of sight, where the narrative's four-dimensional mythological realm emanates into Euclidean pictorial space.

So, the Vortex represents Dante's Primum Mobile—the "locally glued area" between *Paradiso*'s earthly and empyreal spheres, ripped apart in an age of Newtonianism—as available for regluing by messengers willing to traverse the two regions. But progress is arduous. For two spheres that have been "put through" each other remain independent spaces throughout and touch only at their common three-dimensional surface or Satanic "Outside." An earthling might notice a remote event "in the zenith" of his perspective (*M* 15:48; E 110), much as the Blake poet glimpses Milton, but he "cannot know / What passes in his members till periods of Space & Time / Reveal the secrets of Eternity" (21:8–10; E 115). Ololon's journey earthward as "a sweet River, of milk & liquid pearl" (21:15; E 115) is distinctly sluggish. Meantime, Milton during his descent peels off no fewer than four different personas corresponding to the four different mythological levels or States of Blake's narrative: his dead and redeemed self asleep on its Death Couch in Eternity, another persona striving to reform Urizen in a standoff still unresolved by the poem's end, a "vehicular" third one shooting through the sky to land on "Blake"'s left foot, and a fourth who finally manifests in Felpham garden as a historical puritan. Thereby, *Milton* acknowledges the "thick," quantum nature of its redemptive Moment.

Henry More: "Spissitude," the Multidimensional Ogdoaz, and Blake's Vortex

Central to *Milton*'s unfolding of Blakean vision is Henry More. More's claim, "That besides those THREE dimensions which belong to all extended things, a FOURTH is also to be admitted, which belongs properly to SPIRITS,"[53] plausibly showed Blake the way to a more immanent and organic emanationism than could his Neoplatonist friend, Thomas Taylor, a radical dualist. In *The Immortality of the Soul* (1659), More dubs this fourth dimension "Essential spissitude" (from L. *spissitudo*, thickness):

> This Mode or Property of a Substance, that is able to receive one part of it self into another. Which fourth Mode is as easy and familiar to my Understanding, as that of the Three dimensions to my Sense or Phansy. For I mean nothing else by Spissitude, but the redoubling or contracting of Substance into less space then it does sometimes occupy. And Analogous to this is the lying of two Substances of several kinds in the same place at once.[54]

Stripped of its metaphysics, the idea resembles the Kleinian "redoubling" and "coiling-over" of the flesh described in Merleau-Ponty's late phenomenology. As examples, More gives the folding of a string or the "globulation" (balling-up) of a long piece of wax (a likely allusion to Descartes's Second Meditation), where loss of longitude entails no diminishment in substance. Since spirits, for More, were dilatable, contractable, and penetrable, spissitude ensured that any loss of three-dimensional spatial volume in a contracted immaterial substance would be made up for by a proportional increase in its four-dimensional volume. One wonders if he was thinking of the co-location of spirits in three-dimensional space by analogy with the mapping of two hemispheres onto each other, as when the projection of a globe onto a plane superimposes opposite points on the globe's surface. Admittedly, More tends to conflate "inspissation"—in medical terms, the thickening of a liquid almost to a solid—with "conspissation," namely, condensation or congealing—and also "saturation," when "a spirit will not easily admit of further redoubling" (*Enchiridion Metaphysicum* 204, 216). He seems to run together his higher-dimensional geometrical analogy with a more materialist and aetherous idea of spissitude as a simple increase in a body's viscosity or density.[55] Despite this—or because of it—inspissation in the sense of a dynamic physical process supplies the very glue of Blake's Vortex, which reconstructs Dante's God-given four-dimensional cosmos through the thought experiment of two spheres being progressively put through each other to create conical regions of increased depth and volume. A case in point is the hourglass shape of Blake's painting, *The Vision of the Last Judgment*, examined in my conclusion (figure C.3). Blake's greatest student, W. B. Yeats, describes this shape in *A Vision* as a double gyre or pair of interlocking cones (albeit his purely spatial, theosophical perspective was much less relativistic than Blake's).[56] At the heart of More's system stands his "universall Ogdoas,"[57] a Vortex-like image of God's immanency throughout the cosmos. The Ogdoas forms an eight-tiered, symbolical Cone of emanations, which More eventually came to conceive as physical (figure 4.15). The center is the Divine One while the circumference is Hyle/matter, an infinite array of barely vital atoms.

The ideas of Psyche/Mind acquire extension as they ray out through Physis/Nature to the Cone's material cusp. Jasper Reid explains: "Psyche's role was to take the point at the Cone's apex, to multiply it and lay the resulting atoms out in a three-dimensional juxtaposition. Psyche would thus be directly responsible for the generation of Tasis [extension] out of Hyle, i.e. the extraction of an actual bulk out of a state of mere potency, and she would then clothe herself in the perfectly homogeneous matters that she had thus produced. Tasis would constitute '*Psyche's* out-array.' "[58] Reid's last phrase quotes More's early poem, *Psychozoia*. In *Milton*, the naturalizing tendency of what *Psychozoia* calls Psyche's "stole aethereall, / Which . . . down to the earth doth fall" (1.39)—that is, the particles of Tasis/extension as combined to produce sensible bodies—is represented as phenomenal images of nature woven by the seductive Daughters of Beulah. Like More, Blake views Nature as a symbol of Mind.

For, in More's monistic, proto-Priestleyan immaterialism of matter, "body" is defined as "nothing but a fixt spirit, the conspissation or coagulation of the cuspidall particles of the Cone, which are indeed the Centrall Tasis, or inward essence of the sensible world."[59] Since body and space are distinguished only according to whether their atoms are congealed together or not, and since spirits can contract themselves to the size of atoms, it is

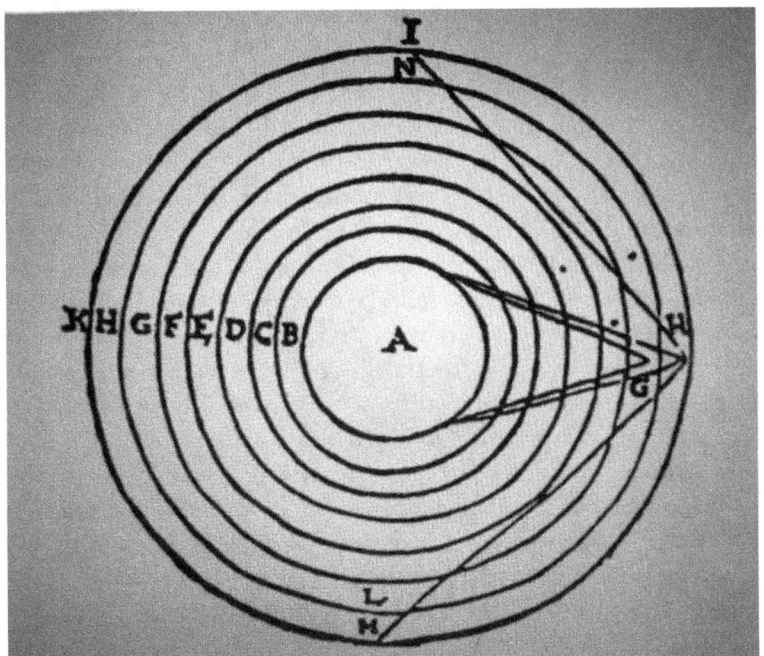

Figure 4.15. Henry More, "Ogdoaz," from *A Platonick Song of the Soul* (1647), ed. Alexander Jacob (Lewisburg: Bucknell University Press, 1998), 110. More refers to this diagram throughout his poems; see Grosart, 13–21, 54, 108. A-D represents spheres of the intelligible world, E-H realms of sense. When Ahad/The One at A projects outward to any particular atom-point at H, it forms a cone with a three-dimensional focus. Jacob comments: "The crucial elements in this representation of the transformation of the intelligible world into the sensible are Psyche [C] and Physis [F], or the Soul and the Spirit of Nature . . . The world of phenomena is formed by Psyche's outward movement, which causes the Ideas contained in her consort, Aeon's mind to take on the appearance of physical extension or Tasis [G]."

possible for two separate bodies to penetrate each other and coexist in the same space (contra both Descartes and Berkeley), while spissitude continues to preserve the essential separateness of the two spirits. We appear to have here a basis for the spiritual mechanics or visionary physics of the vehicular Milton's entrance into the Blake poet's left foot in *Milton*.

The importance of vehicles in More's system sprang from his belief that every soul requires an animating body, without which it is unable to exert its powers or even exist in any effective sense: "For where there is no union with bodie, there is no operation of the Soul."⁶⁰ Since (for the later More, at least) the conspissated masses of cuspidal particles were, just barely, spiritual—constituting, in Jasper Reid's words, "God the Father's point of active contact with the created world" (*Metaphysics of More* 244)—More therefore refused all his life to accept mechanical explanations without additional, non-mechanical ones. Blake similarly insists, "every Natural Effect has a Spiritual Cause, and Not / A Natural: for a Natural Cause only seems" (*M* 26:44–45; E 124). Much as Ault's *Visionary Physics* demonstrates, Blake considered Newton not flat wrong or "insane" like the Satanic Spectre (*J* 33:4; E 179) so much as dangerously blinded by his mechanistic perspective. If Milton can be redeemed from error, so too can Newton at *Jerusalem* 98:9.

More's idea of spissitude followed from his argument that spiritual substance combines the power of interpenetrability with "indiscerpibility," that is, untearability, the topological property by which an extended spirit's total metaphysical unity is preserved throughout all portions of its volume analogously to the indivisibility of atoms. The historical Milton also held this doctrine, as exemplified by the shape shifting of *Paradise Lost*'s angels.⁶¹ Similarly, in *Milton* the threat of violent penetration based on Newtonian mechanics again and again gives way to erotic interfusion arising from the "fleshly" interpermeability of human spirits. Milton's entrance into the Blake poet's foot is painless, and when the poet mistakenly fears a divine rape by Milton-Los at 22:5–12 (figure 4.12), not only does Los's surrender gesture in the design convey reassurance,⁶² but the juxtaposition of the poet's head with the bullseye of Los's groin implies an energetic expansion of vision similarly to Oothoon's solar copulations in *Visions of the Daughters of Albion*. Like her, the poet is about to fall into a Vortex of vision. So, whereas Milton's entrance into the poet's foot seems on plate 15 a mere incidental bit of descriptive narrative, it eventually proves to be a complex, cosmic "put[ting] on" of "new flesh . . . new clay" (Eph. 4:22–24), a rebaptism that is also an inspissation. Milton strives to sculpt the clay of Urizen's body of death into a "spiritual body" while Urizen contrariously assists, despite himself, by "pouring on / To Miltons brain the icy fluid" from the river Jordan (*M* 19:6–14; E 112). Another inspissation is Ololon's gradual emergence from an oozy multitude of Sons and Daughters into a sharply etched "Virgin of twelve years" speaking "words . . . more distinct than any earthly" (36:17, 4–5; E 137). (Blake ironically links the historical Milton's concern for chastity with puritanism's ideological enemy, Mariolatry, as embodied by Dante's Beatrice.) Still another inspissation is the parallel compaction by which Milton, "collecting all his fibres into impregnable strength," casts off his extrinsic Satanic "false Body" (40:35; E 142) and steps into Felpham garden "clothed in black, severe & silent" (38:5–8; E 138). What results from Milton and Ololon's mutual self-annihilation is a thoroughly inspissated (and largely androgynous) being: "One Man Jesus . . . round his limbs / The Clouds of Ololon folded as a Garment dipped in blood" (42:8–12; E 143).

For More, such alterations in the soul's "Perceptive Faculties" are enabled by its responsive "Aiery Vehicle," whose "temper" and "external shape" the soul "has a marvellous power of . . . changing" (*Immortality of the Soul* 337–38). The vehicle was a rarified body that the immaterial soul rode like a carriage, by which the soul's movements were conveyed to the gross matter of the human body. Eventually, the concept was recruited to explain not only alterations of mood and complexion but also how affects travel between different persons. Vehicular physiology was grafted onto a doctrine of perception that extended it into interpersonal situations. For, in eighteenth-century moral theory, grounded as it was in an analogy between gravitational attraction and the mind's association of ideas, a main requirement of "the vehicular hypothesis" was to explain how gravity's action at a distance became transmitted within the world of social feelings to hold that world together the way gravity did the physical world. The answer, according to Hume, was "sympathy," by which the usual sequence leading from sense "impressions" to fainter, second-order "ideas" is reversed in consequence of the fact that humans resemble, and hence identify with, one another more closely than with any of the other objects of perception. "The ideas of the impressions of others, [which] appear at first in *our* mind as mere ideas, and are conceiv'd to belong to another person, as we conceive any matter of fact, . . . are converted into the very impressions they represent."[63] Once Adam Smith's *Theory of Moral Sentiments* (1759) developed Hume's psychologizing into a full-blown doctrine of moral Sympathy, More's self-caused "emanative effect" gained the power to reduplicate itself across separate human bodies.[64]

And yet, *Milton*'s vision lasts no more than a moment. When the Blake poet faints upon the garden path like Saul of Tarsus and then awakens beside his wife, what he experiences is not a climbing back up to consciousness through memory, as Locke's analysis of personal identity would have it, but a radically discontinuous event: a spontaneous resurrection following an instant of self-annihilation when he had gone out of his senses, become a spirit, and received Judgment. The poet's apprehension of Jesus occurs in the interval *between* Moments, as he exits the Vortex of *Milton*'s narrative without yet having entered a new one; for, "every thing has its / Own Vortex" (*M* 15:21–22; E 109). His awakening illustrates that "heaven [is] a vortex passd already, and the earth / A vortex not yet pass'd by the traveller thro' Eternity" (15:34–35; E 109). Since heaven in its eternity precedes earthly existence, earthly spirits necessarily envision it as a vortex already passed, much as Blake implied when he signed Upcott's autograph album the year before his death: "Born 28 Novr 1757 in London & has died several times since" (E 698).

If, as Whitehead says, "a duration is a concrete slab of nature limited by simultaneity," it follows that travel toward the instant, which has no thickness, is not passage from imperfect to perfect knowledge.[65] It is passage beyond knowledge. So, instead of leading beyond time to Eternity and knowing God "face to face" (1 Cor. 13:12), *Milton*'s Moment lands the Blake poet back where he began in pastoral Felpham—a place now much more distinctly realized than it was in the Bard's Song—there to await the next durative "slab of nature" which he will have to envision all over again in the form of *Jerusalem*. Evidently, the doors of perception are revolving. They push you out even as the next one swings round to suck you back in. Said Jesus, "I am the door: by me if any man enter in, he shall be saved, and shall go in and out, and find pasture" (John 10:9).

The Blakean Vortex-Moment, then, is an abstract quantum of the unpicturable energy, change, or action at the base of space-time. It is that to which the solar axis or "burning axletree"[66] of the revolving doors of perception is attached. Real time is the now, an endlessly dividing, eternally returning Moment. Thus, *Milton* and *Jerusalem* are both dated "1804"; the poems were deliberate anachronisms even before they were published. For Newton, the continuity of movement within homogeneous space entailed time's infinite divisibility, as in Zeno's Paradoxes. But Blake refuses to grant durationless mathematical instants any physical existence, so the structure of his space-time is therefore discontinuous. "The Poets Work is Done . . . / . . . / Within a Moment: a Pulsation of the Artery" (*M* 29:1–3; E 127) because man's confinement to the Euclidean "zone of the middle dimensions" grants him access to the Moment only from within "the two Limits: first of Opacity, then of Contraction" (13:20: E 107), whereby divine Providence originally set three-dimensional boundaries to the polypy Satanic formlessness of infinite divisibility. Blake's puns capture the Moment's Vortex-like transitivity from inside to outside. The poet's "Work" "is Done" "Within a Moment" in a double sense: the past participle indicates not only completion of a single, spontaneously inspired artwork but also the labor of a lifetime done and redone continuously in a series of such inspirations.[67]

This middle zone of recurring Moments resembles William James's "specious present" comprised of discrete "drops or buds of experience" lasting roughly between one-half and two seconds (estimates vary).[68] If the Euclidean time model rests on the existence of point-like instantaneous boundaries dividing the successive moments of duration, conversely Blake's psychophysiological model organizes the continuous chronotopic pulsations of duration by the qualitative difference between *gestalts* of the specious present. Each extended now establishes a simultaneity of events within consciousness, and the resultant scenario of presence constitutes the subjective experience of inwardness together with all its ineffable qualia. In contrast, the past and future lack such inwardness. That is the message of the *Visions of the Daughters of Albion* frontispiece (figure 1.1), whose specious "cavern" of ordinary isolated consciousness forms a frozen three-dimensional slice of ongoing four-dimensional reality. Accordingly, when Blake says, "I question not my Corporeal or Vegetative Eye any more than I would Question a Window concerning a Sight I look thro it & not with it" (*VLJ*; E 566), he means that the world he sees like everyone else is a mental content, a representation or model constructed in the mind. Since man cannot see without corporeal eyes, representationalism is unavoidable, though we can still look past it. On the other hand, seeing "with" the eye—as if the eye were not a sight organ but a mechanism transmitting independent objects of vision—is not seeing at all.

James's specious present calls attention to the habitually ignored element of duration that underpins ordinary three-dimensional vision, without which visual objects would vanish the instant they were perceived. If chronological time is a product of the solidity of objects as viewed in Euclidean absolute space, then its origin would seem to reside in a fourth dimension of indefinite duration. A similar geometrical idealization underpins the mistaken supposition that empirical perception of objects is three-dimensional. Strictly speaking, the objects of visual perception exist partway between two and three dimensions since we necessarily always see them from some limited point of view. That is what the Klein-bottle space of the *Daughters of Albion* frontispiece demonstrates. To view the backs of objects simultaneously with their other sides would require "Four-fold Vision" (*M* 35:23; E 135), a divine perspective that the frontispiece represents as having collapsed, in an age of materialism, into the eye of Satan-Urizen and his abstract view from nowhere.

Hence, the curiously poignant sense of loss that suffuses *Milton*'s explication of the Vortex. In the lines cited earlier, the "traveller thro Eternity" sees the Vortex "roll backward / Behind his path, into a globe itself infolding . . . / . . . / like a human form, a friend with whom he livd benevolent" (*M* 15: 21–27; E 109)—perhaps a rueful allusion to William Hayley, the well-intentioned patron of Blake's disastrous Felpham interlude. Travel through the Vortex illustrates the coalescence and collapse—or, to use Whitehead's term, the "concrescence"—of a drop of experience: an open cone of internal sense perceptions on the way in (time as passage) but a closed-off, solid globe ever after (time as history). Through this process, private subjective experiences enter the world as public facts available to other entities and events within their local causal neighborhoods. As Whitehead grandly puts it, "The world is always becoming, and as it becomes, it passes away and perishes."[69] The past in perishing ceases to be an open Vortex and is objectified within the present as an impassable solid.[70] To put it another way, the present depends upon the past, but the past is independent of the present.

So regarded, the Vortex depicts "the ultimate metaphysical principle" behind Whitehead's organic Platonism: "The many become one, and are increased by one" (*Process and Reality* 21). The immanent and cumulative nature of organic unification leaves the totality of its lived forms just beyond reach. Even as their passing feels all but immediate, it painfully lacks the instantaneousness of disembodied conceptual knowledge (Newton's fluxions, for example). The emotional truth of the Vortex is illustrated in the myth of Orpheus and Eurydice: so close; so far away. Eurydice, a Thel figure in Blake's early artwork, enters *Milton* as the Blake poet's wife, who has hitherto remained indoors, "sick with fatigue" (*M* 36:32; E 137). He regains his "sweet shadow of delight," yet their reunion remains no more than a mild earthly reprise of Milton's transcendental embracing of *his* emanation. Blake here captures something of the mood of Abraham Tucker's sentimental encounter with his own dead wife, nicknamed Eurydice in "The Vision" chapter of his *Light of Nature Pursued* examined in my next chapter. Exclaims Tucker, "O! that it were permitted to take you down with me to make a paradise again upon earth."[71] At *Milton*'s close, earthly paradise is evidently what the reunited Blakes can finally anticipate in Felpham-Beulah. However, Blake, if not Tucker, is surely remembering the poignancy of Milton's "Methought I saw My Late Espoused Saint": his Orphean sonnet to his wife, dead and departed but briefly, agonizingly, restored to him in a dream. *Milton* gives the story a happy ending, turning Catherine into Penelope, yet the poet still awaits a full spiritual unification like that of Milton with Ololon in accordance with Whitehead's tantalizing principle of ongoingness: "The many become one, and are increased by one."

In this way, *Milton*'s Vortex introduces the fourth dimension of time into the concept of the horizon Blake had deconstructed in *The Book of Urizen*. If the earlier poem demonstrated "the" visible horizon to be a fallacy of misplaced concreteness whose fixity is a function of the viewer's immobility, *Milton*'s Vortex offers a more fully spatiotemporal image of movement, freedom, and change within limits. Beulah's mobile, flexible horizon along earth's "infinite plane" becomes at 15:28–33 and 29:4–14 (both cited earlier) a source of Edenic repose and comfort: nest-like containment rather than constriction. That is, Beulah is a local Galilean "inertial frame" based on the mutual equivalence of rest and continuous steady motion, whereby motion itself offers a kind of stability and so can become a local residence apart from Eternity, whose cosmic space-time is not amenable to such equivalence. The Vortex is the form of a transcendence that leads over the edge of the known three-dimensional world, even as that world's ungraspable horizon continually recedes from view.

Milton is a prophecy largely because it recognizes how the extended present of human consciousness includes within itself retentions of the past and protensions of the future, all coexisting within an eternally returning Moment that remains fresh and novel because its continual peeling-apart to form the now reflects the differential, quantal structure of time itself. Imagine Buddhism's ever-blooming Lotus or the fractal zoom of a Mandelbrot set. Just as every Blakean atom is a Vortex opening into infinite space, so the Moment offers a window onto Eternal duration. It "encloses and is enclosed by" linear, chronological time, being simultaneously one more of time's passing units and the *last* such unit: the heterogeneous present not yet bounded by the Euclidean timeline, the now whose apocalyptic unfolding redefines the shape and total meaning of history by altering, however slightly, the prior structure of relations between all the moments of time. As a novel physical event whose cosmic "thickness" or "transversal width" includes oneself as witness, Blake's Moment renders every local here-now a universal call to conscience and Judgment. It is a historical burden impossible to bear, he believed, without divine assistance from an extra dimension both within and without.

CHAPTER FIVE

The Neoplatonism of Blake's Mundane Soul

Let us explore further the origins of Blake's cosmology in the "plastic" Spirit of Nature expounded, with often malleable logic, by the Cambridge Platonist Henry More. In addition to influencing Blake directly, More's Neoplatonic Spirit of Nature pervades an unlikely trio whose influence on Blake ran deep: George Berkeley, Joseph Priestley, and Abraham Tucker. Eighteenth-century Neoplatonism was a broader religious current than is commonly supposed, and it often served as a backstop for moderate, non-Calvinist defenses against atheism and what Blake called "Unbelief."

Priestley's argument that "the immateriality of matter" supports Christian revelation has roots in the Neoplatonism of More, whose repudiation of Cartesian "nullibism" a century earlier paved the way for the received English view of Descartes as atheist. Since Descartes's world practically excludes all spirit, his answer to More's question of *where* spirits, souls, and God are located can only be nowhere, *nullibi*.[1] A big problem with Donald Ault's insistence, in *Visionary Physics*, on the unextended and dimensionless nature of Blake's Eternity is that it tends to make Blake a Cartesian nullibist, which seems impossible, not least because Priestley's early version of field theory, which Blake embraced, implies space is not empty. Blake's mythology was greatly shaped by More's claim that space itself offers the best argument for the extended nature of immaterial bodies, and by his corresponding doctrine of emanative "aery vehicles." More says, "It is very manifest that *the Soul has Dimensions*, and yet *not infinite*, and therefore that she is necessarily bounded in some *Figure* or other . . . ," albeit that figure is highly "plastic."[2]

What I am calling the Neoplatonism of Blake therefore does not reflect his absorption of ancient esoteric wisdom, pace Kathleen Raine and, to an extent, George Mills Harper.[3] Far less does it reflect the beliefs of Blake's sometime friend, Thomas Taylor "the Platonist," as these critics supposed.[4] Rather, Blake's Neoplatonism attests a diverse set of influences whose common thread is their shared concern about the atheistic implications of Newtonian mechanism and its besetting problem of attraction at a distance.[5]

This is not to deny the significance for Blake of the fifth-century Pythagorean Neoplatonist, Proclus, as translated by Taylor and as emphasized by Raine and Harper. I've already cited Proclus repeatedly, and more is in store. But Blake wasn't some hermetic antiquarian mystic harkening back to timeless psychological truths of late-Platonic myth and thaumaturgy. Rather, he recognized there existed a strong undercurrent of Neoplatonism running through Newtonian physics and its Enlightened metaphysics. It was Proclus's *contemporary* relevance that mattered. Raine and Harper could reply that everything I am about to claim Blake learned from Henry More concerning vehicles, emanative causes, the physical nature of light, and the soul's need of a body by which to act, he could also have found in Proclus (and, perhaps, Iamblichus, whom Taylor also translated and revered). But if Blake was interested in Proclus, it was for giving Morean insight into the relation of four-dimensional vision to Euclidean geometry and Newtonian physics. Blake's Neoplatonism derived less from Proclus, Iamblichus, and Taylor than it sprang from his deconstructive insight into the organic, nondualist possibilities afforded by Newtonianism's Neoplatonist strain, which vibrates across eighteenth-century rationalist philosophy and especially Priestley's doctrine of "the immateriality of matter."[6]

In their own day, More and his friend Ralph Cudworth were important fellow travelers of Newton. For both, a great virtue of the mechanical philosophy, despite its atheistic tendency, was that its development would eventually disclose insurmountable spiritual deficiencies and so reinforce their own quite different metaphysics. Like Blake, More held the millennial view that error's progressive consolidation would eventually render any further temporizing with materialism impossible, thereby leading to the epiphany of Antichrist and the Last Judgment, as in *The Four Zoas,* Night Nine. The darkness must thicken to become visible, seen as such, and cast out. That said, More was no antinomian, nor did he anticipate an overthrow of existing institutions. His interpretations of Daniel and Revelation led him to believe the Millennium was far from imminent, though still worth preparing for.

On the other hand, Blake's millennialism was urgent. This chapter argues that Blake exploited the traditional concept of the Mundane Soul, derived partly from its grandfather Plato but also from Plato's inheritors More and

Abraham Tucker (whose novella-length visionary satire of the Afterlife is an important source for *Milton*), to develop a cosmology of "plastic" space-time that treats Creation, Fall, and Apocalypse as ongoing events. Time is ending *all* the time. The lastness of the Last Judgment is its recency; what makes each Judgment everlasting is its uniqueness and irrevocability as an event. Creation and Fall are, therefore, productive processes continually recapitulated within the living present. The physics of this cosmology I trace forward to the revised relativity theory Alfred North Whitehead developed in *Process and Reality*, grounded in a manifold he termed "the extensive continuum." Blake's critique of Newton does not miraculously anticipate modern particle theory. But his connection to relativity physics is more than analogical, as I hope the preceding chapter's account of his various treatments of four-dimensional space has indicated. There is a real intellectual history here beyond all *ex post facto* analogizing.

More's Materialism

Henry More is a Janus figure who crossed paths, and sometimes swords, with Newton in ways that complicated Blake's responses not only to Newton but Berkeley and Priestley. On one hand, Blake's lifelong emphasis on toleration and the unity of all religions directly reflects the influence of the Unitarian Priestley, despite his cool deistic rationalism. Like Blake, Priestley supported repeal of the Test and Corporation Acts and promoted Catholic emancipation, a cause he deemed crucial to the success of Unitarianism during the 1780s. On the other hand, radical as those positions appeared at the time, their wellspring goes back to the tradition of Latitudinarian moderates like More a century earlier and their culture of wide-ranging metaphysical speculation. The oft-heard charge of Coleridge's "apostasy" from revolutionary democrat to constitutional moderate ignores the pull of this history. So does the claim that Blake's later mythology reflects his abandonment of radicalism.[7] Blake would have despised More's post-Restoration attempt to ground the apocalypse in ancient institutions like monarchy and episcopacy. And yet, *Jerusalem* concedes that the beauty of George Herbert's Anglican Church might offer a stay against violence and atheism (*J* 12:28–44; E 155–56); so perhaps Blake recognized parallels between More's situation after the Great Rebellion and his own after the Napoleonic Wars. *Jerusalem* portrays the floor of Anglicanism's allegorically abstracted "building of pity and compassion"—an image of Beulah's aesthetic escapism—as smoothing over "the rough basement" of verbal truth that Blake-Los strives to hammer out of "the stubborn structure of the language" (*J* 36:58–60; E 183), chiefly its sinews of syntax as opposed to imagery. Even so, superstructure and foundation are part of the same building, Jerusalem. More's characteristically puritan-millenarian "increased confidence in the capacity of the human intellect," as one commentator describes it, and his expectation that the End time would see "spectacular advances . . . in all fields of learning,"[8] dovetails with *Jerusalem*'s Enlightened utopia of free-ranging "exemplars of Memory and Intellect" (*J* 98:30; E 258) including "Bacon & Newton & Locke" (98:9; E 257).

At the center of More's system, and underlying his complicated doctrine of "vehicles," lies his anti-Cartesian conviction that soul is not thought but action and change. Therefore, soul needs body to act or even exist. For, "what is simply active of it self, can no more cease to be active than to Be" (*Immortality of the Soul* 42). In the most thorough study to date, Jasper Reid remarks: "It clearly followed from this that souls could never become completely disembodied."[9] After death and the dissolution of the body, the soul seeks to occupy another because it has "a very strong Propension, natural Complacency, or *essential* Aptitude always to join with some Body or other."[10] Thanks to the existence of vehicles, "the Soul, speaking in a natural sense loseth nothing by Death . . . For she does not only possess as much Body as before, with as full and solid dimensions, but has . . . this Body more invigorated with Life and Motion then it was formerly" (*Immortality of the Soul* 350). In *Milton*, the restlessness of Milton's soul in Heaven and its eventual entrance into the Blake poet's body demonstrate this point, vindicating Blake's Morean claim twenty years earlier, "Energy is the only life and is from the Body" (*MHH* 4; E 34).

Given its inability to act without a body, More argued against Calvin that the soul's separation from a material vehicle would cause it to lapse into a "psychopannychite" sleep on the verge of annihilation (*Grand Mystery of Godliness* chaps. 6–7). Such is the darkened and ineffectual state of the Blakean Milton asleep on his Couch of Death in Eternity while his vehicle travels heroically toward earth. "For where there is no union with bodie," More emphasizes, "there is no operation of the Soul."[11] More's critique of Descartes's *cogito* as insufficiently founded in action, movement, and body implies a Blakean anti-dualism similarly to Whitehead's when he argues: "Descartes writes (*Meditation II*): 'I am, I exist, is necessarily true each time that I pronounce it, or that I mentally conceive it.' Descartes in his own philosophy conceives the thinker as creating the occasional thought. The philosophy of organism inverts the order, and conceives the thought as a constituent operation in the creation of the occasional thinker."[12] Behind this inversion lies Whitehead's field theory of the "extensive continuum," to be examined shortly. Meanwhile, consider Alexander Koyré's observation in 1957: "The fundamental entity of contemporary science, the 'field,' is something that possesses location and extension, penetrability and indiscerpibility. . . . So that, anachronistically, of course, one could assimilate More's 'spirits,' at least the lowest, unconscious degrees of them, to some kinds of fields."[13]

It has been noted that More's philosophy descends at times into a contradictory materialism.[14] Very likely, that was part of its appeal for Blake (and for Newton, Priestley, and the young Coleridge). The transparent bodysuit or skintight tunic with which Blake clothes spirits throughout his art attests his materialist belief in the existence of a unitary body-soul possessed of "Spiritual Sensation" (E 703), similarly to More's claim that "the Soul . . . cannot act but in dependence on *Matter*, and that her Operations are some way or other always modified thereby" (*Immortality of the Soul* 329). The bodysuit further demonstrates that for Blake, as for More, spirits necessarily have extension in space (a crucial point of disagreement between Blake and Berkeley, as we'll see in chapter 6). As Jasper Reid points out, many apparent inconsistencies in More's thought arise because he gradually repudiated his early Plotinean view of Hyle/matter as nonsubstance, nothing but pure, formless privation and restriction unshaped by Nature's "plastic spirit." In his later writings, More sees body and space as differing, not in essence, but only according to whether their atoms are compounded or not. In the 1647 edition of the *Philosophical Poems*, his earlier graph of "the universall Ogdoaz" (figure 4.15) is reinterpreted to define body as "nothing but a fixt spirit, the conspissation or coagulation of the cuspidall particles of the Cone, which are indeed the Centrall Tasis, or inward essence of the sensible world."[15] Whereas most classical atomists regarded atoms and space as opposites, More viewed space itself as atomic, independent of any bodies that might occupy it. So, the path lay open for him eventually to absolutize space and, thence, to spatialize time similarly to Newton.[16] In a well-known passage, More deems space as "not only something real . . . but even something Divine," and he goes on to assign it no fewer than twenty "divine names or titles," including "*One, Simple, Immobile, External, complete, Independent, . . . Necessary, . . . Uncreated, Uncircumscribed, Incomprehensible.*"[17]

Therefore, space for More was no longer associated with the incorporeal and almost unreal Receptacle of Plato's *Timaeus* as the bare possibility of a place for things. So far from being merely "imaginary," space was now—said More sarcastically, perhaps parodying Anselm's ontological proof of God—"so imaginary that it cannot possibly be dis-imagined by humane understanding. Which methinks should be no small earnest that there is a more than imaginary Being there."[18] We are on the verge of Newton's controversial suggestion in *Opticks,* Query 31 that space be considered as the "sensorium" of God, an idea Blake mocks in *Europe*'s "Five windows light the cavern'd Man" passage (*E* iii:1–5; E 60, also *M* 5:28–37; E 99). In More's later writings, the Mundane Soul, anciently derived from Plato's Receptacle, transforms into a Spirit of Nature operating beyond the material aether as the elusive source of gravitation and attraction at a distance: "There is a *Spirit of Nature* which is the Vicarious Power of God upon the *Motion* of the *Matter* of the Universe," "the great *Quartermaster-General* of Divine Providence," "the Vicarious Power of God upon this great *Automaton*, the World."[19] By contrast, "dis-imagining" space as conceived by the later More and Newton is exactly what Blake does when he reimagines it as Eternity's four-dimensional space-time manifold grounded in an emergent continuum of heterogeneous event-objects or "minute particulars."

More's influence on Newton—or at least, the two men's occasionally shared tendency of thought—is also strikingly evident in Newton's several speculations about Creation. The best known of these, to which Locke alludes in his *Essay*, I examined in chapter 2. Suffice it to restate Alexander Campbell Fraser's summary: "Newton, it seems, suggested that 'creation of matter' means, God causing in sentient beings the sense-perception of resistance, in an otherwise pure space—a theory akin to Berkeleyanism in its recognition of the Supreme Power, and to Boscovich in its conception of the effect."[20] Thus, Newton and Locke were not impervious to intimations of a Whiteheadian (and, we shall see, Agambenian) view of Plato's Receptacle of Becoming. Newton-Locke's "dim and seeming" speculation also recalls More's aforementioned account of his Ogdoaz, where material creation occurs through "the conspissation or coagulation of the cuspidall particles of the Cone, which are indeed the . . . inward essence of the sensible world" (*Poems* 160). A similar emanative process seems to account for Urizen's production of a solid world out of a "Self-clos'd, all-repelling . . . / . . . abominable void / . . . soul-shudd'ring Vacuum" (*BU* 3:3–5; E 70). More's alleged impossibility of "dis-imagining" space is echoed in Newton's claim in *De gravitatione* that, "although we can possibly imagine that there is nothing in space, yet we cannot think that space does not exist" (26). In other words, "space is an emanative effect of the first existent being, for if any being whatsoever is posited, space is posited" (25); spatial extension is "an emanative effect of God" (21–22). Therefore, he implies, God did not create space freely through an act of will. Rather, space arose as an immediate consequence of God's omnipresence, which necessitated he be someplace.[21] In this way, Newton managed to uphold the dictum John Yolton sees as central to eighteenth-century thinking about matter: "Nothing can be or act where it is not; no action at a distance," including cognitive action.[22]

As several scholars observe, Newton here has adopted More's notion of an "emanative cause" in *Immortality of the Soul*, a work Newton owned.[23] Says More, an emanative cause is a cause which, "merely by Being, no other activity or causality interposed, produces an effect." The "Emanative Effect" is thus "coexistent with the very Substance" that caused it: "That very Substance which is said to be the Cause . . . wants nothing to be adjoyned to its bare essence for the production of the Effect; and by the same reason the Effect is at any time, it must be at all times, or so long as that Substance does exist" (*Immortality of the Soul* 33–35). By this means, the extended soul manages and controls its body, including, Jasper Reid points out, the movement and direction of intellectual thought itself. Since spirits are self-moved, the ontological priority of the emanative cause enables it to operate with "immediate" effect and without any intervening mechanisms. The metaleptic transpositions of cause and effect that pervade Blake's myth-making reflect a similar idea. However, the "emanative" power that Newton reserves for God, and More for certain privileged spirits, Blake extends to all of Creation: "every Natural Effect has a Spiritual Cause, and Not / A Natural: for a Natural Cause only seems, it is a Delusion" (*M* 26:44–45; E 124). If there is no mechanical cause and effect occurring within serial time—if divine creation is occurring everywhere all the time—then spatial extension becomes dynamic and emergent similarly to Whitehead's extensive space-time continuum. So, the true reason for the mechanist rule, no action at a distance, is that homogeneous metric distance can exist only after the fact of the heterogeneous events that supply its ground; topology precedes geometry. Since Blake accepts that a spirit's power is inseparable from its substance—like More, unlike Newton—the principle "no action at a distance" therefore requires spirit to be present in the place where it operates.

It is not difficult to see how the doctrine of the "emanative cause" ("coexistent with the very Substance" which causes it) is of a piece with the "redoubling" that produces spissitude, with indiscerpibility, and with More's overall stance of "holenmerianism" ("whole in every part," per the Scholastic principle of the nature of spiritual presence: *tota in tot, et tota in qualibet parte*). Jasper Reid observes More's "favourite analogy" for the indiscerpibility of a spirit in a particular space was that of an orb of light (165). A central light source, assumed to be indivisible, sends out rays to produce a sphere of illumination about itself. Likewise, More thought a "created spirit" with a "central life" would radiate a "secondary substance" inseparable from the center. As he put it in a celebrated phrase, "Parts of a *spirit* can be no more separated, though they be dilated, than you can cut off the *Rayes* of the *Sun* by a pair of Scissors made of pullucid Crystal" (*Antidote against Atheism* 16; his italics). Blake's view that minute particulars are irreducible because each harbors Eternity at its center is "holenmerian": "Every thing in Eternity shines by its own internal light" (*M* 10:16; E 104). Reid points out that More leaves it ambiguous whether the central spirit fills the orb with the whole of itself, or rather irradiates the orb by "reduplicating" its images, as in Plotinus (notably, *Enneads* 4.2.1). If the first option is analogous to the space-creating omnipresence of Newton's God, it is hardly surprising that Blake should embrace the second one, as illustrated by the recursively repeated, even fractally multiplied figures throughout his art. In his illustration to Milton's Nativity Ode (figure 4.5), the Holy Spirit's area of illumination—"A globe of circular light," the sphere within a sphere that "surrounds" the shepherds' sight—discloses both spirit's indiscerpibility and its physical extension by means of angelical reduplication.

Before we turn to cosmology in the next section, it's worth noticing that More's position was not farfetched. Just as there is a Neoplatonic Newton who anticipates Blake's Priestleyan immateriality of matter, there is also a holenmerian Descartes who in at least one well-known instance anticipates Blake's Morean idea of a composite body-soul. More repudiated Descartes—and yet Descartes, in his problem Sixth Meditation, similarly argued the body lacks functional indivisibility or even numerical unity without a soul. Martial Gueroult demonstrated that for Descartes the functional unity supplied by the soul pervades the body and alters the relation of its parts to one another from mechanical to teleological, without however changing the body's overall structure.[24] As in More and Blake, the embodied soul's "internal finality" operates as a kind of intermediary or third entity between corporeal and spiritual realms. In this sense, soul and mind have extension. So, even though Descartes employs the asceticism of "clear and distinct" ideas to prove the distinction between thinking and extended substance, thereby destroying scholastic "substantial forms" and establishing the mathematical character of physics, in *Meditations* Six he finds it necessary—like More—to have recourse to sensation to prove the substantial union of body and soul and, indeed, the reality of material existence itself. We saw earlier how More's reinterpretation of his Ogdoaz likewise concedes cuspidal matter is not mere dearth of spiritual substance but something sensible and real.

As in Newton, sensible matter, though rare and tenuous, then provides for both More and Descartes alike a "bastard" object for physics as opposed to geometry, the Cartesian *esprit géométrique* notwithstanding.

Plato's Receptacle and the Nonsensation of *Anaisthesis*

At the heart of More's many contradictions lie the two most important cosmological theories in Western thought, in Whitehead's estimation: Plato's *Timaeus* and Newton's Scholium to *Principia Mathematica*. Not that Whitehead thought the Scholium contains much metaphysical depth of thought. Its great deficiency, he says, is that "nature is merely, and completely, *there*, externally designed and obedient . . . [S]pace and time, with all their current mathematical properties, are ready-made for the material masses," which in turn "are ready-made for the 'forces' which constitute their action and reaction," and all "are alike ready-made for the initial motions which the Deity impresses throughout the universe" (*Process and Reality* 93–94; his italics). Since the origin of all this Deistic "fitting & fitted," as Blake called it (E 667), remains mysterious, Hume's annihilative *Dialogues concerning Natural Religion* follows inescapably. *Timaeus*, however, offered Whitehead's "philosophy of organism" a crucial model of reconciliation between natural law as divinely imposed and natural law as immanent, the two together forming, in his opinion, the chief antinomy of the four great world religions.

As Lisa Landoe Hedrick recently pointed out, it was Judeo-Christianity's Creation *ex nihilo* that Whitehead held responsible for the longstanding misappropriation of Plato as an idealist. What the *Timaeus* actually describes, he thought, is a "realist" theory of the evolution of matter out of the disorder (in Christian terms, "chaos") of the Receptacle.[25] Based on this erroneous understanding of Plato, Newton's mathematical approach further reduced the reality of "passage"—nature's extensive series of overlapping events—to permanent bits of matter suspended in empty space (*Process and Reality* 95). Whitehead's reading of *Timaeus* follows that of his friend, the classicist A. E. Taylor, whose arguably distortive *Commentary* of 1928 claimed that "the *Timaeus* is wholly free from any form of the doctrine which Professor Whitehead calls the 'bifurcation' of nature. In Aristotle we find 'bifurcation' beginning in the distinction between 'substances,' which are imperceptible and the 'accidents' of the substances, which are what we perceive. But the very distinction between 'substance' and 'accident' plays no part in the *Naturephilosophie* of Timaeus."[26] As Hedrick stresses, Taylor sharply rejected the standard association of Timaeus's "timeless space" of mathematics with a transcendent realm of Ideas beyond all temporal process. According to Taylor, Timaeus designates the "eternal" as that which does not "pass" because he "asserts that it is not 'in' anything [namely, absolute space] except in the sense that it is 'in' itself, it 'fills' itself" (*Commentary* 351). Thus, Taylor remarks "an almost exact equivalence" between Whitehead's "[eternal] objects" and Plato's "ideas," even as he further emphasizes that "the 'metaphysics of nature' which runs through the dialogue and that of Professor Whitehead and others among our contemporaries throw light on one another" (*Commentary* x).

For Whitehead, then, that which is "given"—what Timaeus calls Necessity—is not traditional monotheism's "wholly transcendent God creating out of nothing an accidental universe" (*Process and Reality* 95). Rather, the given occurs when indefinite potentiality (which arises from the ongoing ingression of the past into the present) becomes restricted to the actual ("clear and distinct ideas" in the perceptual mode of "presentational immediacy," as epitomized by Hume). The given is therefore a product of activity and is essentially relational. Says Hedrick, this "is the sort of givennesss [Whitehead] credits Plato as recognizing as the limit of 'theory,' insofar as any theory at all presupposes something given to be theorized" (10). At the far end from Newton, Whitehead's critique of Einstein's general relativity is based on the same "corrected," nonbifurcated Platonic realism. The biological uniformity of our spatial and temporal intuitions demands correspondingly uniform measurement practices. Otherwise, Einstein's mathematical theory of a contingently curved but smooth space-time generates incoherence between what we know and what we feel. Whitehead's position thus corresponds with Blake's similarly epistemological criticism of Newton for failing to see that limited human perception requires space to be measured atomistically according to "a Globule of Mans blood," and time by "a Pulsation of the Artery" (*M* 29:1–27; E 127).

In this way, both Blake's and Whitehead's thought runs parallel to Timaeus's difficult "bastard kind of reasoning" (*logismo tini notho*).[27] In Plato, the product of such reasoning, the Timaean Receptacle of Becoming, constitutes an ontologically necessary "third kind" beyond intelligible pattern and its sensible imitation, by means of which mind and matter are rendered capable of real intercommunication. As the ancient Neoplatonists recognized, the Receptacle's all-but-unimaginable formlessness signifies a limit of the pure potentiality that exists in matter before the entrance of form into it. By means of this formlessness, the Receptacle can be molded by the Ideas into a substrate or matrix for the production of actual bodies. Since visible objects (copies of the Forms) neither exist in themselves nor in the Forms, they must exist "in another," Plato reasons, much as reflected images exist within the medium of a mirror.[28] Whitehead remarks, "The Receptacle . . . is the way in which Plato conceived the many actualities of the physical world as components in each other's natures. It is the doctrine of Law, derived from the mutual immanence of actualities."[29]

Elsewhere, Taylor elaborates the similarity of Whitehead's own doctrine to this "corrected," realist view of the *Timaeus*:

> His account of the "ingredience of objects into events" corresponds almost exactly with that given by Timaeus of the determination of the various regions of the "receptacle" by the "ingress" and "egress" of the impresses of the forms. The "receptacle" itself only differs from "passage" in being called "space" and not "space-time." If we try to picture "passage" as it would be if there were only "events" and no "objects" ingredient in them, we get precisely the sort of account Timaeus gives of the condition of the "receptacle" before God introduced order and structure into it.[30]

Admittedly, Taylor reductively maps Whiteheadian concepts onto Plato here. Nevertheless, Taylor's Whiteheadian interpretation wasn't *merely* self-ratifying; it did bring about a significant development in Whitehead's own metaphysics, if not in the received classical view of *Timaeus*. Whereas Plato's Receptacle is mere passive space, the "room" where objects acquire visibility as bodies, the Receptacle Taylor and Whitehead imagine is the very ground of becoming, the basis of the cosmic principle of creative change.[31] In other words, Whitehead subsumes Plato's creative Demiurge into the Receptacle, turning their offspring, Plato's World Soul or Cosmic Animal, into a dipolar creature of both being and becoming able to mediate "the relation of reality as permanent with reality as fluent" (*Adventures of Ideas* 130).[32] According to Whitehead, the Cosmic Animal "is not a static organism" but rather "an incompletion in process of production" (*Process and Reality* 214–15).

For, this is a World Soul "whose active grasp of ideas conditions impartially the whole process of the Universe." Its reason expresses the skillfulness of the Demiurge "on whom depends that degree of orderliness which the world exhibits," and without whose "living intelligence" the Ideas would be "static, frozen, and lifeless" (*Adventures of Ideas* 147). In Whitehead's summary: "Plato's Receptacle may be conceived as the necessary community within which the course of history is set, in abstraction from all the particular historical facts. . . . There is one all-embracing fact which is the advancing history of the one Universe. This community of the world, which is the matrix for all begetting, and whose essence is process with retention of connectedness,—this community is what Plato terms the Receptacle" (*Adventures of Ideas* 150). As community, the Receptacle already includes a potentiality to become Plato's World Soul, here reconceived under Whitehead's modified relativity principle as the intelligence behind the extensive continuum and, more generally, "the togetherness of things." "The extensive continuum is that general relational element in experience whereby the actual entities experienced, and that unit experience itself, are united in the solidarity of one common world" (*Process and Reality* 72). Whitehead's extensive continuum is the materialized form of Plato's Receptacle.

Blake's name for this combined Receptacle/World Soul is Albion, the human totality behind the three-dimensional world or Mundane Egg whose outer surface the Daughters of Beulah weave and reweave into a solid-seeming "fabric of existence." Blake's exploration of Albion begins in *The Four Zoas*, whose opening Circle of Destiny episode describes how the Daughters of Beulah create provisional "Spaces" to prevent the fallen sleepers in Eden from falling further into "Eternal Death" (*FZ* 5:35; E 303). Thereby, they make it possible for mankind to have reassuringly beautiful perceptions of time and space, albeit the sleepers risk mistaking Beulah, a mere way station, for the fully awakened consciousness of Eden. (Perhaps Blake was exploring his own temptation to turn art and aesthetic perception into a consolatory escape from dire political reality after the government crackdown of late 1794, when he stopped publishing for over a decade.) We noted in chapter 1 a wider consciousness at the back of the relatively early *Visions of the Daughters of Albion* frontispiece (figure 1.1). If that design's curvilinear perspective indicates a collective awareness that includes not only the three seemingly encaverned figures but viewers whose stance of detachment renders them part of the alienation they behold, then it is now possible to locate the source of this irony in the underlying presence of the Mundane Soul, Albion.

Better than the *Daughters of Albion* frontispiece, *Jerusalem*'s opening paragraphs give a notion of what it would be like to recognize Albion the Mundane Soul for what he is, namely, a form or mode of global human being that includes one's own self while extending immeasurably beyond it. Such is the "theme" *Jerusalem*'s invocation announces to the poet's torpid audience at daybreak:

> Of the passage through
> Eternal Death! and of the awaking to Eternal Life
> Awake! awake O sleeper of the land of shadows, wake! expand!
> I am in you and you in me, mutual in love divine:
> Fibres of love from man to man thro Albions pleasant land.
> . . .
> . . . return Albion! return!
> Thy brethren call thee . . .
> . . .
> Ye are my members O ye sleepers of Beulah, land of shades!
> (*J* 4:1–12; E 146)

The speaker here is Jesus, the element of fellow feeling or solidarity that holds Albion the Divine Humanity together, the Imagination by which one sees oneself as and in another. "[S]elf-enjoyment in the enjoyment of something other" is how Hans Robert Jauss describes the aesthetic pleasure of reading; a "pendulum movement in which the self enjoys not only its real object, the aesthetic object, but also its correlate, the equally irrealised subject which has been released from its always already given reality."[33] It seems an apt account of the kind of union Blake's Jesus invites us to join. This is further to say that the Blakean Jesus is an anthropomorphization of Plato's epistemic "third kind" whose immanent, higher-dimensional presence beyond mother and child we noticed earlier in "A Cradle Song" (chapter 2). Accordingly, the *Timaeus* suggests an answer to our question: What would it be like to recognize Albion in person? Plato says that apprehension of the Receptacle—a "room" (*khora*) that yet is not a material "place"—comes about "as in a dream" through "the aid of a non-sensation" or "lack of sensation" (*met'anaisthesias*, 52b). As Giorgio Agamben explains, what is apprehended here is the pure potentiality for sensible impression in the absence of any of the actual sensible determinations in matter. Plato's "matter" is not in itself sensible, rather its receptiveness allows the sensible to be formed in the image of the divine Architect's uncreated model or blueprint. Therefore, perception of the "third kind" is perception of something beyond being or experience. Its very glimpse is loss.

In *Potentialities*, Agamben describes Plato's *khora* as "the perception of an imperception, the sensation of an *anaisthesis*, a pure taking-place (in which truly nothing takes place other than the place)."[34] This "room" is the very opposite of absolute space made up, Newton claims, of portions or "spaces" that cannot be "moved out of their places" without being "moved (if the expression may be allowed) out of themselves."³⁵ Agamben compares Plato's *anaisthesis* to Aristotle's "cardinal secret" of *dynamis*, and he remarks that Plotinus went on to locate it in an understanding that the eye in darkness "can sense its own lack of sensation" and so sees "the self-affection of potentiality." Blake throughout his work evokes a similar idea when he symbolizes sunrise as God's actualization of potentiality hidden in the darkness not of pre-existent space but, evidently, the Receptacle, perception of which constitutes, in Coleridge's phrase, "a repetition within the finite mind of the eternal act of creation in the infinite I AM" (consider, again, figure 4.13).³⁶ Hence, Blake insists, the "pernicious idea" of a Creation *ex nihilo* "takes away all sublimity from the Bible & Limits All Existence to Creation & to Chaos [i.e., unorganized matter] To the Time & Space fixed by the Corporeal Vegetative Eye . . . Eternity Exists and All things in Eternity Independent of Creation which was an act of Mercy" (*VLJ*; E 563). If "Mercy" here signifies limits suitable to "the middle zone" of human perception, "Eternity" is synonymous with potentiality.

In *Timaeus*, Agamben says:

> Potential thought (the Neoplatonists speak of two matters, one sensible and one intelligible), the writing tablet on which nothing is written, can thus think itself. It thinks its own potentiality and, in this way, makes itself into the trace of its own formlessness, writes its own writtenness while letting itself take place in separating itself. . . . The potential to think, experiencing itself and being capable of itself as potential not to think, makes itself into the trace of its own formlessness, a trace that no one has traced—pure matter. In this sense, the trace is the passion of thought and matter; far from being the inert substratum of a form, it is, on the contrary, the result of a process of materialization. (*Potentialities* 218)

Potential thought in all its formlessness, and the materialized trace of its almost perfectly negative thought of its own potentiality: Agamben's deconstruction is probably as close as we are going to get to what Blake means when he speaks of the vision of an unconditioned Imagination which "is the Human Existence itself" (*M* 32:32; E 132). Such statements look like a Plotinean affirmation of a superordinate One at the back of the Forms whose multiplicity, if disarrayed, might preclude the recognition of necessary truths (for example, the claim that two things are equal or the same, such as 2+2 is 4). So petrific was the impact of Hume, not until Reid and Stewart in the later eighteenth century were Locke's innate faculties of sensation and reflection recuperated as *capabilities* by which man *learns* what he needs to know. Since Locke's God would not have put man into the world without endowing him with means to thrive, Hume's atomistic skepticism loses its paralyzing effect. Even today, it tends to be forgotten that Locke's *tabula rasa* derived, ultimately, from the "wax tablet" model of thought and perception in Plato's *Theaetatus* 190e—a model closely related not only to *Timaeus* but Aristotle's discussion of potentiality in the *Metaphysics* in terms of wax "capable of being built" into a beehive. According to Aristotle, potentiality is "an originative source of change in another thing or in the thing itself *qua* other," and it underpins his concepts of motion and action.[37]

Plotinus asserts that at the limit of signifying language, thought touches the "abyss" or, as Taylor's translation has it, the "profundity" (*bathos*) of "matter,"[38] namely, the *khora*, the pure taking place of each entity. For, Plato's "bastard reasoning" (*logismos*) remains, as Agamben says in *What Is Philosophy?* "an experience of language" (*logos*). What Plotinus describes is therefore a mystical encounter of "the pure dwelling of language at the limit of signification, . . . language's bare giving of itself" as "a pure quantum of signification, which, however, does not signify a thing or a concept, but only the giving itself, the pure 'taking place' of something" ("quantum" is also Thomas Taylor's term in his translation of *Enneads* 2.4.9).[39] At this limit, the Platonic opposition between the intelligible and the sensible collapses, yielding an idea "which does not take place in heaven or earth [but] in the taking place of bodies, with which it coincides." In Agamben's summary,

> the [*khora*]—the space and the taking place of each thing—is what appears when we take away, one after the other, the *semantic* elements of discourse, and move toward a purely *semiotic* dimension of language, not in the direction of writing but in that of a voice. [For, "Language takes place in the non-place of the voice" (sect. 11)]. In other words, the [*khora*] is the threshold at which the semiotic and the semantic, the sensible and the intelligible, numbers and ideas seem to coincide for an instant." (*What Is Philosophy?* sect. 21; his italics)

To put it another way, the *khora* provides a basis for the step-by-step, "quantal" aspect of logical reasoning by which Zeno's paradoxes of motion are defeated—what Descartes famously called "*l'ordre des raisons*" and Blake, quite differently, "the Building up of Jerusalem" by means of irreducible minute particulars (*J* 77: E 232).[40]

Specifically, Plato's bastard fusion of sensible and intelligible constitutes the form of geometrical reasoning, which concerns not sensible bodies directly but rather their pure taking place in space. That is why Plato follows his account of the Receptacle with an explanation of how the Demiurge fashions the elements within it by means of isosceles and scalene triangles constructed according to exact numerical ratios (as seen in Blake's *Newton* print, figure I.1). Agamben thus considers it "decisive" that the opening definition in Euclid's *Elements* speaks of "a point which has no part," where a literal translation of the Greek *semeion* would be not "point" but "sign": namely, says Agamben, "not a material entity but a quantum of signification" without any concrete denotation. It is, then, geometry that enables us to bridge the gap between the intelligible and the sensible. For, a point is not real; it is what results from the sheer signifying relation between word and thing, language and the world, when all semantic content has been removed. One cannot imagine a point without extension or some space within which the point is located. Accordingly, Plato's sixth-century commentator Simplicius argued that in the case of astronomy, the task of Platonic science was to "save the appearances" through geometrical artifices that do not claim literal or qualitative truth but merely serve as hypotheses by which to explain the errant movements and epicycles of the stars as circular, uniform, and regular. Blake's abhorrence of "experiment"—"People flatter themselves that there will be No Last Judgment & . . . That Error or Experiment will make a Part of Truth (VLJ; E 565)—reflects his appreciation that by means of experiment, Newtonian science had translated its mathematical discourse into natural language, forgetting its confrontation of limits by substituting controlled conditions for real ones and thereby abandoning the attempt to save the appearances. Blake's "God forbid that Truth should be Confined to Mathematical Demonstration" (E 659) seeks, conversely, to preserve the Receptacle from reduction to a concept. For, he says next, "beside real, there is also apparent truth." As my introduction indicated with respect to the *Newton* color print, Blake seeks to straddle "Ideal Beauty" (E 658, 661) and perform it at the temporal threshold of the *khora*.

The Four Zoas as Plotinean Contemplation, Whitehead's "Extensive Continuum"

Here we might turn to *The Four Zoas*, the epic that preceded *Milton* and *Jerusalem* and whose relation to Plato's Receptacle differs significantly from theirs. In *Narrative Unbound* (1987), Donald Ault argued that Blake's goal in this unfinished masterpiece was construction of "a transformational process at the service of (and brought into existence by) the temporally unfolding surface narrative itself," a "continuously originary process" of "inexhaustible revision" and "self-differing" whose "heterogeneity" "requires constant retroactive reconstitution of 'facts' or reader 'events.'"[41] So far, I've been implying that *Milton* and *Jerusalem* approach the same utopian goal, but they indicate the impossibility of its realization, inasmuch as Earth and Eternity remain Contraries whose dialectic cannot be subsumed, sublated, or reconciled. More than one critic has objected that a poem conceived along the lines Ault describes seems unpublishable, and hence an abrogation of the prophet's public duty. It's as if Blake in *The Four Zoas* was, on Ault's representation, trying experimentally to merge with that entity Whitehead designates as God, "the underlying energy of realisation" and "creativity" that generates "the advance from disjunction to conjunction" that constitutes the extensive continuum and its operating principle: "The many become one, and are increased by one" (*Process and Reality* 21). To be sure, a poem may "contain" revisions to itself; that is, it may represent or gesture toward them. But you don't have to be a historicist to see that a poem spontaneously able to render itself ever fresh and new—to "increase itself by one" through revision, thereby pre-empting interpretation—cannot have been conceived with any living, earthly audience in mind. Since narrative is "extensive," it can never quite catch up to the time of its telling and become, in Ault's words, a "continuously originary process" of self-differing revision.

I therefore side with those who view Blake's project of writing *The Four Zoas* "in the intransitive" as a glorious failure.[42] Ault acknowledges that his rigorously narrow reader-response approach, which excludes all reference to the poem's intellectual and historical context, repeatedly "enacted" the difficulties of Blake's text: "The revisions of *Narrative Unbound* have at times uncannily paralleled those of *The Four Zoas*" (xxiii). Ironic to say, this echoes the difficulties David Erdman had already encountered when he tried to historicize the poem in opposing positivistic terms, by interpreting it as an allegory of breaking political developments in Europe based on newspapers and contemporary reportage. The part of *Prophet against Empire* Erdman revised most across three editions from 1954 to 1977 was, by far, his analysis of *The Four Zoas*; so he, too, wound up "enacting" Blake's revisions. It seems to me the true value of Ault's study is that it highlights the deficiency that Blake succeeded, by great effort, in making good in the two epics that came after. By looking at the Vortex concept Blake first introduced in the

Four Zoas, Night 6, but that he then extended into the very structures of *Milton* and *Jerusalem,* we can see how his approach to narrative composition changed and with it his idea of the prophet's role and place in the world.

The first point to make here is that vortexes in *The Four Zoas* are not objective physical event-objects like *Milton*'s. Rather, they are projected by Urizen as the path he is forced (he believes) to travel to explore his universe, in parody of the Road to Hell opened by Satan in *Paradise Lost.* Whether they are Cartesian or Newtonian (critics disagree) or an amalgam of both (as Ault argues[43]), or whether they exist in a dense ethereal medium or in a vacuum, these vortexes remain primarily spatial effects produced by gravitation. Significantly, the passage that introduces the Vortex concept on *Four Zoas* page 72 seems to describe the author's own participation in Urizen's self-conflict. We are told that Urizen's clothes rot during his protracted life-in-death,

> But still his books he bore . . .
> . . .
> . . . the books remaind still unconsumd
> Still to be written & interleavd with brass & iron & gold
> Time after time for such a journey none but iron pens
> Can write and adamantine leaves recieve nor can the man
>   who goes
> The journey obstinate refuse to write time after time.
> (*FZ* 71:35–72:1; E 348–49)

This looks like an allegory, not of writer's block, but prophet's block arising from Blake's inability or unwillingness to publish by means of illuminated printing. Urizen evidently "bears" (carries, but also suffers and endures) his books as unrealized ideas "still unconsumd [undecayed, but also unread] / Still to be written." On the other hand, "the *man* who goes / The journey" is obliged or compelled to write at all costs, stuck in a private no-man's-land without real progress, as instanced by the repetition of the already redundant phrase, "time after time." The double negative, "nor can the man . . . refuse to write," conveys the oppressiveness of his task. For, the journey he goes is Urizen's journey, and yet Urizen's "poor ruind world" lacks the attainments of relief etching: "books & instruments of song & pictures of delight" (*FZ* 72:35–73:3; E 349–50). (Relief etching is not musical per se, but it does qualify as an "instrument" of song, as the "Introduction" to *Songs of Innocence* emphasizes.) It is unclear if the cause of Blake's "obstinate refus[al] to write" for publication after *Urizen* in November 1794, when Pitt launched his clampdown on dissent, was the specific threat of jailtime or sheer nervous dread. Then again, it's rarely possible to distinguish cleanly between heavy government censorship and self-censorship, its intended effect.

The Four Zoas illustrates this cause-effect conundrum through Urizen's inability to grasp his own role in generating the vortexes that obstruct his journey and render it a contradiction leading nowhere. It seems he is confused about how to integrate space with movement in time.[44] Is it possible Blake faced a parallel uncertainty over how to integrate his cosmology with the confused historical conditions of its imagining, uncertainty that resulted in a narrative endlessly rewritten and never finished? The frenzied scribbler portrayed on the *Book of Urizen* title page (figure 3.4) begins to look like a prophecy of the poet himself under repression. To be sure, Blake's published works *always* satirize their own inability to communicate and the social complicity this reflects, from *The Marriage*'s only half-true Blake-Devil and the alienated auditing of Tom the Sweep's seemingly natural, birdlike expression of pain and grief, "weep weep weep weep" (*SI*; E 10), to the visibly deleted appeal for the reader's "[love] and [friendship]" in *Jerusalem*'s Public Address (*J* 3; E 145). But *The Four Zoas* goes further still and deliberately instantiates the crippling of Blakean public prophecy during a period of (so-perceived) government thought control.[45] The poem's writtenness—its markedly drafty aspect—makes it a verbal equivalent of Blake's *Newton* (figure I.1) and the large color prints of 1795, which, as Robert Essick points out, merge themes of fallenness with the relatively heavy, occluded, and, for Blake, ideologically inimical graphic medium of color printing (as mentioned in my introduction). Blake's later condemnation of the heavily worked chiaroscuro paintings he made while composing *The Four Zoas* as "Blots & Blurs" (*PA*; E 576) applies also to the manuscript. Viewed in hindsight of the evangelical *Milton*, his frantic revisionary activity in composing the poem looks like an evasion of Judgment, a "Satanic" failure to face up to the temporal limitations of his cosmic epic—such limitations being, increasingly for Blake, the main reason for believing human mortality needs forgiveness and mercy. To say, as I am saying, that this failure constituted an "objective" response to dire sociopolitical conditions is simply to acknowledge the prophet had reached a limit beyond which he could do little more than look out for himself.[46]

To put it another way, Blake's decade-plus silence following *The Book of Urizen* seems to have corresponded with a despondent period of Plotinean rationalism when he abandoned belief in a personal God. Stressing Plotinus's Greek-rationalist indifference to the Christian God of personality, J. H. Randall, Jr. explains that for Plotinus, "there must be a Reason Why Being is unified. There must be an [*arche*, or principle of order] of Being itself. This [*arche*] must be logically prior to Being, as its Logical Source. . . . This is the ultimate potentiality of there being an intelligible universe: it makes [*nous*], the Intelligible Realm, itself an actualized potentiality."[47] Crudely simplified, this sounds like every prophet's fantasy: sheer thought can actualize potentiality. As Leopold Damrosch, Jr., emphasizes, throughout his work Blake identifies Jesus as an abstracted symbolic "function"[48]: namely, "the Poetic Genius" (*ARO*; E 1–2) or "the Imagination [that] is not a [human emotional] State: it is the Human Existence itself" (*M* 32:32; E 132). Blake's evangelicalism in *Milton* and *Jerusalem* evidently reflects awareness after *The Four Zoas* that if his poetry was again to take material, public form, it would be through incarnation like Jesus, not an embodiment but an instantiation or "*image* of the Invisible God" (*M* 2:12; E 96; my italics), and thence through a less dualist, more Agambenian reading of Plotinus than either J. H. Randall or Blake's friend Thomas Taylor describes. *Milton*'s invocation, with its many echoes of *Paradise Lost*, can be seen to associate the compositional strategy behind *The Four Zoas* with the Forbidden Fruit: the poem was a "False Tongue" whose "sacrifices" made "even Jesus . . . its prey" (*M* 2:10–13; E 96).

At any rate, Urizen's "world of Cumbrous wheels" without traction (*FZ* 72:22; E 349)—a parody of Ezekiel's chariot—resembles Blake's own career following *The Book of Urizen* when he was spinning his wheels on *The Four Zoas*. "Creating many a Vortex fixing many a Science" (72:13; E 349) is no more than a stopgap against falling since Urizen continues to pass through each new vortex back into the voids between them. Since these voids lack any up or down, his impression of falling persists as a kind of free-floating anxiety:

> For when he came to where a Vortex ceasd to operate
> Nor up nor down remaind then if he turnd & lookd back
> From whence he came twas upward all. & if he turnd and viewd
> The unpassd void upward was still his mighty wandring.
> (*FZ* 72:16–19; E 349)

Urizen confounds creation and falling because he lacks a way to bring the moment to its crisis. Not until he arrives at the putative queen of the sciences, Euclidean geometry, does his sense of falling end because geometry puts him in a position to *measure* the psychophysiological form of his anxiety. "Here will I fix my foot," he boasts, compass-like, and so "the Sciences were fixd

& the Vortexes began to operate / On all the sons of men" (73:14–22; E 350). That is, the vortexes operate on created mankind but no longer on Urizen for he has finally achieved his goal of pure dislocation and exemption from the space-time continuum replaced, now, by a purely spatial world: "a void / Where self sustaining I may view all things beneath my feet" (72:23–24; E 349). He gains "a New Dominion" as "King / Of all & all futurity . . . bound in his vast chain" (73:19–24; E 350). In *The Four Zoas*, then, the vortex form emerges as a false universal. It is fixed in an absolute space conceived in abstraction from Urizen's actual experience of journeying cyclically, "Time after time" (72:1; E 349), through a multiplicity of local spatiotemporal vortexes. The poem's vortexes exemplify the self-contradictory phenomenology of Newtonianism's "fallacy of simple location," as Whitehead termed it, based on the separation of space from time. Just why Urizen feels he is always falling downwards into a vortex even as he struggles upward to pass through another one is subsequently shown in *Milton*. Namely, there is just one Vortex—the psychophysiological form of the Moment, the eternally recurring now of time and motion in space, the experience of passage—but Urizen's fearful insistence upon fixity prevents him from recognizing it.

To return to the *Timaeus*, Plotinus, and Agamben, we are in position to see how Ault's revision-centric interpretation of *The Four Zoas* looks like an attempt to track the Receptacle's material traces of thinking without assimilating them, in Urizenic-Newtonian fashion, to Agamben's "inert substratum of a form" (*Potentialities* 218). Narratively, this inert substratum corresponds to stable descriptive realism as opposed to Spenserian romance (the latter a space-expanding, phantasmatic *khora* and "a field for thoughts to take place," according to one recent critic[49]). Physically, it equates to the engraved copperplate required to print Blake's poem as a minimum public reality; cognitively, to the activity of the Demiurge in *Timaeus* 30c as he contemplates the Form of the Cosmic Animal while in a state of virtual identity with it. In the passage that so impresses Agamben, Plotinus explains that "the indefinite consist[s] in a certain negation in conjunction with a certain affirmation; and [it is] like darkness to the eye" (*Enneads* 2.4.5, *Select Works* 33–34). The soul, "taking away whatever in sensibles resembles light, and not being able to bound what remains, is similar to the eye placed in darkness, and then becomes in a certain respect the same with that which, as it were, it sees." Nevertheless, Plotinus stresses the soul does not thereby "understand nothing." Rather, "when she beholds matter, she suffers such a passion as when she receives the resemblance of that which is formless."

Yet, this feeling is fleeting because the soul "immediately impresses [matter] with the form of things, being pained with the indefinite, as if afraid of being placed out of the order of beings, and not enduring to stop any longer at nonentity." Hence, the existential anxiety that drives Urizen's search "for a solid without fluctuation" (*BU* 4:11; E 71). If, as Plotinus asserts, "the profundity [abyss, *bathos*] . . . of each body is matter," which remains "dark" because "reason [*logos*, language] is light and intellect is reason" (22), then what Urizen seeks to consolidate is a perception of the being of the indefinite. Like Newton, he wants to measure and objectify potentiality. Just as Blake recognized Milton in *Paradise Lost*'s Satan, so perhaps he came to see himself as Urizen seeking to prolong and stabilize his mystical experience of the Receptacle.

Ault's reading of *The Four Zoas* conclusion, his interpretation of which he expressly says he finds dissatisfying, further helps to explain why Blake in *Milton* widens his mythological heterocosm to include direct first-person narration and the constrained, subjective figure of the author. Ault notes that by *Four Zoas*, Night 9, there remain "few legitimate options for halting the narrative's seemingly infinite capacity for repetition/transformation of events under ever new and intertwined delusive guises." Nevertheless, he emphasizes that the "poem is *not* a self-consuming artifact. . . . Instead, the narrative reveals the narrator to be, at the poem's close—in order for there to *be* a close—the hidden ally of the Newtonian reader . . . The reader constructs a narrator whose wish-fulfillment enacts the reader's own desire for closure . . . the hallucination of a narrator who has the authority to close the poem, by speaking its 'End' " (*Narrative Unbound* 447, 467–68; his italics). In other words, Ault's "narrator"—an artefact of the textual formalism entailed by his reader-response phenomenology—serves to mediate the relation between the hitherto self-contained cosmic narrative and the actual, earthly world about to subsume it as the poem reaches its close. But in *Milton*, this is exactly the role played *within* the poem by the Blake poet. The same irony Ault feels forced to read into *The Four Zoas* conclusion with frankly uncomfortable self-consciousness, is provided in *Milton* by the narrative's final shift to Felpham's local perspective. In place of *The Four Zoas*'s ostensibly universal but manifestly imaginary Apocalypse, *Milton* offers a fictional but potentially real apocalypse wherein living the limited and temporary Judgment of "Blake" is juxtaposed against his precursor's permanent and transfiguring one.

Conversely to *The Four Zoas*, *Milton* tells the story of how a functioning authorial self gradually coalesced out of an indefinite, almost voiceless awareness that had been drifting confusedly within a mythic cosmos without individuation: that of the Mundane Soul. In retrospect, we can say the narrative restores "Blake" to embodied first-person perspective following a separation between himself as cognitive agent and his physical body, as seen in the largely self-enclosed preliminary allegory of the Bard's Song. From an initial field of forces dispersed confusingly across multiple worlds, personifications, and personas, the poem puts together a unitary Blake poet and, at the Felpham climax, delivers him into the commonsense, low-energy universe of Newtonian naïve realism. Thereby, the Newtonian universe crucially provides a concrete location wherein the poet can experience a specific, nonimaginary vision of Redemption and Judgment. The terrifyingly decentered energies seen in the archaic Bard's Song as sweeping through Eternity beyond all rational perspective, and which are subsequently glimpsed sidelong in *Milton*'s episodic cutaways to its underlying myth (for example, plates 18–19 and 19–20), become naturalized as the persisting vague background to a newly immediate and motivated subjective foreground.

I am suggesting that *The Four Zoas* narrative of "continuously originary process," as Ault calls it, constitutes a Plotinean raising-up of contemplation to the status of a productive principle in compensation for, or reaction against, the poem's drastic narrative of collapse and disintegration. Interestingly, Agamben's *khorus*-based, semiotic rather than semantic "quant[a] of signification" appear close kin to "inner speech," the fragmented stream of self-directed verbal behavior that humans generate throughout waking hours to regulate perception, consciousness, and agency. The example of *The Four Zoas* indicates that Blake wanted to close the distance between the conscious self and its constant spontaneous subvocalizations, disabling the cognitive "source monitoring" that ordinarily serves to discriminate between internally generated sensations (thoughts, memories, imagery) and sensations originating from the external world. Inner speech could thus become hallucination of a voice, without the "belief evaluation" or epistemic normalization otherwise enabled by background context.[50] Neurologically, the eye's Plotinean power to "sense its own lack of sensation" closely resembles "the self-illumination . . . of the visual sense" generated spontaneously within the eye and brain along the visual pathway by phenomenal "phosphenes" within the retina in the absence of external optical stimuli.[51]

Blake's grand experiment of turning *The Four Zoas* into an essentially incomplete work-in-progress looks, therefore, like an attempt to gain access to the mind's "default mode network" or "minimal phenomenal self," the substrate neural source of the human capacity for dreaming, daydreaming, mind wandering, unfocused rumination, and states of reverie in which narrative fragments combine and recombine semi-spontaneously (as, for example, in *Songs of Innocence*). All of these activities involve "simulation": the manufacture or confabulation of story elements that provide the raw material of autobiographical narrative by allowing us, as in dreams, "to escape the stimulus-bound present, review the past and preview the future by projecting ourselves into different scenarios."[52] If *The Four Zoas* was automatic writing that served as a spiritual meditation or prophetic practice by which Blake "kept the Divine Vision in time of trouble" (*J* 44:15: E 193) during a period of intense censorship, then the manuscript's unpublishability would have been (or became) part of its *raison d'être*.

In his foreword to Ault's *Narrative Unbound*, George Quasha speaks of "laying eyes on" *The Four Zoas* manuscript during its exhibition at the New York Public Library, a tactual experience he compares to "seeing through (not with) the words to the incomplete erasures . . . like moving through veils to touch the body of the 'real poem' " (n.p.). Thus, Quasha, like Plotinus, seems to gain mystical access to the Agambenian semiotic, nonsemantic "quant[a] of signification" of Plato's Receptacle. Yet, Quasha's "mov[ement]" can surely never reach its object, for the palimpsestic "body of the 'real poem' " remains as untouchable as the unextended reflections in the Receptacle's mirror (an auditory equivalent might be the sound of one hand clapping). By contrast, *Milton* and *Jerusalem* portray the poet as directly participating in the temporality of his narrative and its several falls and emanations. The result is not a timeless vision of Being as the untouchable source of all making-possible but a brief personal apprehension of indefinite duration as the transcendental potentiality that imbues being with becoming, thereby generating the experience of temporal passage and change. And the form of this becoming-of-being is that of the Vortex-Moment discussed in chapter 4. The Vortex, then, is the geometrical form of Blake's prophetic travel to the Receptacle of potentiality, travel that renders him the Demiurge's mortal vehicle even as he falls back into three-dimensional existence, thence to enter a new Vortex. Blake's retreat into mythology and quietism came, not in *Milton* and *Jerusalem*, as is often claimed, but in the Neoplatonizing manuscript poem that preceded them, *The Four Zoas*, whose self-cycling private contemplation of pure duration the later works resubmitted to the world of time, change, and public prophecy.

Consider, therefore, how *Jerusalem*'s introductory verses, cited eleven or twelve pages back, trace Eternity's emanation into time as a nonlinear feedback loop that complicates any *Four Zoas*-like Plotinean union with the

Demiurge. Blake subverts the necessary logical equality between the statements "God is man" and "Man is God" by temporalizing them as concrete, context-bound speech acts. For if it is true that "God becomes as we are, that we may be as he is" (*NNR* [b]; E 3), so is the obverse: man, his eyes opened to notions of superiority and moral judgment, himself "becomes as God" (Gen. 3:5), who thus ceases to be seen except as Urizen. *Jerusalem*'s invocation begins in the indefinite present:

> Of the sleep of Ulro! and of the passage through
> Eternal Death! and of the awaking to Eternal Life.

A paragraph break, and then a voice is heard, presumably the poet's:

> This theme calls me in sleep night after night, & ev'ry morn
> Awakes me at sun-rise, then I see the Saviour over me
> Spreading his beams of love, & dictating the words of this
>   mild song.

Then another paragraph break, followed by the words of the song:

> Awake! awake O sleeper of the land of shadows, wake! expand!
> I am in you and you in me, mutual in love divine:
> Fibres of love from man to man thro Albions pleasant land.
> . . .
> Lo! we are One; forgiving all Evil; Not seeking recompense!
> Ye are my members O ye sleepers of Beulah, land of shades!

But disaster ensues, and the next verse paragraph shifts into the main narrative's definite past tense:

> But the perturbed Man away turns down the valleys dark;
> [*Saying. We are not One: we are Many, thou most simulative*]
> (*J* 4:1–23; E 146)

And from Albion's reaction here ensues the whole of *Jerusalem*'s vast tale of fall and fragmentation. How did this happen?

Through the reader's inattention. One had assumed the Saviour's rousing words were traditional epic boilerplate preliminary to the main tale. Yet Albion "the perturbed Man"—a character in the impending narrative—reacts to them directly, as though the Saviour had spoken to him as well as the poet. (Compare the shift, in "Nurse's Song," from the generalized, moody first-person voice of a naturalistic Browningesque character in the first two stanzas to disruptive immediate vocalizations in the last two.) Evidently, *Jerusalem*'s narrative was underway from the outset. The invocation was not a partitioned-off set piece but already part of the story—as was the reader. Apparently, the representational-realistic conventions of what Ault calls "out-there," "Newtonian narrative" merely travesty the inspiring interconnectedness proffered by the Saviour a moment before in the (only seemingly formal) invocation. The Saviour's words were not a fictional dramatization of song framed within the author's lyrical account of a private dream. They were the very song itself in all its unifying self-presence, much as the indefinite present tense would suggest.[53] Poet and Saviour are "in" each other, such that Albion's fall into disunity is precipitated by the self-ignorant reader's deafness to the urgency of their joint call for community. So, the grammatical shift from inspired invocational present to narrative past tense enacts the reader's membership in Jesus only after the fact, as a betrayal.[54] Henceforth, readers will be "in" miserable Albion. This is a narrativized version of the turning outside-in that Blake dramatizes in the shift from the text of "A Cradle Song" text to its realistic design, or in his suggestion that the tableau seen in the *Visions of the Daughters of Albion* frontispiece is really a Minkowski time slice of a wider being who contains the viewer, namely, fallen Albion the Mundane Soul. It seems the Saviour's initial invitation to unity was not secure and unconditioned but carried a potential to be changed by subsequent events: the present reacts upon the past and alters its meaning, just as the way we think about the future ends up shaping it. Apparently, Blake then extended this organic revisionary process back into the poem's opening address, "To the Public," whose deleted passages (*J* 3; E 145) bear prophetic witness to the reader's imminent obtuseness. Too many deletions, however, and there would be no new poem at all but just another over-revised and unpublishable *Four Zoas*.

In *Jerusalem*, therefore, the experience of duration arises as an essential element in sense awareness, specifically in hearing and listening. If the running-in-place revisionism of *The Four Zoas* privileges contemplation as giving insight into the metaphysical absolute of continuous, indivisible duration (as in Bergson's famous example of the melody whose successive notes have no before or after because it is grasped as a *gestalt*), then in *Jerusalem* Blake affirms a more Whiteheadian, more empiricist recognition that "the quality of passage in durations is a particular exhibition in nature of a quality which extends beyond nature."⁵⁵ Passage, which Whitehead's *The Concept of Nature* designates as "the ultimate metaphysical reality," is only measurable so far as it appears in nature, where it yields the perception of a serial time that seems to extend beyond nature.⁵⁶ But what actually extends beyond nature is passage itself, which is even more fundamental than space, time, and matter. Since every perception is a grasping of sense data into a unity, its specific here-now constitutes no more than one perspective on a spatiotemporal surround of more distant and indefinite sensory images whose backdrop of "beyond" eventually fades to black. *Milton*'s Vortex expresses, topologically, how any punctual here-now emerges noninstantaneously from a "transcendent" beyond of raw, imperfectly apprehended impressions. This emergence—the realization of the Moment's indwelling potential for movement and change—is a function of the spatiotemporal element in all perception, from which we derive the concept of time itself and its distinctions of before and after. One reason why *The Book of Urizen*, *Milton*, and *Jerusalem* all start so estrangingly is to dramatize the crossing of a threshold—the open lip of a Vortex. Thereby, some small fragment of Eternity's prediscursive logical concept space begins to be retrieved and brought into earthly orbit, thence to become realized temporally within earthly minds through an act of imaginative perception, however small and limited. Conversely, we've already seen in chapter 2 the crucial role played by endings in several Songs of Innocence, where the heavenly vision granted within the poetic heterocosm is self-reflexively seen or heard to fade away as the reader grasps its sensory basis into a completed event-perception occurring in the public world beyond subjective reading.

Implicit in *Milton*'s concluding first-person perspective, then, is a view of the poem's Vortex-shaped "drop of experience" as a unitary local event with universal implications, a process of becoming spread at first across a broad mythic region but concrescing, eventually, into the garden outside "Blake's Cottage at Felpham" in "1804" (*M* pls. 36, i). The dynamism of this universe, where each new event not only redounds upon the meaning of history as given but also shapes the very geometry of the space-time within which future events will occur, bears close resemblance to Whitehead's "extensive continuum" as set forth in *Process and Reality* (61–82). This was Whitehead's quantized generalization of Einstein's special relativity, and it differed significantly from other interpretations of general relativity on offer at the time. Whitehead's concern was that Einstein, in saying space is curved, even variably curved, and in attributing this differential curvature to the physical characteristics of space itself, seemed to imply space is something substantive with causal power to affect mass. He reasoned, "The whole theory of the physical field is the interweaving of the individual peculiarities of actual occasions [the topology of event-objects] against the background of systematic [i.e., uniform, Euclidean] geometry" (*Process and Reality* 333). Accordingly, Whitehead renounced his earlier revised relativity principle of 1919–24, based on a view of space as a continuous abstraction from events, for an atomic view of motion or change as a series of "occasions" or events the sum of which we call, for example, "distance" along the "path" of the particle. Compare how "Miltons Track," which appears as continuous in the diagram so titled (figure 4.11), manifests in *Milton*'s narrative as a series of temporally discrete but connected vectorial phase states: falling star, close-up Los-sun, black-clad puritan, Jesus. Since infinitesimals are not real, says Whitehead, "a comparison of finite segments is thus required"; accordingly, "it would be better to abandon the term 'distance' for this integral, and to call it by some such name as 'impetus,' suggestive of its physical import" (*Process and Reality* 332–33). In this way, Whitehead's theory stakes out a middle ground between Berkeley's rejection of infinitesimals and, on the other hand, his "fallacy of simple location" rooted in the claim that distance exists purely in the mind as an inference drawn from differently sensed lengths (see chapter 6).

To Whitehead's disappointment, the extensive continuum was never much accepted by physicists, who found it too complex for problem solving even as it yielded results nearly identical to the Einsteinian equations. His reasons for the theory were metaphysical rather than practical, and Einstein, who was not persuaded to accept the early version when the two men met in June 1921,[57] likely would have deemed the 1929 version a mere psychologizing extension of the quantum mechanics he was already beginning to despise. Undeterred, Whitehead went on to install the extensive continuum at the very foundation of his cosmology. It appeared to supply the number-crunching operationalism of quantum mechanics with a sorely needed theory of space. Recently, it has been recruited to explain findings of microphysical research into quantum gravity.

The distinctive feature of the extensive continuum is that its space-time emerges from the collective character of the event-objects that constitute its physical structure. These processes supply the ground of extension rather than the other way around. The geometry of extended space-time derives from metaphysically prior "occasions of experience" (to use Whitehead's term), instead of pre-existing them. Einstein's continuous metric space-time is therefore seen to be a macroscopic abstraction from a pregeometric quantum substrate dominated by other interactions than gravity. Whitehead's space-time does possess a metric, but it is one described by the uniform relatedness of events by which new occurrences achieve "solidarity" or "togetherness" and so become facts. We might hazard an analogy. As Newton's low-energy, long-range gravitational system was to Einstein's general relativity—namely, its limit—so general relativity is to Whitehead's quantum gravity space-time.

In contrast to Whitehead, Einstein's theory of relativity, which does not allow the definition of a global past, present, and future, notoriously treats the felt human experience of time's flow from a historical past to an open future as an illusion. To quote the ending of Einstein's great letter of condolence to the family of his lifelong friend Michele Besso: "Now he has again preceded me a little in departing this strange world. This has no importance. For those of us who believe in physics, the separation between past, present, and future has only the significance of a tenacious illusion."[58] Though Einstein associated relativity's time-symmetrical "block" universe with Schopenhauer's argument of the nonexistence of free will—a position Blake shared[59]—Blake likely would have linked the "block" with determinism, predestinarianism, and the tenseless Christian God. Indeed, Einstein insisted that philosophical problems of the subjective now, time's arrow, and the cosmology of the Big Bang all had nothing to do with the mathematics of relativity. In contrast, Blake and Whitehead anticipate recent anthropic arguments, based on the observer's dynamic role in quantum mechanics, to the effect that time is the animal inner sense that brings the spatial world to life by uniting successive snapshot pictures of spatial states in a colligation or compilation, much as repeating frames of a film reel are perceived to show objects in motion. This is to say, further, that colligation selectively *creates* enduring "objects" from changes in background field. The preliminary snapshots may be produced by the eyes through saccades or by a double diffraction slit on special paper in a laboratory: either way, consciousness synthesizes reality as we know it. That last statement may be an idealist tautology but it's also neuroscientific fact. Thanks to the brain's colligations, we can anticipate and track changes in position of objects or events—a valuable evolutionary adaptation—and measure their "momentum."

Blake's version of this idea appears in *Milton*'s doctrine of States:

Distinguish therefore States from Individuals in those States.
States Change: but Individual Identities never change nor cease:
You cannot go to Eternal Death in that which can never Die.
(*M* 32:22–24; E 132)

"Satan & Adam are States," but "The Imagination is not a State: it is the Human Existence itself" (*M* 32:25–32; E 132). Blake's "States" are phase states of matter washing over the Imagination like waves over a lighthouse (the idea recalls Wordsworth's apostrophe to Imagination in *The Prelude*, book 6). *A Vision of the Last Judgment* reverses the terms of the opposition in seeming endorsement of Einstein's block universe, yet the meaning remains the same: "These States Exist now Man Passes on but States remain for Ever he passes thro them like a traveller who may as well suppose that the places he has passed thro exist no more" (*VLJ*; E 556). States change from the shifting viewpoint of the traveler who passes through and leaves them behind on pilgrimage to Eternity, but in actuality, States are static, inert, inorganic;[1] it is the traveler who undergoes change and growth.[60] In sum, the Imagination generates time, movement, and momentum in an otherwise fixed universe of instantaneous changes in spatial position of adventitious bits of matter: differences that make no qualitative difference, change without movement, direction, or meaning. This is to say, per chapter 1, that reality is analogical all the way down. For Whitehead as for Blake, the present lies in the eternal now of thinking itself—including not only dreams and memories, as my next chapter shows by way of a striking example provided by Whitehead, but also daydreams, lucid dreams, mind wandering, and enthusiastic Vision. Even so, an important function of the Whiteheadian extensive continuum, for Blake, was its physical basis. This prevented a slide back down induction's slippery slope of merely odd psychological phenomena which the Lockean epistemology tended to assimilate, bloblike, to its underlying mind/body dualism.[61]

Traditional Newtonian-Cartesian physics had defined space as homogeneous extension without process and time as a serial process without spatial extension. As Whitehead put it in a lecture, if Descartes "conceived extension as essentially a quality of matter," then for Einstein "extension is essentially a quality of events and so is process." But then Whitehead insists, further, that "nature's becomingness cannot be restricted within one serial linear procession of time. It requires an indefinite number of such processions to express the complete vision."[62] Whitehead envisions not a single, lawfully heterogeneous space-time but rather one whose heterogeneity is indefinite because it changes, slightly, with the introduction of each new event. In Blakean terms, the reason nothing is lost—"not lost not lost nor vanishd, & every little act, / Word, work, & wish, that has existed . . . / . . . every thing exists & not one sigh nor smile nor tear, / One hair nor particle of dust, not one can pass away" (*J* 13:60–66; E 157–8)—is not because the minute particulars are all smoothly destined for salvation but because reality's advance is uneven, contingent, and irregular, being a cumulative product of the particulars that go to make it up. If Blake's "not . . . / One hair" is lost alludes to Luke 12:7 and its reassurance that "not one" sparrow is "forgotten before God" because "the very hairs of your head are all numbered," then "particle of dust" recalls atoms. Like Whitehead, Blake affirms the particularity of the individual particles as against their mathematic uniformity as points.[63] For Whitehead, this is a consequence of Faraday's electromagnetic field theory: "The modification of the electromagnetic field at every point of space at each instant owing to the past history of each electron is another way of stating . . . that in a sense an electric charge is everywhere."[64] The past is gone, but the *being* of the past persists since settled facts supply the potentiality that allows new entities to arise. Hence, becoming and extension are united in a single comprehensive process of emergence. As Isabelle Stengers points out, Whitehead's extensive continuum "relegates to the 'empirical case' many of the properties that Einstein's general relativity defines *a priori* as belonging to the space-time metrics."[65] Since this continuum "is not a fact prior to the world" (*Process and Reality* 66), it lacks the transcendental status of space and time in Kant.

Proclus's commentary on Plato's *Timaeus* again proves helpful in explicating theological implications of the Receptacle-like condition of betweenness that Whitehead imagines:

> Though the world exists through the whole of time, yet the existence of it consists in its coming to be, and is in a part of time. This however is *once*, and is not simultaneously in all time, but is always once. For the eternal is always in the whole of eternity; but the temporal, in a certain time, is always at a different time in another time. As with reference therefore, to the eternally existing God, the world is very properly called, *the God that would once exist*.[66]

Christoph Helmig and Carlos Steel explain that, for Proclus, there are two measures of duration. Eternity "measures at once the whole duration of a being," whereas time "measures piecemeal the extension of a being that continually passes from one state to another." Yet, some beings partake in both eternity and time. Souls "are immortal and indestructible; nevertheless, they are continually undergoing change."[67] Thus, *Milton* suggests the world—"*the God that would once exist*"—is a Cosmic Animal sustained, ultimately, by the perpetually perishing divinity of Jesus, the change principle, Blake's human symbol of the becoming process behind Whitehead's "extensive continuum," where all exists once only and yet forever after. As Blake-Los asserts, "not one Moment / Of Time is lost, nor one Event of Space unpermanent / But all remain . . . / . . . / The generations of men run on in the tide of Time / But leave their destind lineaments permanent for ever & ever" (*M* 22:18–25; E 117).

On this view, the initial narrative sequences of *The Book of Urizen*, *Milton*, and *Jerusalem* appear so bewildering because they express the central paradox in Proclus and Whitehead alike: "There is a becoming of continuity, but no continuity of becoming. . . . In other words, extensiveness becomes, but 'becoming' is not itself extensive." For "continuity concerns what is potential; whereas actuality is incurably atomic" (*Process and Reality* 35, 61). In these poems, one sees how the narrative extends out of language's self-circling mythopoeic potentiality through series of perspective scene shifts. Each Illuminated Book represents a step forward in the progressive unveiling of an Eternal ur-myth that it verbalizes and reperforms in response to changing conditions on the ground. Consider the classicist Gregory Nagy on the three-dimensional nature of oral-epic transmission:

> From the standpoint of oral poetics, each occurrence of a theme (on the level of content) or of a formula (on the level of form) in a given composition-in-performance refers not only to its immediate context but also to all other analogous contexts remembered by the performer or by any member of the audience. . . . Whatever we admire in our two-dimensional text did not just happen one time in one performance—but countless times in countless reperformances within the three-dimensional continuum of a specialized oral tradition.[68]

The individual Homeric text is the product of "an underlying system, which is Homeric poetry," but the text "is not the same thing as the system." Blake's mythology forms, likewise, an underlying four-dimensional system—the poetic equivalent of Whitehead's extensive continuum and its attendant "solidarity" or "togetherness of things"—from which each new poem must birth itself through a difficult "Striving with [against yet by means of] Systems" (*J* 11:5; E 154).[69]

This difficult birth process is plainly shown in *Milton*'s opening plates, which repeat passages from several of Blake's earlier works to dramatize the poem's gradual emergence out of the past into the present. *Milton*'s narrative doesn't get underway until the Bard's Song on plate 7, yet even then it remains for a while temporally indefinite since it continues to be interrupted twice by the injunction, previously repeated no fewer than six times: "Mark well my words! they are of your eternal salvation" (*M* 7:16; E 100, and *M* 7:48; E 101). Is this the epic Bard exhorting his ancient audience of Eternals or a sporadic voice-over by the author Blake addressing present-day readers in a prolongation of his invocation on plate 2? The effect is a kind of static, a partially garbled transmission like that of the rationalist-sounding Voice of the Devil in *The Marriage of Heaven and Hell*, plate 4; or the literary-pastoral artificiality of "Laughing Song"'s innocent "Ha, Ha, He," an allusion to Shakespeare's bawdy Pandarus (*SI*; E 11)[70]; or the crude, unmended deletions Blake perpetrated upon *Jerusalem*'s address "To the Public," and then went ahead and printed in direct contradiction of its claim that "every word and every letter is studied and put into its fit place" (*J* 3; E 146). In Whiteheadian terms, one could say Blake plunges readers into the quantum discontinuity of becoming. After all, *we* are the audience of Eternals, *Blake* is the ancient Bard, and that Bard's Song is *Milton*: we, Blake, and Blake's poem are all mutually "vehicles" of Eternity. In terms of Plato's *Timaeus*, the experience is akin to contemplating the indefinite potentiality of the Receptacle as it begins to become accessible through the nonsensation of Plotinean *anaisthesis*.

The Influence of Abraham Tucker and "The Vision"

The third main source I see for Blake's Mundane Soul is considerably less exalted than More or Plato, and even more unsystematic. It's hard to imagine Blake ever read all six long volumes of Abraham Tucker's *Light of Nature Pursued* (1768–77), but he evidently did peruse at least part of volume 2 around the time he wrote *Milton*, likely in Hazlitt's abridged edition brought out in 1807 by Johnson.[71] Tucker would have appealed to Blake on several grounds. As opposed to Hartley's mechanical theory of volition, he argued human action also has ideal causes. Reason, however, operates too slowly to supply the ideas that prompt spontaneous human behavior; rather, these are given by the imagination, whose "instantaneous motives" reason directs toward long-range goals able to provide what Tucker calls "satisfaction" (1.97–209). In his letter to Trusler, Blake links a similarly idealist view of the imagination with the very lodestar of British empiricism: "Consider what

Lord Bacon says: 'Sense sends over to Imagination before Reason have judged & Reason sends over to Imagination before the Decree can be acted'" (E 703). Thus, the *gestalt* of Hartley's "decomplex" ideas is raised to the level of an epistemological principle, as in one of Blake's notes to Berkeley's *Siris*: "Knowledge is not by deduction but Immediate by Perception or Sense at once" (E 664). "Satisfaction" is Tucker's modest utilitarian version of Plato's eros. It replaces Locke's purely goal-oriented description of desire in terms of uneasiness. Tucker apparently never read Hume, but his argument that the formation of habit is promoted by ideal causes beyond discrete sensations accords with Hartley and anticipates Whitehead, who critiqued Hume's doctrine of association of ideas for tacitly assuming the survival of the past within the present contrary to Hume's professed mental atomism.

So, one sees why Hazlitt probably regarded Tucker as anticipating his own *Essay on Human Action* (1805), which makes imagination the source of voluntary action. In the *Essay*, disinterested imagination of the future moves the mind by metalepsis toward qualitative change and "compassion" beyond Adam Smith's self-centered doctrine of vicarious moral sympathy.[72] Similarly to Hazlitt, in Blake's *Milton* the dead poet's disinterested imagination of the future—namely, now: the dire condition of Europe in "1804" which includes "Blake," who is in correspondingly bad shape—constitutes a "self-annihilation" based on prophetic compassion for alienated Ololon. Though Blake is commonly viewed as a religious enthusiast, he would have seconded Tucker's condemnation of Evangelical enthusiasm for perverting human solidarity through its gloominess, fearful dwelling upon the horrors of death, and artificial ratcheting-up of devotional feeling. Tucker's attacks on the element of self-regard in Wesleyan "virtue" and its liability to become "a species of intemperance" rooted in the notion that morality has "a distinct essence of its own" beyond ordinary human pains and pleasures[73]: Blake criticized these religious attitudes all his life. On the other hand, Tucker's *via media* of rational toleration and modesty combined with his robust openness to intellectual speculation make him an inheritor of More's Cambridge-Platonist Latitudinarianism and a key source for the peaceable syncretism of Blake's Mundane Soul.

Tucker's Mundane Soul—"the receptacle for particular spirits as they can disengage themselves from their vital union with matter" (2.107), an Albion-like "god or animal or glorified man containing all men" (2.90), a powerful figure of near-telepathic human unity—significantly influenced Blake's ideas on expanded sense perception. And notice how Tucker, like Whitehead, tends to conflate his Mundane Soul with the Platonic Receptacle.

Since "all space not occupied by matter is replete with spiritual substance called the mundane soul," explains Tucker, "it will follow that by the mutual communication of perceptions every one [individual member] may have those arising in every other" (2.70). He concedes, saying that "all perception must begin by the action of matter." But

> this need make no difficulty, for when we consider how the stars with their several systems of planets are dispersed up and down, how light, ether, and perhaps many other subtil fluids we know nothing of, are diffused every where, and that all these bodies, great and small, must lie contiguous to some parts of the mundane soul, we shall see there will not want objects for it to perceive. . . . the variety of bodies floating about in the mundane soul may exhibit a greater variety of ideas thereto, whereby it may discern them all, their combinations and modifications, together with the comparisons and other relations resulting therefrom. (2.71)

Interestingly, the passage begins by placing matter "contiguous to" the Mundane Soul, but it goes on to speak of celestial and ethereal bodies "floating about *in*" the Soul, which therefore must contain them as Berkeleyan ideas of sense similarly to the starry sun- and moon-containing Albion on *Jerusalem,* plate 25 (figure 4.5). In Blake's design, Albion's daughters are disemboweling him to create an external object world. Although Tucker says he finds Berkeley's immaterialism ridiculous, he evidently imagines a humanized, "inside-out universe" like the one in George Gamow's illustration (figure 4.2). Indeed, Tucker's aetherous Mundane Soul may owe something to Newton himself. In his letter to Henry Oldenburg of 9 December 1675, Newton puts forward a corpuscular theory of light that anticipates the modern idea of light quanta and speculates in passing:

> The gravitating attraction of the earth [may] be caused by the continual condensation of some . . . aethereal spirit . . . ; and bearing much the same relations to aether which the vital aerial spirit, requisite for the conservation of flame and vital motions, does to air. . . . And, as the earth, so may perhaps the sun imbibe this spirit copiously, to conserve his shining, and keep the planets from receding further from him. And they, that will, may also suppose, that this spirit affords or carries with

it thither the solary fewel and material principle of light: and that the vast aethereal spaces between us and the stars are for a sufficient repository for this food of the sun and planets."[74]

Insofar as Tucker's Mundane Soul takes the form of Gamow's inside-out gastrula, Newton led the way by supplying its astral nourishment in that last sentence. As "a perpetual worker," says Newton, nature conserves the activity of the cosmos by "making a circulation." The physiology of this is further evident in another letter to Oldenburg at this time which speculates: *"the frame of nature may be nothing but aether condensed by a fermental principle* . . . nothing but various contextures of some certain aethereal spirits or vapours condensed, as it were, by praecipitation . . . [and] *condensed in fermenting or burning bodies, or otherwise inspissated in the pores of the earth to a tender matter, which may be, as it were, the succus nutritius of the earth, or primary substance, out of which things generable grow. . . ."*[75] In P. M. Heimann's summary, "[c]hemical, electrical and aethereal spirits thus acted as active principles in nature."[76]

Tucker's celebration of the Mundane Soul and Newton's similar idea of a universal "aethereal spirit" that feeds nature lay the ground for Albion's apotheosis as God in *Jerusalem*'s concluding celebration of Plato's Great Year: "all / Human Forms identified, living going forth & returning wearied / Into the Planetary lives . . . / And then Awaking into his Bosom" (*J* 99:1–4; E 258). In *Milton*'s account of "the nature of infinity," the reason the "traveller thro eternity" perceives the conglobed Vortex that "roll[s] backward behind / His path" as being "like a moon, or like a universe of starry majesty, / . . . / Or like a human form, a friend with whom he livd benevolent" (*M* 15:21–27; E 109), is because that globe is really Albion the Mundane Soul, who a Moment ago contained the traveler within himself.[77] The Mundane Soul constitutes, Tucker says, "one Mind, and therefore cannot, like compound bodies, consist of distinguishable and separable parts discerped [i.e., torn off] from it." Individual souls are all "the same person with . . . [this] one created mind in nature . . . yet upon their immersion into matter we see they are distinct persons and things" (2.61). "Discerped," "immersion into matter": this is the language of Henry More and Neoplatonism. Newton's "aerial spirit" in the preceding passage is, likewise, a central term in More's system of vehicles.

From this it follows, claims Tucker sounding like a combination of More and Berkeley, that what we call the material world is the "sensory" of the Mundane Soul "exciting sensations and reflections and exhibiting ideas" in its "spiritual part," which operates "as a percipient to receive them and a vivifying principle to invigorate and actuate the motions of the other [i.e., the sensory part]" (2.89–90). In Whiteheadian terms, the Soul's sensations and ideas actualize the potential in passive matter to produce creative advance. Tucker says if "it will appear impossible that so vast an infinitude of knowledge can be contained in any created mind," that is simply because we are unaccustomed to the experience of expanded sense perception. "In our present condition 'tis our organs that set the limits to our understanding, nor do we know what our mental capacity is. . . . We may possibly be capable of twenty senses, but being provided with inlets for only five, have no more conception of the others than a blind man has of light" (2.73). It seems likely that Blake had already come across this passage when *The Marriage of Heaven and Hell* referred to the "enlarged & numerous senses" of "the ancient poets" (*MHH* 11; E 38). In contrast to the abrupt and polemical *Marriage*, however, *Milton*'s narrative sequence demonstrates how such enlarged understanding takes place—namely, as a micro-event within a symbiotic universe whose interactions are simultaneously more intensive and more far-reaching than either Tucker or the youthful *Marriage* poet could imagine.

Newton had proposed the aether, a spiritual substance, as the source of such different phenomena as gravitation and cohesion. As others did, Tucker pseudo-scientifically adds "fire, heat, muscular motion and sensation" to the list of phenomena that might be so explained. Yet, inasmuch as these various "effects" are all different, the aether must operate differently to produce them, "which indicates a kind of choice and discernment not to be found in the motions of matter unless where directed by some understanding; and this direction it may receive from the mundane soul" (2.79).[78] Notwithstanding Newton's rule 1 for scientific explanations in the *Principia*'s Rules of Reasoning, the so-called "principle of parsimony," Tucker deems his suggestion entirely plausible: "Since then we experience in ourselves a power of giving impulse to matter, and there is none of it [i.e., no matter within the Mundane Soul] but must lie within the reach of some spirit contiguous thereto, why need we scruple to believe [such matter] liable to the like action as we exert upon our motory fibres?" Thus, "the repulsion of ether, whence all other material agents derive their vigour, begins" in the Mundane Soul, "by whose ministry the laws of nature are executed, the continual decay of motion repaired, [and] the world and all things therein are kept in order" (2.81). Like Henry More (also like A. E. Taylor in his *Timaeus* commentary, who was roundly censured for it), Tucker goes on to associate biblical creation with this demiurgic Mundane Soul: "The very expression commonly used that God made all things by his word warrants our supposition of an intelligent agent who should understand and obey the word when spoken" (2.99).

Nevertheless, Tucker's Mundane Soul is only half Blakean. If his vision of it previews *Jerusalem*'s finale, the Soul's material grandeur equally recalls the superhuman Pantokrator of Newton's Scholium: "By his strength he rolls the huge planets along the boundless sky; by his agility he dashes the light on all sides with inconceivable velocity; by his energy he produces gravitation, cohesion, heat, explosion, fluidity, contraction and dilatation of the circulating vessels in plant and animals" (2.91–92). Thereby, Tucker's Newtonian Mundane Soul performs the maintenance work Newton had admitted was necessary to keep the planets in their proper orbits and hold the universe together. Blake's Providence and Divine Hand, mentioned in passing in *Milton* and *Jerusalem*, serve a similar function in accordance with the eighteenth-century occasionalist doctrine, "Conservation is but continuous creation."[79] At the same time, unlike Newton but like Blake, Tucker emphasizes the Mundane Soul's dependency upon the limited human souls it contains. "It may be presumed that we are likewise of some use to him," including our sufferings (2.94). For, it is just because the Mundane Soul "may have nothing to desire for himself, and nothing to do but exert his powers and contrivance in lessening the burthens and enhancing the enjoyments of animal life as much as possible," that "the most glorified Beings may be constantly attendant upon the services of man, not for the greatness of his [man's] importance, but because he is the only poor creature that wants their cares" (2.96). Indeed, "The existence of evil, which proves to us a stumbling block, would teach him [the Mundane Soul] a usefull lesson," since, being innately harmonious, the Mundane Soul has no other way of discovering it (2.106). Thus, Tucker blithely anticipates Blake's more strenuous paradox: "God becomes as we are, that we may be as he is (*NNR* [b]; E 3). As the Christly agent of a supreme God who remains largely beyond discussion, the Mundane Soul's dynamic aspect resembles the "consequent nature" of the dipolar Whiteheadian deity whose "lure for feeling, the eternal urge of desire," causes "a reaction of the world on God." With respect to his becomingness, says Whitehead in words that well apply to Blake if not quite the pragmatic Tucker, God is best conceived under the image—"and it is only an image"—of "infinite patience . . . a tender care that nothing be lost" (*Process and Reality* 344–46). The mutual redemption-in-weakness of the aging Albion-Jehovah and youthful Jerusalem-Orc in *Jerusalem*'s full-page design to plate 99 images this dipolar reciprocity, leaving it unclear who is saving whom.

But further, since Tucker's Mundane Soul also serves as "the receptacle" (2.107) that receives the vehicles of human souls after death, it necessarily grows over time. Tucker imagines it as becoming an increasingly intricate and harmonious network of human relations, to the point where this self-unified "host of happy spirits" (2.107) eventually achieves apotheosis. In this way, Tucker offers an eighteenth-century "optimistic" anticipation of Whitehead's doctrine of creative advance. The progress of his Mundane Soul is a metaphysical version of the physiological process whereby Hartley argued increasingly "decomplex" or recursively synthesized associations of ideas promote sympathy and theopathy, and so must "reduce the State of those who have eaten of the Tree of the Knowledge of Good and Evil, back again to a paradisiacal one," thereby producing a society of "*Members of the mystical Body of Christ*" possessed of "new Sets of Senses, and perceptive Powers . . . so as to increase each other's Happiness without Limits."[80] Tucker says, "Let us consider that since creation is currently esteemed so common as to be practiced every day in furnishing souls for children in the womb, we may as well suppose the same creative power constantly employed in producing new spaces, extending the bounds of the universe, and giving room for the mundane soul to expand according to the new members it continually receives" (2.66). The "new Expanses" Blake prophesizes at the end of *Jerusalem*, "Creating Space, Creating Time . . . / . . . / . . . according to the Expansion or Contraction, the Translucence or / Opakeness of Nervous fibres" (*J* 98:30–37; E 258), appear to combine the visions of Hartley and Tucker both. Significantly, however, Blake adds time to Tucker's expanding spaces. As a four-dimensional space-time, *Jerusalem*'s "new heaven and new earth" (Rev. 21:1) are not merely something new under the same old sun; the sun, too, is recreated within time and nature, as we've seen in figure 4.13.

Jerusalem's conclusion is triggered when Albion throws himself into the Furnaces of Affliction. Thereby, the indefinite collectivity that makes up the mythic Mundane Soul attains the concrete particularity of living human kindness epitomized by Jesus. For, "every kindness to another is a little Death / In the Divine Image" (*J* 96:27–28; E 256), an unselving by which God recreates his vehicle, man, after his own image. And so Albion awakens into Eternity. Albion's Godlike realization of the permanence of passage itself—a fusion of the ontological Contraries, being and becoming, or in Whitehead's terms, the "everlasting" and the "perishing"—constitutes a transcendence of "the five Senses. the chief inlets of Soul in this age"

(*MHH* 4; E 34). Albion's imaginative dream come true briefly realizes the coming face-to-face with God that Tucker likewise contemplates: "We may be capable of new senses, higher faculties and sublime reflections than our present organization can exercise. When totally disengaged from the veil of matter enwrapping us we may be able to see even as also we are seen and discern sensibly that glorious object which no man can behold and live" (2.104–05). In Tucker's redemption scenario, the Mundane Soul, purified through intensive interrelationships between its parts, constitutes an "intimate communion of saints," so it "may be reckoned a resurrection of the body." Since "the vehicular state is a resurrection too, therefore that [vehicular state] may be reckoned the first, or resurrection into the kingdom of Christ, and this of the mundane state the second, when he shall deliver up all power to the Father" (2.114). *Jerusalem*'s ending develops Tucker's speculation into the sadly grand vision of a single cyclical scheme of fall and redemption, passage and permanency, whose underlying unity is God.

At the heart of *Jerusalem*'s wistful, valedictory conclusion lies an apprehension of time similar to Bergson's when he observes that "it is impossible to imagine or conceive a connecting link between the before and after without an element of memory and, consequently, of consciousness" (*Duration and Simultaneity* 48).[81] Bergson's notorious panpsychism finds echoes in the various "proto-mental" entities of Whitehead, Herbert Spencer, and David Bohm, all ultimately derivations from Plato's World Soul. After all, without some intercommunication arising from the past's immanence within the present, two successive "snapshot" moments would lack heterogeneity, thereby depriving the present of its distinguishing character of novelty and undermining the processive nature of duration itself. Blake's Mundane Soul, Albion, houses the prophet's reluctant acknowledgment of the past's necessary drag on the present, without which the future would lack any real connection to time and passage. His position resembles that of the phenomenologist Mark Currie when he argues that if it is correct to regard the present as the future of the past (as historians have recognized at least since R. G. Collingwood, not to mention Joseph Priestley as quoted in the introduction), there can be no real *self*-presence without the distancing of self-reflection.[82] Indeed, the whole thrust of *Milton*'s Moment is to give pause, so that the present may be opened by "Self-examination" (*M* 40:37; E 142) into fuller consciousness beyond the impersonal collective memory of the Mundane Soul.[83] After telling how Milton entered his left foot, the Blake poet ruefully acknowledges:

> But I knew not that it was Milton, for man cannot know
> What passes in his members till periods of Space & Time
> Reveal the secrets of Eternity: for more extensive
> Than any other earthly things, are Mans earthly lineaments.
> (*M* 21:8–11; E 115)

Milton's Moment serves to disclose the full meaning of the past before the present passes, and it becomes too late.

An overlooked source both for the dilatoriness of *Milton*'s narrative and its first-person perspective is Tucker's novella-length chapter 23, "The Vision," reprinted in full by Hazlitt and Johnson. Tucker's satirical "Vision" deploys a Henry More–like theory of vehicles to explode received ideas about the size and shape of the human body per Blake's *Marriage*, plate 4. His fabulous first-person narrative supplied a key model for *Milton* no less than Dante's *Commedia* did. Indeed, Tucker's own narrative is partly modeled on Dante. The author journeys through a purgatorial afterlife or bardo of history's greatest thinkers while guided by a sternly pedagogical John Locke. His body laid asleep in a lucid dream—like Milton in *Milton* 15:1–16—the author gains access to "the vehicular state" and visits the land of "the vehicular people," "little diminutive men" (2.19). When his vehicle eventually "bursts," he becomes "instantly absorbed into the Mundane Soul" (2.300):

> My body was immense yet I could manage it without trouble, my understanding extensive yet without confusion or perplexity. For the material universe was my body, the several systems my limbs, the subtile fluids my circulating juices, and the face of nature my sensory. . . . I rolled the bulky planets in their courses, and held them down to their orbits by my strong attraction: I pressed heavy bodies to the earth, squeezed together the particles of metals in firm cohesion, and darted the beams of light thro' the expanse of innumerable heavens. (2.300–01)

In the end, Tucker drops back down into London from "the zenith" like "a falling star"—like Milton at *Milton* 15:47, word for word, when he begins his own fall to earth "thro Albions [the Mundane Soul's] heart" (*M* 20:41; E 114). Clearly, the "Vision" chapter's Google Earth zoom-in on home from outer space provided a model for the Blake poet's safe return to Felpham cottage in *Milton*'s closing plates after wandering lost in mythology's outer reaches. Tucker's depiction of the spiritual vehicle is not a mere gratification of Smithean Sympathy and its mystified, *in situ* transports of received feeling, as in Sterne and other sentimental novelists, but a genuinely estranging thought experiment—perhaps even a "Fiery Chariot of Contemplative Thought" (*VLJ*; E 566) by which to explore lucid dreaming as a prelude to vision (as in Andrew Baxter, discussed in the next chapter).[84] For Blake as for Tucker, objects are not self-moved, so they lack power to achieve presence to the mind. Rather, presence is achieved when the mind travels, shamanically, to seemingly freestanding objects where the body as we know it is not.

Indeed, Tucker speculates repeatedly that now might be time for a change of perspective: "That we are out of heaven, is not owing to any distance we stand at from thence, but to our being pent up in walls of flesh which cut off our communication with the blessed spirits. . . . We are like persons inclosed each in a sentry-box having all the chinks and crannies stopped that might let in the least light or sound" (2.100). The passage alludes to Locke's famous comparison of the understanding to "a closet wholly shut from light, with only some little openings left [the five senses], to let in external visible resemblances, or ideas of things without."[85] By contrast, Tucker wonders if "heaven is not local but everywhere, all around, above, below, on each side, and within us" (2.100). Inasmuch as Tucker's little vehicle gets back inside his bedchamber by slipping through "the *chink* under the door" (2.330; my italics)—an echo of the earlier passage—"The Vision" demonstrates that sorties from the Lockean closet can and do occur. As Blake puts it in *The Marriage*, if "man has closed himself up, till he sees all things thro' narrow chinks of his cavern" (*MHH* 14; E 39), it is merely because he has limited "the chief inlets of Soul" to "the five senses" (pl. 4; E 34).[86]

What enables Tucker's hypothetical change of perspective is the vehicle's almost infinitesimal tininess, by which it gains intimate access to "Queen Mind" (2.60) and, thence, extrasensory—in Plotinean terms, nonsensible—perception of heaven within. In a recollection of More's principle of indiscerpibility, Tucker explains that a kind of fractal vision is required to grasp just how intensive so much tininess can be. "The great divisibility of matter" entails that "the least conceivable [vehicular] particle is capable of containing as great a variety of parts and machinery as the whole human body. But what clogs our comprehension of these minute divisions is that we commonly think of making them by dividing the whole without dividing the parts, which must certainly spoil the composition" (2.14). Compactness also yields surprising improvements in the vehicle's athletic abilities because "the finer parts a body contains, the fewer atoms they must severally consist of (for these cannot be divided,) the less of pore there will be among them, and consequently its nerves and sinews will be so much tougher and stronger . . . more agile and pliant . . . more alert and vigorous . . . consisting all of muscle and fibre . . . extremely flexible and obedient to the Will, susceptible of any shape" (2.55–56). Accordingly, once the vehicular Milton "collect[s] all his fibres into impregnable strength" (*M* 39:35; E 142), his high-res "outline of Identity" (37:10; E 137) allows him to become a preternaturally clear and distinct idea in the Blake poet's mind. (The combination of muscular compression and flexibility seems paradoxical only if you're still thinking in Democritean-Euclidean terms: ask any athlete.)

Since time in the vehicle is measured by the succession of ideas, which because of its miniscule size occur more rapidly than in geophysical time as measured by the sun (2.164), Tucker's vehicle travels 40,000 miles in one minute (2.183). Here Tucker follows Berkeley, whose *Principles of Human Knowledge* argues: "Motion is not without the mind, since if the succession of ideas in the mind become swifter the motion, it is acknowledged, shall appear slower without any alteration in any external object."[87] Yet, as perceived time slows, movement for Tucker ceases to be Euclidean or even metric. His account of how the "little diminutive men" (2.19) transport themselves through the stream of light corpuscles by pushing off against them like skaters, or tacking back and forth between them like sailboats, renders them as Lucretian atoms whose deviatory movements defy geometric rule.

This is a telling departure from Berkeley, who had performed several striking thought experiments (such as traveling toward the moon) to demonstrate that distance exists only in the mind and that the visible image alone is the true object of sense. In a passage that recalls the experience of looking through a microscope, Berkeley argues:

> In proportion . . . as the Sense is rendered more acute, it perceives a greater Number of Parts in the Object, that is, the object appears greater, and its Figure varies, those arts in its Extremities which were before unperceivable, appearing now to bound it in very different Lines and Angles from those perceived by an obtuser Sense. And at length, after various Changes of Size and Shape, when the Sense becomes infinitely acute, the Body shall seem infinite. (*Principles of Knowledge* 24)

But for Tucker, Berkeley's idea of infinitely expanded sight remains too intellectualistic and rational. He seems on the verge of transcending Berkeley's purely spatial idea of location: "What is there room for optics among those [vehicular bodies] to whom the corpuscles of light are so gross as to be objects of touch rather than of sight? Of mathematical lines and angles among bodies continually moving?" (2.65). Blake says, "The Microscope" and "Telescope" "alter / The ratio of the Spectator's Organs but leave Objects untouchd" (*M* 29:17–18; E 127). Pace Locke, perception cannot be reduced to sense: "Mans perceptions are not bounded by organs of perception. He percieves more than sense (tho' ever so acute) can discover" (*NNR* [b]; E 2). That is why the Blake poet's initial glimpse of Milton as a falling star takes so long to resolve into a visible object. As we shall see, it was in defense of plain, ordinary sense that Berkeley opposed Newton's attempt to solve the problem of locating "bodies continually moving" by means of mathematical "fluxions" or infinitesimals. Notwithstanding, when Berkeley extrapolates infinite vision from existing sense in the above passage, he takes the same approach as Locke does when he naturalizes ideas of eternity and angels through incremental, ratiocinative reasoning (*Essay* 1.251–55). Tucker would have known that in 1727, James Bradley, with Molyneaux's assistance, calculated that light took 8'12" to travel from the sun to the earth: strong evidence of its materiality.[88] Unlike Berkeley but compatibly with Bradley's finding, both Tucker and Blake seek to explain the movements of barely visible bodies by locating them within a physical, specifically psychophysiological space of expanding or contracting sense perceptions.

At the far horizon of these imaginings lies Einstein's famous thought experiment of what would happen if he caught up to a light beam: "I should observe such a beam of light as an electromagnetic field at rest."[89] Bradley's experiment plainly indicated a speed limit for light. Thus, the Einsteinian theory was already foretold, which places light at the center of space and combines space and time by defining them both in terms of light signals sent from one part of the universe to another. Proclus's claim, "Space is nothing other than the finest light,"[90] underpins a similarly unified doctrine where light engenders space as a container of event-objects. Thereby, light creates matter, which, in turn, gives the event-objects spatial extension. Throughout his work, Blake's celebration of daybreak as a new Creation-Fall dramatizes a similar unification of light, space, and matter, while crucially adding time as the dimension that makes this revelation momentarily visible to sense (consider *Milton*'s dawn conclusion, *Jerusalem*'s sunrise invocation, and figure 4.13). The original source for both Blake and Proclus appears to be *Timaeus* 58c, where Plato treats light as a special kind of body, an emanation of "flame" that does not "burn" and possesses the power of brightness without heat.[91] Isidoros Katsos comments: "In a loose analogy, one may even suggest that Plato's light particles fulfill a function similar to our light quanta or photons. They mark the limits of the sensible, corporeal dimension."[92] In his *Timaeus* commentary, Proclus seems to understand Plato's light as marking a limit of the sensible, and so does Blake, whether his source was Plato, Proclus, Henry More, or Priestley, whose *History and Present State of Discoveries Relating to Vision, Light, and Colours* describes Bradley's experiment in detail.

Broadly speaking, Tucker's "vehicular hypothesis" stands in playful reproof of Newton's austere "*hypotheses non fingo*." So does his reasoned defense of "Hypothesis" in the chapter that begins *Light of Nature*, volume 2. In the preface to *The Wisdom of the Ancients*, Bacon had explored fable, parable, and allegory as mythic prototypes of the new science. Tucker evidently wanted to add hypothesis to Bacon's list to engender a more inclusive empiricism than Newton's. He defines hypothesis as "a kind of continued allegory" by which men represent "another state under such veils as may render it discernible":

> Fable represents an action impossible to have been performed, parable one that is possible and similar to those frequently happening but not proposed to our belief as an historical fact; whereas hypothesis exhibits such representation of things as may be the real case for anything that can be shown to the contrary. It requires no positive evidence to build it upon, . . . the burden of proof lying upon him that would overthrow it: but its strength lies in the consistency and mutual dependence of its several branches upon one another, its not contradicting any known phenomena or received principles, its helping to join into a regular body those which before were detached and independent. (2.5–6)

Hypothesis, then, is a chain of analogies whose persuasive power lies in their interconnectedness. Tucker emphasizes its importance to empirical science, not unlike the Blake-Devil when he quotes the Proverb of Hell: "What is now proved was once, only imagin'd" (*MHH* 8:33; E 36). Unsurprisingly, "continued allegory" whose "veils" represent "another state" describes *Milton* quite as aptly as Blake's own definition of "Sublime Allegory" as "Allegory addressd to the Intellectual powers while it is altogether hidden from the Corporeal Understanding" (E 730).[93]

Yet, as strange as Tucker's long, elaborate dream vision is, its utter lack of repercussions in the waking world is stranger still. Once Tucker abruptly passes "through the atmosphere" to arrive "at the top of mine own house" (2.326–27), his mentor, Locke, is forgotten, and he goes straight back to bed. Absent a space-time able to lift his experience of expanded sense perception to the level of prophecy by uniting it to physical reality, the shapeshifting "bag" of organic sensitivities that is Tucker's vehicle ultimately reduces to a corpuscle operating within a container of external, three-dimensional space. "The Vision" remains, therefore, a *jeu d'esprit*. Despite the Tuckerian narrator's lengthily described early struggles to gain control of his new amorphous body bag, it is apparent his cognitive "I" persists unshaken throughout. Indeed, this Cartesian "I" makes prolix comedy out of the spectacle of the narrator's befuddlement. The reintegration of Tucker's vehicle with the author in bed is, then, a private, subjective experience: the naturalistic termination of a lucid dream, not a full-blown annihilation and reconstitution of the self as in *Milton*.[94]

This is simply to observe that Tucker's educational Locke figure is a far cry from the dislocated identity-in-difference of Blake's Milton-Los-Jesus, whose action transcends the collective, hypostatic operations of the Mundane Soul. Blake's is an anonymous Jesus who "may be" (*NNR* [b]; E 3) our own selves unwittingly transfigured into a blessing or inspiration in the eyes of a needy recipient through our doing of some casual, small act of kindness, and without our even knowing it.[95] Milton remains throughout the poem quite oblivious of his timid Romantic successor, even as the latter joins him gratefully in victory: "I also stood in Satans bosom & beheld its desolations!" (*M* 38:15; E 139). In this sense, the Blakean vehicle bears witness to the inseparability of Berkeley's *esse est percipi* and *esse est aut percipere* ("to be is to be perceived" and "to be is to perceive"), quite as Northrop Frye suspected.[96] Milton's heroic, high-powered trajectory through wide swathes of mythology into the eye and brain of the homely Blake poet is matched, conversely, by the earthbound poet's slow coming to consciousness of the cosmic magnitude of his present moment. Since the two viewpoints are interdependent and mutually constitutive, there can be no external perspective from which to judge or measure their relation. Milton's Judgment is final, and so is the Blake poet's small-scale, local vision of it.

In other words, Milton-Jesus represents a caring and connectedness that transcends "Felpham," "1804," and all specific time and place. He is the opposite of Smithean sympathy's imitative self-duplications based, ultimately, on a literalistic Renaissance view of the Mundane Soul as the source of sympathetic magic. You might say Blake's Jesus principle of interdependence constitutes relativity within the world of moral relations. If self is not a substance, one becomes a person nonautonomously through other people, whose diminishment or growth therefore affects one's own development. Still, let's not feel too pleased about such human interconnectedness. If the dispersive Blakean self is also *internally* modal and conditionalized, then in the words of the psychoanalyst Adam Phillips, "We share our lives with the people we have failed to be."[97] There is a dark Contrary even to the redemptive claim that "nothing is lost," as shown by the Blake poet's poignant failure to follow his precursor into full union with Jesus. Milton becomes forever a "new man" (Eph. 4:24); Blake, only for a moment of vision no sooner seen than it has passed. Not every potentiality can become realized or it wouldn't *be* potential but merely a delayed predetermination. Necessarily, many fail of their purpose.

CHAPTER SIX

Berkeley

Very Close, but No Cigar

In a late essay, "The Unity of Human Knowledge" (1960), Niels Bohr summarized his descriptivistic or "operationalist" understanding of relativity: "Physics is to be regarded not so much as the study of something a priori given, but rather as the development of methods for ordering and surveying human experience."[1] The classical concepts of space, time, causation, and continuity are (Kantian) idealizations, he argued, insofar as they assume the existence of an independent physical world with inherent properties of its own. And yet it is difficult to discard these concepts entirely since they are adapted to the language of sensory experience and serve to connect the formalism of quantum mechanical theory with actual experimental observations.

Bohr's statement sounds like Berkeley, and not by coincidence. Berkeley wanted to reduce the mechanists' search for occult efficient causes to the study of "phenomena" whose regularities and "uniformness," as reflected in nature's observed "analogies, harmonies, and agreements," constitute the "signs" by which to formulate "probable conjectures" and thereby "predict things to come" in proper scientific fashion.[2] At the same time, Bohr's position—and Berkeley's, too—sounds compatible, *prima facie*, with the positivistic approach of Ernst Mach, whose elevation of empirical sense data over mathematics, physical modeling, and the alleged "pseudo-problems" of traditional philosophy was thought, in his day, to ally him with the neo-Kantians' reduction of knowledge to subjective mental ideas. Karl Popper even went so far as to argue that Berkeley's critique of Newton, based on his rejection of a material world behind the physical appearances, makes him Mach's "precursor."³ But this distorts Mach's profoundly relational view of mental and physical phenomena alike as "neutral" events or "elements," neither subjective and private nor independently "out there," which continually enter and exit existence without any substantial basis. What Mach, like Blake, critiqued was not metaphysics per se but rather mechanism's assumption that the real world must correspond to our intuitive visualizations of it.⁴ Arguably, he anticipated Whitehead's critique of the "bifurcation of nature" in attempting to fuse together phenomenal and material worlds. Popper's misreading illustrates how difficult it still was, decades after the triumph of relativity and quantum theory, to give up the materialism/idealism binary based on the opposition between a knowing subject and substantial objects with "properties" independent of the interactions that disclose them to an observer.

This chapter investigates the close relation between Blake's and Berkeley's respective idealisms, while asking how Blake was able to embrace Priestley's materialism at the same time. My argument is that, for all his indebtedness to these two opposing philosophies, Blake rejected their shared underlying rationalism, and this explains why he could learn so much from both without falling into contradiction. Blake understood "understanding" as an interactive, two-way process, unlike Berkeley and Priestley but like Plato in the *Timaeus* and, in various ways and degrees, More, Tucker, Mach, Whitehead, and Bohr himself. Much as Mach argued, ever-changing event-like relations form the substrate reality that constitutes subjects and objects alike. Blake similarly sought for a "third way" able to provide empirical explanations of psychological phenomena: explanations beyond subjective visualization, which in an epistemological context is bound to be circular. While he shared in the general Enlightened abhorrence of Mystery, he also loathed the bland cruelty with which Enlightened Reason operated to reduce everything to a solution—as if logic-chopping could explain away "thick," "vague," often miserable situatedness in the world through some preordained Manichaean pattern of black-and-white *concordia discors*.⁵ Such self-ratifying Ratios find their Contrary in Blake's nonlinear metalepses of cause and effect grounded in "the phenomenological performative," whereby "what the poet predicts will happen *is* happening in and through his writing, and vice versa."⁶ "As none by traveling over known lands can find out the unknown. So from already acquired knowledge Man could not acquire more. therefore an universal Poetic Genius exists" (*ARO*; E 1). The mystery of "more"—the ontological problems of change and potentiality, and the corresponding concern that nothing genuinely existent should be overlooked or "lost," not even a bare potentiality doomed never to become realized—is what Blake, like Whitehead but unlike Berkeley, denominated as God.

Berkeley rejected not only Henry More's and Newton's shared idea of space as absolute and independent of sense perception, but also More's idea of spirit as spatially extended—an idea Blake embraced, as we've seen. These, Berkeley argued, are more of "the errors arising from the doctrine of abstract general ideas, and the existence of objects without the mind," to which "the mathematicians" are especially prone because "their first principles are limited by the consideration of quantity" (*Principles of Knowledge* 1.217–18). If Blake's mythology deliberately exploits "inconsistencies" in Newton's thought, as Donald Ault demonstrates, then Berkeley led the way by attacking the mathematical side of Newton's experimentalism as unempirical and impious. Nevertheless, Blake was not a Berkeleyan—far from it.

The Canadian critic F. E. L. Priestley pointed out some time ago that Berkeley's *Principles of Human Knowledge* (1710) "is a continuation of the long series of attacks on 'Hobbists, Atheists, Epicureans, and the like' mounted in the late seventeenth century by the Cambridge Platonists and their followers, including Newtonians like Bentley and Clarke."[7] Blake surely appreciated that Berkeley's denial of independent material substance comprised an attack on atheism's materialist underpinnings. For Berkeley, motion, space, and extension are not qualities of material bodies but ideas of relation. Newton's absolute space therefore leads to "thinking either that Real Space is God, or else that there is something besides God which is eternal, uncreated, infinite, indivisible, immutable" (*Principles of Knowledge* 1.217). Berkeley here mockingly recalls More's previously quoted list of space's twenty "divine names or titles."[8] Really, he says, extension, color, motion, and so on are simple qualities of particular objects. Only by abstraction can these qualities be conceived as existing without any particularity—for example, "an idea of colour in abstract which is neither red, nor blue, nor white, nor any other determinate colour" (1.140). As Blake put it, "Deduct from a rose its redness. from a lilly its whiteness . . . rectify every thing in Nature as the Philosophers do. & then we shall return to Chaos & God will be compelld to be Excentric if he Creates" (E 595). "Excentric," because in a spatial chaos devoid of God's four-dimensional presence, Creation is necessarily decentered and arbitrary.

Accordingly, Blake endorsed Berkeley's attack on Newton's fluxions as "the Ghosts of departed Qualities"[9]: abstractions from sense experience which, in the view of both men, is alone what supplies the basis for science once gathered and ordered by reason. *Principles of Knowledge* explains that geometers do indeed abstract lines or figures from their magnitude, thereby enabling them to stand as universals for other lines of different sizes. Thus, an inch line "must be spoken of, as though it contained ten thousand parts, since it is regarded not in itself, but as it is universal . . . in its signification, whereby it represents innumerable lines greater than it self" (1.222). Nevertheless, "it will not be found that in any instance it is necessary to make use of or conceive infinitesimal parts of finite lines, or even quantities less than the *minimum sensible*. . . . And whatever mathematicians may think of fluxions or the differential calculus and the like, a little reflection will shew them that, in working by those methods, they do not conceive or imagine lines or surfaces less than what are perceivable to Sense" (1.225).

When Blake asserts, "They say there is no Strait Line in Nature this Is a Lie like all that they say, For there is Every Line in Nature" (*PA*; E 575), he affirms like Berkeley geometry's basis in sense. Hence, his anti-metric assertion: "A Line is a Line in its Minutest Subdivision Strait or Crooked It is Itself & Not Intermeasurable with or by any Thing Else" (E 783). Consider, therefore, Gilchrist's anecdote of the youthful Blake's mathematics tutorial by Thomas Taylor in 1784. The two had reportedly gotten "as far as the 5th propositn [of Euclid] which proves that any two angles at the base of an isosceles triangle must be equal. Taylor was going thro' the demonstration, but was interrupted by Blake, exclaiming 'ah never mind that—what's the use of going to prove it, why I see with my eyes that it is so, & do not require any proof to make it clearer.' "[10] Amusingly, the visionary styles himself as a pragmatic empiricist in contrast to his teacher, Taylor "the Platonist." His position even resembles Hume's in criticizing Euclid's Parallel Postulate as an example of "the fallacy of geometrical demonstrations, when carry'd beyond a certain degree of minuteness. . . . The original standard of a right line is in reality nothing but a certain general appearance."[11] What Blake, like both Hume and Berkeley, rejects is not geometry but the reduction of sense to mathematics and Descartes's *esprit géometrique*. He remained open to the non-Euclidean concept of a space-time experientially accessible through event-objects at sense's very limit—namely, as I've been trying to show, "drops of experience" whose form is that of the Vortex-Moment: indefinitely topological on the way in, geometric in retrospect. At the opposite end of his career from Taylor's tutorials, Blake's exasperated appeals in the *Public Address* to the "Man of Sense" (E 577, 578, 579) who cannot possibly support the caverned solipsism of chiaroscuro, Lockean epistemology, or the *Examiner*'s scurrilous charge of madness, reveal his rage at the contradiction thrust upon him by polite society. He is being forced to *explain* his self-evident, commonsense perspective as if it were just another deductive mathematical abstraction.

I am suggesting Blake recognized Berkeley as a fellow empiricist for whom the true measure of things is man's "middle zone of perception": neither macroscopically all-swallowing like the omnipotent Newtonian Pantokrator's view from nowhere, nor infinitesimally small to the point of nonexistence like his fluxions. As Whitehead concedes, "the practical atomicity of the physical . . . is essential for the intelligibility of the apparent world to a finite mind with only partial perception. Without atomicity we could not isolate our problems; every statement would require a detailed expression of all the facts of nature."[12] Thus, "substance" is a macroscopic effect of limited human sense. And since geometric lines are sensible quantities, there are no infinitely small atoms of space. At least since Proclus, mathematical objects had been deemed as immaterial but also as extended and divisible and, hence, as known through the understanding, in contrast to sensible things on one hand and objects of pure intellect or being on the other. The Zenonian paradoxes of motion supposedly resolved by infinitesimals are therefore misleading abstractions from the start. Achilles doesn't need to take an infinite number of steps to reach the tortoise because atomic space, like time, is discontinuous (much as More's idea of space implied). So, action or energy is itself quantized. Achilles is just behind the tortoise; then instantly he is past it. In the same way, *Milton*'s Blake poet is just about to attain union with Milton-Jesus, then suddenly he has overshot his target only to awaken outstretched in the dirt of his garden path.

Significantly, Whitehead was at some pains to recruit Berkeley to his cause. Berkeley's implication that, as Whitehead puts it, "the realization of natural entities is the being perceived within the unity of mind" from different individual perspectives is found to be an immaterialist anticipation of Whitehead's own theory of the extensive continuum.[13] In *Three Dialogues* (1713), Berkeley writes that, unlike God,

> We are chained to a body, that is to say, our perceptions are connected with corporeal motions. By the law of our nature, we are affected upon every alteration in the nervous parts of our sensible body; which sensible body, rightly considered, is nothing but a complexion of such qualities or ideas as have no existence distinct from being perceived by a mind: so that this

connexion of sensations with corporeal motions means no more than a correspondence in the order of nature between two sets of ideas, or things immediately perceivable.[14]

Priestley comments on this passage: "What we call 'physical events' are events we perceive, that is, patterns of related ideas, as are all the 'things' the mind constructs from sensations. . . . [T]he ideas we call 'corporeal motions' and those we call 'perceptions' . . . all together form a unity of relations which we call 'the order of nature'" (66). For Berkeley, nature "bundles" our sensations in a regular and predictable fashion.[15] Contrast this position with Locke's brutally pragmatic correspondence theory of knowledge, defined subjectivistically as *"nothing but the perception of the connexion and agreement, or disagreement and repugnancy of any of our ideas.* In this alone it consists. Where this perception is, there is knowledge, and where it is not, there, though we may fancy, guess, or believe, yet we always come short of knowledge" (2.167–8; his italics). Berkeley's opposing view is that nature's unified order offers a much stronger argument for God's existence from Design than can Newton, to say nothing of those who imagine, like Locke, "any unthinking *substratum* of the objects of sense" (*Three Dialogues* 1.333). Other philosophers "acknowledge all corporeal beings to be perceived by God, yet they attribute to them an absolute subsistence distinct from their being perceived by any mind whatever, which I do not." Others say, "*There is a God, therefore he perceives all things*," but Berkeley says: "*Sensible things do really exist; and, if they really exist, they are necessarily perceived by an infinite mind: therefore there is an infinite Mind, or God.* This furnishes you with a direct and immediate demonstration, from a most evident principle, of the *being of a God*" (*Three Dialogues* 1.304). Berkeley here approaches the self-evidence of Blake's divine Imagination based in immediate "Spiritual Sensation" (E 703).

Indeed, Creation, Berkeley writes, means nothing more than making things perceptible. His mouthpiece Philonous declares, "I imagine that if I had been present at the creation, I should have seen things produced into being—that is become perceptible—in the order prescribed by [Moses]" (*Three Dialogues* 1.348). Blake perhaps recalls this passage in his annotation to Lavater: "Let it be rememberd that creation is. God descending according to the weakness of man" (E 599). Berkeley thus solves the problem of "creation out of nothing" more convincingly than Locke's Newton-inspired "dim and seeming conception" in the *Essay concerning Human Understanding*. Thomas

Reid, at least, thought so. For Reid, Berkeley's system "followed from Mr. Locke's, by very obvious consequences" that Locke himself failed to see.[16]

More's editor, Alexander Jacob, compares Berkeley's position to More's. Unlike More, Berkeley denied the reality of spatial extension, which he considered a main culprit behind the mistaken notion of a mind-independent material substance. On the other hand,

> The extension that More had granted to spirit as well as to matter . . . allowed him to show how Mind and matter were equally extended, the latter imperfectly reflecting the former, so that Nature is basically an unconscious shadow of Reason. In fact, the Plotinian model of the universe as formulated by More reveals not only the way in which Mind motivates matter through Soul, which ranges between the intelligible and sensible realms as Divine energy, but also what Berkeley himself was at pains to establish throughout his career as a philosophical immaterialist, namely that Nature or the *phenomenal* world is but the final result of the Mind's positing, or *perceptio*, of Itself and, therefore, must be as incorporeal as its original.[17]

Jacob's Morean unconscious Nature is the Whiteheadian vague "togetherness" and "solidarity" of the Mundane Soul, in contrast to Rationalism's "clear and distinct ideas." Jacob's analysis suggests that Blake and Berkeley were likely following More when they regarded the notion of independent matter as Reason's greatest failure to recognize its own participation in the ideas of sense. In the above-mentioned passage from Berkeley's *Three Dialogues*, Philonous is replying to Hylas's question of how, if existence is mind-dependent, as Philonous claims, everything can also exist eternally in the mind of God. Philonous explains: "May we not understand it [the Creation] to have been entirely in respect of finite spirits; so that things, with regard to us, may properly be said to begin their existence, or be created, when God decreed they should become perceptible to intelligent creatures, in that order and manner which He then established, and we now call the laws of Nature? You may call this a *relative*, or *hypothetical existence* if you please" (1.350). Indeed, this is much the same reasoning as underpins Tucker's vehicular hypothesis, which, we have seen, is an explicitly relativized construction based on his definition of "Hypothesis" as a kind of analogy whose purpose is to regularize observed phenomena: "such representation of things as may be the real case for anything that can be shown to the contrary. . . . [I]ts strength lies in the consistency and mutual dependence of its several branches upon one another, its not contradicting any known phenomena or received principles, its helping to join into a regular body those which before were detached and independent."[18]

Conversely, Urizen's emanative Creation in *The Book of Urizen* is the mistaking of "a *relative*, or *hypothetical existence*" for absolute "laws of Nature." In a Frankensteinian travesty of Blake's dynamic redemption principle, "God becomes as we are, that we may be as he is" (*NNR* [b]; E 3), Urizen simply plays God *tout court*. He collapses the Contrary, metaleptic relation between God and man so that instead of transforming man into God, he merely drives fallen man—man as he already is—to worship his own selfishness while rationalizing this *status quo ante* as divinely decreed. Urizen resembles the subjective or cultural relativist who emphasizes the limited nature of all viewpoints but his own. Blake's position is clarified by another marginalium to Lavater: "Man can have no idea of any thing greater than Man as a cup cannot contain more than its capaciousness But God is a Man not because he is so percievd [*sic*] by man but because he is the creator of man" (E 603). God's humanity is not a mere subjective effect of man's anthropomorphizing mind; it is a consequence of God's having freely created man in his own image. In turn, man's perception of God after *his* own image—man's apotheosis in Jesus—is a logical and correspondingly creative reciprocation of that original event. So the loop is closed nonlinearly, like a Möbius strip. There *is* a distance between man and God. It exists in error, in the delusory mind of fallen man.

Blake's visionary deconstruction builds on Berkeley's attempt to explode the illusion of "outness," man's chronic habit of belief in a mind-independent material substance. Berkeley accordingly reduced the concept of distance to visible, phenomenal objects. Yet he never denied the existence of extension or a real, sensible distance between two points on a line. Simply, he claimed that distance does not exist "without the mind." Since "objects of sense exist only when they are perceived" (*Principles of Knowledge* 1.59), the inside/outside dualism proves to be a Lockean abstraction from immediate experience. So far, so Blakean.

What Blake nevertheless rejects is Berkeley's God, whose independent presence conserves the reality of objects when there is nobody to perceive them. For Blake, objects derive from an organic Cosmic Animal or Mundane Soul whose perceptivity grounds the perceptions of the humans who partake in him. My view therefore differs somewhat from Chris Townsend in a recent article laying out the extent of Blake's indebtedness to Berkeley, largely taken for granted since Northrop Frye but never investigated so carefully. Townsend points out that Berkeley's rejection of Locke's "abstract general ideas" and his emphasis on the particularity of perception are the likely source of Blake's term "minute particulars."[19] He goes on to examine Blake's only known remarks on Berkeley, the marginalia to the late *Siris* (1744). When Berkeley attacks Newton's divine sensorium, writing, "God knoweth all things, as pure mind or intellect, but nothing by sense, nor in nor through a sensory," Blake comments, "Imagination or the Human Eternal Body in Every Man" (E 663). And when Berkeley goes on to imply that God's perfect immateriality precludes his being attached to a Cosmic Animal or Mundane Soul—"Nor hath he any Body: Nor is the supreme being united to the world, as the soul of an animal is to its body"—Blake practically repeats himself: "Imagination or the Divine Body in Every Man" (E 663). From these comments, Townsend concludes:

> Spirits [for Blake] are not imprisoned in or contained by bodies [contrary to Berkeley's "We are chained to a Body . . . ," quoted earlier], but they exist *as* bodies. As [Simon] Jarvis puts it, this is the difference "between the subjective and the objective body: between the body which I am and the body which I have." Thus, Blake's account of embodiment asks us to reconceive of "spirit" as something that necessarily entails sensory perception. (368; Townsend's italics)[20]

Yes! My own account of the Blakean body is similar, though I've traced it back to Henry More. Since it is perception that creates objects from sense, and sensory perception requires a body, spirits lacking a body would be unable to perceive the divine object. God would be the inaccessible, superhuman entity of Calvinism or mainstream Neoplatonism like Thomas Taylor's, an empty abstraction of self-mutilating reason. Humans, on this view, would be incapable of spiritual vision and so could never imagine God except "outwardly." Elsewhere, Townsend demonstrates the importance of Berkeley's theory of divine language to Coleridge and Wordsworth.[21] It seems almost too obvious to mention in the case of Blake. In *Alciphron* (1732), Berkeley claims that nature is the language of God: "You have as much reason to think the Universal Agent or God speaks to your eyes, as you can have for thinking that any particular person speaks to your ears."[22] Since spirits alone can be causal agents, motion not our own must be the sign of another agent. So, where the motion belongs to insentient nature, it can only be a sign of God making his presence known to man. How? Blake answers, by means of the collective, transpersonal Mundane Soul.

On the other hand, what enables Blakean "Spiritual Sensation" (E 703) in the first place is the organic power of sense perception to "expand"—an intuitive capability quite unlike Berkeley's idea of perception as a rational, God-given faculty applied to sensations to convert them into objects of consciousness. As we've seen, it is this capacity for expanded vision that renders the Blakean spiritual body a "vehicle" in both More's and Tucker's sense of the term. Blake's "As the Eye—Such the Object" (E 645) implies that close, attentive vision can discriminate minute objects and even detect a subvisible, nonsensory one like Jesus at the Imagination's threshold of identity, that is, self-annihilation; while, conversely, to "become what [you] behold" (*J* 39:32; E 187) implies vision passively conforming itself to corporeal objects as given. The flexibility of such seeing is made possible by the vehicle, whose intermediation maintains separation between perceiver and object and staves off blindness, whether through transcendental union with God or Urizenic collapse into insentience. In *De anima* ("On the Soul"), Aristotle explains that every sense requires a physical medium in which to operate. Far from being seen more clearly, objects right against the eyeball unmediated by air or water obstruct vision. For Blake, it is not corporeal aether but spiritual vehicles that provide this necessary medium. Tucker's "Vision" chapter and the changes undergone by Blake's spiritual Milton demonstrate how the flexibility, minuteness, and compactness of man's "Spiritual Body or Angel" (E 663) operate to dissolve Jarvis's above-mentioned dualism "between the subjective and the objective body: between the body which I am and the body which I have." Berkeley's insistence in *Siris* that "Body is opposite to spirit or mind" only goes to show that his similar dualism holds no room for the Blakean spiritual body.[23] Blake's claim, "that calld Body is a portion of Soul discernd by the five Senses" (*MHH* 4; E 34), underscores the vehicular, Contrarious, and interconnected nature of the body *he* has in mind.

It's not just that Townsend expounds similarities between Blake's metaphysics and Berkeley's whereas I stress differences. Rather, it's the differences-in-similarity between them that explain Blake's need, unlike Berkeley, for intermediaries, a pantheon of spiritual bodies—a mythology. Townsend says of Blake's above-cited marginalia:

> This "Divine Humanity," which Blake elsewhere refers to as the "Human Form Divine" . . . , is the total sum of all human

agents, or spiritual bodies. The members of this community are connected by their minute particularity, which at once preserves individual identities as well as the relationship of these individuals to divinity. In the annotations to *Siris*, Blake makes it clear that it is the imaginative faculty, as in Berkeley, which acts as the nucleus of a minute particular and connects it to the Human Form Divine. . . . He also . . . implies that this structure of the spiritual body's 'minute particularity' might constitute the body of God: "God is Man & exists in us & we in him" (E 664). (376)

But the Divine Humanity—the overall structure of relationships existing between minute particulars—is not static, nor do the minute particulars arrive ready-made for instantaneous inclusion, as in the "fitting & fitted" (E 667) of Wordsworthian divine Design. The salvation Townsend describes is a becoming that takes place in time. If God is Man, it is because "God becomes as we are, that we may be as he is (*NNR* [b]; E 3). The body of this becoming is Albion, the Mundane Soul, the living historical repository of the minute particulars, Blake's humanized counterpart to Newton's divine sensorium of Nature. In Whiteheadian terms, Albion embodies not God as such but God's "consequent nature" resulting from the action of the world upon him: "the record of all achieved fact, a perfect memory of what has been . . . [,] the 'objective immortality' of the world in God."[24]

In *Three Dialogues*, Philonous states it as "a universally received maxim that *Everything which exists, is particular*" (1.283). Blake turns this into a kind of relativity principle, in accordance with his super-empiricist, Leibnizian insight that, as Whitehead says in criticism of Berkeley, "everything is everywhere at all times. For every location involves an aspect of itself in every other location. Thus every spatio-temporal standpoint mirrors the world" (*Science and the Modern World* 91). This doctrine is no paradox, says Whitehead, "but a mere transcript of the obvious facts":

> Your perception takes place where you are, and is entirely dependent on how your body is functioning. But this functioning of the body in one place, exhibits for your cognizance an aspect of the distant environment, fading away into the general knowledge that there are things beyond. If this cognizance conveys knowledge of a transcendent world [i.e., things beyond], it must be because the event which is the bodily life unifies in itself aspects of the universe. (91–92)

For Blake, this unification begins in sensation and moves through perception to imagination. *Milton*'s Vortex-Moment demonstrates how an accident of matter—Milton's remote fall to earth, at first barely noticed—gradually achieves form as a mental object through a concrescence of "the distant environment" or "transcendent world" of things beyond the horizon. The Blake poet's distant glimpse of Milton as a shooting star leads to a vast and overwhelming close-up vision of Milton-Los as the daystar (figure 4.12), and finally to a focused, almost comradely ("clear and distinct") perception of the historical puritan's redemption in Jesus.

The anthropomorphism by which perceptual objects are thus created out of nature's flux resembles Platonic remembrance. In his marginalia, Blake notices without demurral Berkeley's approving observation, "It is a maxim of the Platonic philosophy, that the soul of man was originally furnished with native inbred notions, and stands in need of sensible occasions, not absolutely for producing them, but only for awakening, rousing or exciting, into act [i.e., perception] what was already preexistent, dormant, and latent in the soul" (E 664). Yet, the contrastingly organic, eventful, and sculptural character of Blake's Platonism is exhibited on the very next page. Berkeley mentions Aquinas's view that "all beings are in the soul. For, saith he, the forms are the beings. By the form every thing is what it is. And, he adds, it is the soul that imparteth forms to matter." Blake remarks: "This is my Opinion but Forms must be apprehended by Sense or the Eye of Imagination" (E 664). In other words, he sees Berkeley here has abandoned his earlier stance of strong direct realism that tied the mental forms to the visible appearances. As Alexander Campbell Fraser writes, Berkeley in *Siris* "approached absolute Idealism by making sense absolutely subordinate to constructive reason."[25] Blake's comment registers his protest that once you separate the appearances from the participatory and metacognitive act by which they become recognized as such, they lose concreteness and float heavenward as detached Ideas. (So, for example, one reason Blake's Milton is "Unhappy tho in heav'n" [*M* 2:18; E 96] is that when he looks down on earth, he can see how the idea of him as a paragon of English nationalism has been pried away from his contentious republican politics, notably in Wordsworth's 1802 sonnets which enlisted Milton as a Burkean prop for patriotic reform and moral renewal.)

For Blake, apprehensions originate in and as events. He endows Berkeley's "phenomena" with materiality. The events arise, pass on, and perish; yet the mental objects, which give inner form to the events, endure, thereby endowing nature with order, discreteness, and regularity. But it is just here that Blake parts ways with Berkeley's inside/outside binary, and with the world of nature whose stability that binary operates to preserve. The reason Eternity "expands . . . within [the] Center" (*M* 31:48; E 131), or that "the immortal Eyes / Of Man [open] inwards into the Worlds of Thought: into Eternity / Ever expanding in the Bosom of God: the Human Imagination" (*J* 5:18–20; E 147), is because the mental objects are topologically situated within the events themselves, whose divine potentiality they model analogously to how Blake's revered Michelangelo envisioned and brought forth human forms within chunks of marble.

Thus, the importance for Blake of dreams, especially lucid dreaming. Dreams occur within the body analogously to how every individual exists in Albion, the Mundane Soul. They are not just immaterial ideas in the mind but physical events, hence part of space-time's extensive continuum. Indeed, Whitehead points out that his revised relativity principle carries "a vital connection with the theory of dreams."[26] For, absent Newton's absolute space and time, you cannot date the dream-space and its contents to consciousness of a real night, thence to render them purely imaginary. Relativity requires that the dream-time and dream-space be conceived together: "The dream-world is nowhere at no time, though it has a dream-time and a dream-space of its own." It seems an apt description of *Milton*'s visionary Moment, provided we add that the Moment, like the dream world, is not cut off from ordinary time and space but arises within them. Nor will Hume's argument for distinguishing dream images from sense-impressions according to degree of "force and vivacity" avail to dismiss the dream world as unreal. Whitehead instances his own dream of "hovering"—a lucid dream, I surmise, like Tucker's in "The Vision," or that of Blake's Milton while experiencing "still perceptions of his Sleeping Body; / Which now arose and walk'd . . . / . . . tho walking as one walks / In sleep" (*M* 15:4–7; E 109). Whitehead says, "I remembered that I had had the experience before, and I had subsequently decided that it was a dream. Accordingly, I decided to observe all the circumstances with great exactness, so as not again to be led into disbelief by my vague recollection of the details" (*Essays in Science and Philosophy* 103). Yet, though he remembered all the details upon waking, they still did not fit into Whitehead's empirical, everyday "dominant spacetime continuum." Otherwise, however, he reports that the dream space-time seemed quite as vivid and convincing as his waking one, indeed more so. Thus, this seemingly parallel reality cannot be dismissed on Humean grounds. (In fact, while rapid eye movements in sleep do correlate with the relative spatial locations of reported dream objects and events, there remains no way yet to reality-test for the accuracy of reported dream descriptions themselves.)

But further, Whitehead continues, Hume's whole argument that "necessary connexion" is "the sentiment or impression . . . which we *feel* in the mind" through the "habit" that arises from a "constant conjunction of events," is based on his uncritical acceptance of the three-dimensional spatiotemporal contiguity of events (*Treatise of Human Nature* 155–72; his italics). If Hume finds that our "idea of power" as able to bring diverse events into causal relation has no reality outside the mind, that is because a more thoroughgoing empiricism than his is required to recognize the power of a single "atomic" instance to signify something beyond itself when it flows or "ingresses" into the three-dimensional world from its surrounding regions of space-time as conceived in *four* dimensions. The synchronisms, parallelisms, and interconnections between "the four worlds of Blake's myth" supply this wider perspective.

Whitehead does not pursue the question of where dreams come from, but Andrew Baxter's *Enquiry into the Nature of the Human Soul* (1737) did. The independent reality of the dream world is a central tenet of this widely known book.[27] Baxter's argument resembles Berkeley's in *Alciphron*. Since matter is impassive, and only spirits are self-moved, motion in us not our own—specifically, the co-existence of active and passive states within the dreaming mind—must signal the presence of another spirit. Locke had posited personal identity in continuity of consciousness, giving rise to several bemusing paradoxes such as the question of what happens to the soul in deep sleep or in consequence of memory lapses.[28] Baxter's discussion of dreaming takes the problem of discontinuous personal identity much closer to the surface of consciousness. How do we explain dreams in which we "see persons, who on their being presented for the *very first time*, are familiar to us, and seem to have had former concerns with us"; or dreams where we "become instantly possessed of a *tract of experience*, which we never acquired"?[29] How is it that "the knowledge of things that never were, should appear as belonging to our former consciousness," not as "*bare information*," but as "a *familiar reminiscence*" (2.241–42), not to mention "*that real matters of fact have been discovered in dreams*" (2.244)? Indeed, says Baxter, brandishing his empiricist credentials, "without such a trial and experience in sleep," we could never from our waking state alone have conceived such prior knowledge to be possible.

His answer is that these are likely instances of Platonic remembrance. Certain dreams grant access to the state of *anamnesis* and offer foretaste of the soul's true self-unity and interconnectedness in the Afterlife (in redeemed

Albion, Blake would say). They suggest, "That the soul is capable of a more perfect and ready knowledge of things, than that which it attains to know, by the methods of *sense* and *reflection*" (2.239); and, "That the several parts of our past consciousness (which we are perpetually losing) may be recovered instantly, united together, and become one, by a firmer union, than the having recourse to perishable impressions on a corporeal organ, or our present method of reminiscence" (2.241). Baxter acknowledges he has not proved his case but only demonstrated its plausibility. And yet, "Whatever the Sceptick may say, . . . How delightful is it to think that there is a *world* of spirits; that we are surrounded with intelligent living Beings, rather than in a *lonely, unconscious Universe, a wilderness of matter!* It is a *pledge* given us of *immortality itself*" (2.189). Baxter's non-Rational Berkeleyanism here verges on Blake's own as seen, for example, in *Milton*, where the Blake poet's vision of the living Milton is at the same time the eternal Milton's lucid dream of Blake.

As we saw in chapter 5, a key source of the *Milton* Blake poet's vision is Abraham Tucker, whose resorption into the Mundane Soul doesn't, in his account, transform his individual consciousness but only temporarily suspends it: "As upon a man awaking in the morning out of sleep the dreams and visions of the night vanish away, his senses which had been kept stupefied throw open their windows, his activity that had lain suspended returns . . . : so upon my absorption I found myself, not translated into another species of creatures, but restored to myself again" (2.300). At *Milton*'s outset, Blake echoes this passage while yet undermining Tucker's prosaic realist viewpoint: "As when a man dreams, he reflects not that his body sleeps, / Else he would wake; so seem'd [Milton]" (*M* 5:1–2; E 109). The leisurely grand style ("As when," "reflects not," "Else") appears to be ushering us toward the second half of a well-upholstered epic simile. Then "seem'd" hints at a shift in perspective. One assumes the prior "so" will introduce an analogy but, instead, it unexpectedly puts an end to the grammatical sentence which nevertheless continues with skewed parallelism as a kind of run-on: ". . . so seem'd [Milton] but / With him the Spirits of the Seven Angels of the Presence / Entering; they gave him still perceptions of his Sleeping Body" (*M* 5:2–4; E 109). Blake rejects the sleeping/waking binary and affirms a liminal consciousness akin to lucid dreaming or an out-of-body experience.

For, of course, the true sleeper in *Milton* is the Blake poet himself, whom Milton-Los-Jesus aims to awaken into conscious awareness of his dreaming body. Like other Romantic writers and artists, notably Fuseli and Coleridge, Blake was fascinated by the coexistence of active and passive states in the dreaming mind and how phenomena like apparitions and sleepwalking seem to disclose body as a bound or outward circumference of questing spirit.[30] Interestingly, several watercolors Blake made about the time of *Milton* depict dreaming as the portal leading to a Vortex of vision via fleshly interpenetration with one's surroundings. *Queen Katharine's Dream, Ezekiel's Wheels,* and *Jacob's Dream* convey the lucid dreamer's residual awareness of his or her sleeping body by depicting it in conventional perspective as reposed beneath swirling, sometimes larger-than-life angels in the dream, comparably to Milton as he slumbers on his Death Couch. As Blake wrote Hayley a few months after his return to London:

> My wife joins me in wishing you a merry Christmas. Remembering our happy Christmas at lovely Felpham, our spirits seem still to hover round our sweet cottage and round the beautiful Turret. I have said *seem*, but am persuaded that distance is nothing but a phantasy. We are often sitting by our cottage fire, and often we think we hear your voice calling at the gate. Surely these things are real and eternal in our eternal mind and can never pass away. (E 759; his italics)

If Hogarth's well-known *Satire on False Perspective* (1754; figure 6.1) makes mockery of Berkeley's phenomenal world without distances, conversely Blake's *Milton's Mysterious Dream* (figure 2.1) suggests that abandoning depth perspective and scale might serve to re-enchant the world by imparting somatic immediacy to hoary images of ancient myth.

What Blake found in Priestley's "immateriality of matter," then, was a sufficient weight of physics to save him from the subjectivistic extreme of Berkeley's idealism, which by denying the existence of external bodies implies that dream and waking perception are phenomenologically equivalent. In effect, Blake plays the two philosophies off against each other and progressively splits the difference between them, drawing out the proprioceptive sensory implications of Berkeley's phenomenalism while emphasizing the dynamic, organic aspect of Priestley's idealized universe of atoms. But why did he want to unite these two very different philosophies in the first place? Alexander Jacob's earlier suggestion of a submerged Neoplatonism in Berkeley's thought hints at an answer. For Neoplatonism was a significant influence on the young Priestley, as well. The dissenting academies of Daventry, where he went to school, and Warrington, where he taught, both prominently featured Cambridge Platonists in their history and curriculum.[31]

Figure 6.1. William Hogarth, *Satire on False Perspective* (1754). Wikimedia.

In Priestley, Berkeley's denial of the material reality of distance is effectively inverted. By resolving matter into "nothing but the divine *agency* exerted according to certain rules,"[32] Priestley restored spirit's physical extension in space, which presumably was a main source of his appeal for Blake. Newton had imagined an atomic particle whose solidity was the product of infinitely powerful forces of attraction and repulsion, and Boscovich went on to devise an entire mathematical system of ideal force points. In *Disquisitions Relating to Matter and Spirit* (1777), Priestley misappropriates Boscovich's system to argue that if the cohesive force were removed, the particle's solidity would be, too, because "matter has, in fact, no properties but those of *attraction and repulsion*" (17). Priestley adds that his theory "greatly relieves the difficulty which attends the supposition of the *creation of [matter] out of nothing* . . . by a being who has hitherto been supposed to have no common property with it. For, according to this hypothesis, both the creating mind, and the created substance, are equally destitute of *solidity* or *impenetrability;* so that there can be no difficulty whatever in supposing that the latter may have been the offspring of the former." This looks like the materialist counterpart to Berkeley's theory of Creation as "[seeing] things produced into being—that is become perceptible" (*Three Dialogues* 1.348). Indeed, in 1810 Dugald Stewart, in discussing Locke's "dim and seeming conception how matter might first be made" (*Essay* 2.311), observed that Locke's passage "when considered in connexion with some others in his writings, would almost tempt one to think, that a theory concerning Matter, somewhat analogous to that of Boscovich, had occasionally passed through his mind."[33] Stewart then cites the above Priestley passage in explanation of what Locke must have been thinking. Blake evidently wanted to extend these confused Lockean-materialist approaches toward Berkeleyanism into a deconstruction of the two opposing positions, the better to delve into the creative "third kind" lying at their shared boundary.

In fact, Berkeley and Priestley had already practically deconstructed themselves. Priestleyan matter's elasticity as stretched or compacted by forces of attraction and repulsion recalls More's plastic, passive Spirit of Nature. Priestley emphasizes that even though the behavior of the complex bodies we observe is the product of more fundamental forces of electricity, light, and gravitation, these, too, are ultimately no less passive than matter: "As there is no active force in nature but that of God, this being is the infinite force which unites all the parts of matter, an immense spring which is continual action" (26). This is mechanism but, as John Yolton says, it is "a dynamic or organic version of mechanism."[34] Since therefore no real difference exists between matter and force, Priestley finds it equally acceptable to be called an advocate of "spiritualism" as of materialism: "If they say that on my hypothesis there is no such thing as matter and everything is spirit, I have no objection. . . . The world has been too long amused by mere names" (33). Strangely enough, Priestley in this way joins hands with Berkeley. In

Principles of Knowledge, Berkeley concedes that if "it sounds very harsh to say we eat and drink ideas, and are clothed with ideas," then so long as "you agree with me that we eat and drink, and are clad with the immediate objects of sense, which cannot exist unperceived or without the mind: I shall readily grant it is more proper or conformable to custom that they should be called things rather than ideas" (174–75). In *Three Dialogues*, Philonous likewise throws his hat in with the vulgar: "I am not for changing things into ideas, but rather ideas into things; since those immediate objects of perception, which according to you, are only appearances of things, I take to be the real things themselves" (340). When his interlocutor, Hylas, urges him to avoid shocking the public by "an Innovation in Words"—"What think you, therefore, of retaining the name *Matter*, and applying it to *sensible things*?"—Philonous gives the valedictory reply: "With all my heart: retain the word *Matter*, and apply it to the Objects of Sense, if you please; provided you do not attribute to them any subsistence distinct from their being perceived" (358), since that way lies atheism.

It seems totalizing monistic theories, whether idealist or materialist, tend by their reductivism to dismiss objections as semantics, mere quibbling failures to recognize how dualist concepts have been reframed "analytically." At the limit where explaining becomes explaining away, they thus are liable to trade places with their opposites. As Coleridge was wont to say, "Extremes meet."[35] More than anything else, what the very different victories of Berkeley and Priestley over the two-substance doctrine have in common is their unconvincing effect of sleight of hand.[36]

This may explain why Berkeley, though a metaphysician of the first order (unlike Priestley), and whose criticism of the doctrine of simple location or "outness" anticipates relativity theory and Whitehead, has had so little impact in the history of philosophy. As Hume famously observed, his arguments "*admit of no answer and produce no conviction.*"[37] Priestley himself saw the Berkeleyan argument of Andrew Baxter, whom he "considered as the ablest defender of the strict immaterial system," as an illicit occasionalist collapsing together of matter and spirit. Baxter "acknowledges that *powers of resistance and cohesion* are essential to matter"; but then he claims that since "these powers are the immediate agency of the Deity himself, it necessarily follows that there is not in nature any such thing as *matter*, distinct from *the Deity*, and *his operations*" (8). Says Priestley, this "is in effect to annihilate the substance, and to make the Deity himself to *do*, and to *be* every thing" (9). Matter thus becomes "wholly superfluous," a viewpoint Priestley compares with Berkeley's system of ideas (65). And yet Priestley himself was liable to the same objection in reverse. Increasingly disillusioned with Unitarianism, Coleridge in a letter of early 1796 finds that Priestley collapses spirit into matter with dangerously pantheistic implications: "How is it that Dr. Priestley is not an atheist?—He asserts in three different Places, that God not only *does*, but *is*, every thing. But if God *be* every Thing, every Thing is God—: which is all, the Atheists assert—. An eating, drinking, lustful God—with no unity of *Consciousness*. . . ."[38]

As the critic F. E. L. Priestley summarizes, "It is possible . . . to view the orthodox Newtonianism as poised between a roughly Platonic atomism and a roughly Epicurean atomism. . . . Berkeley isolates in his system the element which can be called 'animist' . . . [but] the balance can tip heavily on either side" (70). Joseph Priestley's theory that matter's extreme porosity and scarcity can provide "*a common property,*" rendering it capable of "*intimate connection* and *mutual action*" (xxxviii) with spirit, looks like a crude Epicurean (or Democritean) attempt to reprise the Platonic Receptacle's intermediate "third kind," which, we have seen, was always liable to be confused with the extended space of his World Soul. If the Receptacle's "near-ineffability is a function of its near-non-entity," as Cornford observed (159), then to the eye of materialist Reason its "mirror" of phenomenal reflections might appear as the immateriality of matter.[39] Indeed, Wayne Glausser notes that "Neoplatonism . . . shares a pool of sustaining metaphors with atomism." At times, Lucretius's wandering simulacra distinctly resemble Plotinus's alienated emanations, and so the two "can easily overlap, and slide toward convergence."[40] Ralph Cudworth, More's ally, recognized as much in his *True Intellectual System of the Universe* (1678), which begins by establishing that Democritean atomism supports, rather than refutes, Christian theism.[41] More's steadily more materialist reinterpretations of his universal Ogdoaz show, further, that once Neoplatonism's elaborate hierarchy of emanations and hypostases was logic chopped finely enough for intercalated stages to blend together—so turning distinct states into transitional phases—the pure effulgence of the Nous could grow spatialized and extensive enough to seem like, in Priestley's words, "divine energy, an energy without which the power of gravitation would cease, and the whole frame of the earth be dissolved."[42] And the backdrop to Priestley's divine energy was established by Newton himself, whose denial of activity to matter is what led, in a return of the repressed, to his lifelong flirtation with Cambridge Platonism's various immaterial and spiritual explanations of gravity and aether in the first place.

Part of Priestley's appeal for Blake was surely his overriding concern, from the *Disquisitions*' first page, to explain the implications of his "divine energy" for "the whole frame" not only of the external universe but man. "I had always taken it for granted, that man had a soul distinct from his body" (2), he writes. Now, though, he thinks man "some *uniform composition*" whose "property of *perception*, as well as the other powers that are termed *mental* is the result of . . . such an organical structure as that of the brain" (xiii–xiv). The *Disquisitions* thus appears a prime source for Blake's pronouncements, "Man has no Body distinct from his soul for that calld Body is a Portion of Soul . . . Energy is the only life and is from the Body" (*MHH* 4; E 34). Blake never reduced mind to brain, but his interest in brain anatomy ran deep; his bookseller, Johnson, was a leading medical publisher.[43] Along similar lines, Robert Schofield has argued that, while Coleridge's German idealist–inspired organicism is often seen as a repudiation of Priestley and materialism, in fact it flowed in no small measure from Priestley, whose Cambridge Platonist background transmitted an idealist element already present in the thought of Locke and especially Hartley.[44] To add an example, Coleridge's celebration in "The Eolian Harp" of "one intellectual breeze . . . Plastic and vast" that sweeps through all of nature is, as the middle adjective signals, straight out of More and Cudworth.[45] Blake's overtly paradoxical fusions of body and spirit can be seen to arise, similarly, out of the strange confluence of Priestley's thought with Berkeley's. The two together exposed longstanding inconsistencies in British empiricism and made the familiar names of things—notably, "that calld Body" (*MHH* 3; E 34)—ripe for late-Enlightened Romantic reimagining.[46] Arguably, the later Blake's cranky mythology of spiritual bodies, vehicles, emanations, Mundane Egg and Mundane Shell springs, for all its quietism, from the Neoplatonic element already present in the early revolutionary *Marriage of Heaven and Hell*.

Each in his own way, Blake and Whitehead understood how the "bastard reasoning" and creative confusions of Plato's *Timaeus* could counteract scientific reason's tendency to reduce real, sensible change to mathematical abstraction. In *Process and Reality*, Whitehead went out of his way to locate his system in British empiricism by exploring contradictions in that tradition's chief mainstays from Bacon, Descartes, and Newton to Locke, Berkeley, and Hume, thence to sketch revisionist readings of potentiality, change, and time in their work. Thereby, he paved the way for Donald Ault's *Visionary Physics* to show how Blake effectively anticipated Whitehead by directly investigating many of the same contradictions in Newton. In this book, I've tried to suggest how Blake's mythmaking constituted a two-front battle both against systematic rational empiricism and the natural linguistic tendency of his own empirico-materialist cosmology to harden into system. The involuted ironies of his cosmology open a "third way" by which to drill down into the shared boundary of monism and dualism and perform their relationship, if only for the space of a Moment (whereas Priestley, Blake came to think, merely wanted to reconcile the two in Bayesian fashion without room for revelation or vision[47]). The roots of this project go back to the overtly specious empiricist reasoning of Blake's first Illuminated Books, *There Is No Natural Religion* and *All Religions Are One*. Companion pieces, the titles themselves indicate the two works' incompatibility in contradiction of their form as logical proofs. How can both be true?

The Marriage's doctrine of Contraries further highlights Reason's paradoxes and inner contradictions. For instance, "The Voice of the Devil," who extols Energy over Reason, speaks in the numbered propositions of a geometer (*MHH* 4; E 34). Even as the Devil designates religious dualism a corruption of Contrary relationship, he pairs its three "Errors" against a matching set of three so-called "Contraries," which he flatly says "are True," thereby reproducing the dualism he purports to reject. For, "the Devil" is, by his own logic, himself a corrupt effect of ideology. Among other things, his self-entanglement can be seen to dramatize the "unconscious hypocrisy"[48] that arises when elite intellectuals follow Berkeley's urbane advice to avoid ridicule by clothing their Enlightened metaphysics in the customary language of natural causes, while privately they "ascribe every thing to the immediate operation of spirits . . . [I]n such things we ought to *think with the learned, and speak with the vulgar*" (*Principles of Knowledge* 51; his italics). Berkeley here concedes his intellectualistic monism must make its way in the commonsense, two-substance world. Blake's paired tractates acknowledge a similar dissonance between the ideal and the real, only they announce it with overt irony. If "all religions are one" in their supernaturalism, then oxymoronic "Natural Religion" is manifestly false; yet, as Hume also recognized, it isn't any less empirically *real* for being a large public delusion.

Like the occasionalist Berkeley, Blake believed, "We who dwell on Earth can do nothing of ourselves, every thing is conducted by Spirits, no less than Digestion or Sleep" (*J* 3; E 145). Indeed, this is the vehicular basis of his whole mythology. At the same time, Blake's bold Swedenborgian visions of the spirit world fly in the face of Berkeley's prudent admonition that "it sounds very harsh to say we eat and drink ideas, and are clothed with ideas" (*Principles of Knowledge* 1.194). Far from smoothing away reception problems like Berkeley, Blake's discourse of invented names, puns, two-way syntax, overlapping voices, satiric double-speak and ironic indeterminacy heightens cognitive dissonance to negotiate a middle path between "*the learned*" and "*the vulgar.*" Poetry that "rouzes the faculties to act" (E 702) overthrows Berkeley's assignment of all causal power to God, by requiring that Berkeley's God-given "language of nature" be interpreted and rewritten in the light of actual human conditions at the time. Over and over in *The Book of Urizen*, *Milton*, and *Jerusalem*, the grammatical conventions of descriptive realism and stable everyday sense-making are assimilated into an extensive continuum whose pliable syntax emerges in symbiosis with the narrative's breaking events. Continually and unevenly, new arrivals are incorporated into "English, the rough basement" whose "stubborn structure" the poet struggles to "buil[d]" (*J* 36:58–60; E 183), somewhat as the displacements of *Jerusalem*'s opening paragraphs showed earlier.[49] Readers willing to embrace the verse's ongoing demands to be parsed and reparsed can therefore recover, bit by bit, linguistic and imaginative agency previously forfeited to God. (Consider how different this structural form of suspension is from the lengthily suspended architectonic sentences of *Paradise Lost*, often regarded as expressions of Milton's predestinarianism.) Thus, the later Prophetic Books extend into metaphysics the Hartleyan psychology of learning to read via "decomplex" and synthetic associations of ideas in *Songs of Innocence*. Too bad Blake's adult re-education program in reading remained, itself, largely unreadable except as a hermetic symbol system with all the temporality shaved off (as in Yeats, S. Foster Damon, and Kathleen Raine), until Einsteinian Relativity filtered into the criticism of Northrop Frye and Donald Ault well over a century later.

Conclusion

The Unified Space-Time of *The Vision of the Last Judgment*

To conclude, let us compare the *Marriage of Heaven and Hell* (1791) title page with Blake's painting of the Last Judgment (1808). The two works together can serve to demonstrate just how systematically Blake went about developing a four-dimensional geometry out of Priestley's airy 1780s rationalist theory of "the immateriality of matter." We might suppose these pictures reflect his shift from a satirically inverted antinomian perspective to a traditional Christian one. But they don't, largely because each only appears to locate heaven above and hell below. Blake did change, but mainly to remain the same during turbulent times when history seemed to be running in circles, as when the Revolution began to devour its own, Bonaparte the liberator became Emperor Napoleon, or King George III—"the Pharaoh of England"[1]—spun himself as defender of the chartered rights of Englishmen against godless French imperialist aggression. While *The Marriage* approvingly identifies hell with "the Body" (*MHH* 4; E 34), the head–shaped *Vision of the Last Judgment* hardly repudiates this, as its dense effect of fleshiness goes to show.

Indeed, a visionary head already pervades the *Marriage* title page (figure C.1). It takes a moment to hit home. Blake constructs an optical illusion joining the Contrary realms, heaven/soul/Reason and hell/body/Energy, within the negative space of a floating head. As commentators have observed, the head's upper half emerges in outline from tree branches suggesting hair and eyebrows.[2] The title's arabesque "and" indicates a mouth. The region beneath can be seen to constitute Energy's hell of inchoate bodily sensations. Some of these have evidently taken on human form as Hartleyan "vibratiuncles"

Figure C.1. William Blake, *The Marriage of Heaven and Hell*, copy D, 1790, 1795. Library of Congress.

or "miniatures" of perception, or perhaps tiny Tuckerian "vehicles," as they drift upward into consciousness. Their counterparts in the aboveground world of Experience appear as "clear and distinct ideas": a properly clad couple strolls on the left, while on the right a different twosome has evidently been separated in death, both pairs in sharp contrast to the passionate naked embrace at the bottom.

So, the marriage of heaven and hell takes place as a Last Judgment inside embodied mind. The title page head's top half spreads across open sky, presumably heaven. The design thus seems a partial illustration of the long, perspectivistic Memorable Fancy where the Blake devil proves how an angel's metaphysics rest on nothing more substantial than "the deep," which however, the angel, a traditional substance dualist, insists on identifying with hell's "infinite Abyss." Perhaps one of the design's snaky tree roots is "the twisted root of an oak" in which, the Blake devil reports, he "remaind . . . sitting" while the angel was "suspended in a fungus which hung with the head downward" (*MHH* 17–19; E 41–42). At any rate, on the *Marriage* title page rational and largely vacant mind presides over "the abyss of the five senses" (6; E 35). Reason's familiar object world begins at the horizon where eternal hellish Energy leaves off. The printing house episode on plate 15 portrays "books" and "knowledge" similarly as the rational, objectified "bound or outward circumference" (4; E 34) of a mental "cave" whose interior is "infinite" (15; E 40).

In a conventional perspective, the V-shaped gray-blue area below the title page's horizon line of trees and human figures would represent a foreshortened foreground. In Blake's design, this area falls away into the *vertical* space of a chthonic underground. Visually conflating top/down and right/left axes, Blake marries the two regions.[3] Still more disturbing, this marriage also includes the forward/back axis, extending the picture's visual space beyond the plane of representation and into the real world. Whereas the aboveground humans and, possibly, the top three of the floating figures beneath them belong inside the visionary head, the couples flying below do not. Indeed, those on the right appear considerably closer than the others. Maybe the whole nether portion of the image is rotating like a vortex, in contrast to the stable Euclidean perspective on top. The dark rocky ledge in the right foreground juts out almost from the very space where the viewer is positioned. Then are we, too, in hell? Blake's title page recalls the chaos glimpsed in the bottom right foreground of his hero James Barry's final mural in his epic series for the Great Room of the Society of Arts, exhibited

222 / A Bastard Kind of Reasoning

in 1784 to great acclaim and widely regarded as Britain's enlightened civic answer to the Sistine Chapel. In *Elysium and Tartarus or the State of Final Retribution* (figure C.2), Barry shockingly undermines decorum by suggesting a Contrary, figure-ground interdependence between the sunlit neoclassical republic of virtuous dead white males and the seething erotic tumult that gapes open beneath them along a ledge at the corner.[4]

One has the impression that the entire space of the *Marriage* title page is an externalization, still taking shape, of some activity occurring even closer-in than the embracing couple or the blackish foreground ledge—that is, on the other side of the eyes in *our* head, whose orderly assumptions the floating head mirrors and mocks (similarly to the eye-sun in the *Daughters of Albion* frontispiece, figure 1.1). Even the text pane of written words and letters is ambiguous since, as we've seen, some of them form elements in the design's scheme of three-dimensional imagery. Arguably, the true locus of Priestley's "immateriality of matter" here is not the negative space of the page's visionary head, as we might have thought, or even the cleverly positioned words of the title, symbolic linguistic signifiers made to double as representations of physical objects. Rather, matter's immateriality is exemplified every day in ordinary written language and verbal symbolism itself. Scholars have recognized how Blake's "autographic" method of etching his books in relief heightens awareness of the text's reliance upon the engraver's activities of writing and seeing and, I would add, by extension the reader's subsensible speaking and hearing.[5] The latter depend, of course, on a certain physical distance, which delimits the legibility of the marked page; too close

Figure C.2. James Barry, *Elysium and Tartarus or the State of Final Retribution*, etching and engraving, 1791. Yale Center for British Art, Paul Mellon Collection.

or far off, and it can't be viewed or read. Blake's undermining of rational perspective creates a continuum joining the imaginary spaces inside his picture to the actual local space extending outward from them to the viewer-reader.

Behind the design to *The Marriage*'s title page lies Newton's "nutshell" theory of matter (see chapter 2). If Blake's visionary head fuses heaven and hell in an inward, personal Last Judgment, then the universal nature of that Judgment implies a universe of such heads. In other words, the three-dimensional material world at the top of the design is really the outward phenomenal effect of a higher-dimensional belowground that wraps around the material world at every point, analogously to how space surrounds the earthly globe. Earth appears as a molecule condensed out of innumerable vortexes swirling in every direction toward hells of infinite Energy. The head shown in the *Visions of the Daughters of Albion* frontispiece (figure 1.1) is, similarly, a negative space formed out of three-dimensional landscape elements. The profile of a skull in cutaway discloses three figures in an unhappy tableau of social oppression. Imagine this enclosed headspace as extended continuously and repeating like a honeycomb, and it will form exactly the sort of recursive pattern by which Newton pictured matter in *Opticks* Query 31: a self-similar lattice in which atoms gradually crystallize to form molecules of the size involved in chemical operations.

Almost twenty years later, Blake's *The Vision of the Last Judgment* pursues the force-field implications of *The Marriage*'s title page design to the very limit of representation (figure C.3). The original five-by-seven-foot tempera painting was apparently intended for an 1810 exhibition Blake decided to scuttle after the humiliation of his 1809 show. After his death in 1827, the painting went missing; *The Last Judgment* is lost. What remains are a pair of diagrammatic ink drawings (one is partially shown in figure 4.6) and the finished watercolor and ink version Blake made in 1808 for Elizabeth Ilive, wife of the Earl of Egremont, along with his lengthy account of the painting for a projected show catalog modeled after Barry's *Account* of his murals.

Most (not all) viewers of this watercolor agree it portrays a kind of visionary head. At top, the cranial space of Christ the King seems to echo Locke's allocation of consciousness to a sensorium where nerves transmit sensations to "their audience in the brain–the mind's presence–room."[6] Below it appear two of the four Major Prophets, emblems of the divine vision as transmitted through reading and writing. Presumably, the other two are in back; those in front serve as metonymies for the eyes. However you look at it, the painting's shifty anthropomorphism plainly stands opposed to the contemporary bourgeois fashion for "face painting," whose

224 / A Bastard Kind of Reasoning

Figure C.3. William Blake, *The Vision of the Last Judgment*, pen and watercolor over pencil, 1808. Petworth House, The Egremont Collection, The National Trust.

"Imbecillity" of adherence to sordid detail Blake attacked at this time for masking "Physiognomic Strength & Power" and "the physiognomies or lineaments of universal human life" (*PA*; E 571).

Blake's Eternity, personified by enthroned Christ at the picture's top, closely corresponds to Bergson's idea of duration as pure succession and continuous transformation—not a series of measurable spatial states, as in Newton's fluxions, but the condition of absolute change, passage as such. And yet, Bergsonian sheer succession seems no less abstract and unimaginable than Newton's mathematical concept of instantaneity. Surely, change presupposes the presence of some *thing* undergoing the change? Therefore, Blake distinguishes Christ's pure duration or passage from the finite set of real, enduring event-objects that make up time and history: namely, the multitude of imperfectly human "States" (*VLJ*; E 557, E 556) portrayed in the rest of his painting. The latter are impure mixtures of space and time, whereas Christ, "The Imagination . . . the Human Existence itself," is eternal (*M* 32:32; E 132). The clockwise rotation of the many small figures indicates each is a Vortex generated by the intersection of two different planes of existence, as eternal duration ingresses perpendicularly into earthly existence to produce lived process. One could say the emanationism of Henry More's universal Ogdoaz is operating organically from inside each human individual or type to create experiential change: growth on the left, decay on the right. Blake's God of Passage achieves perceptibility via quantum "pulses" or "drops" of becoming whereby he "becomes as we are" (*NNR* [b]; E 3)—namely, other people, each a member of Albion, the Mundane Soul. This is Blake's version of Whitehead's extensive continuum. He calls it the "fabric of Six Thousand Years" which weaves itself into and out of humanity's "lineaments" (their words and deeds) and which, by embedding local acts and behaviors in a wider, universal pattern, also defines "permanent for ever & ever" the individual identities behind life's passing show (*M* 22:18–25; E 117).

Implicit here is awareness that the three-dimensional objects we see already do possess duration, though we habitually disregard the fact. A purely three-dimensional object would exist only for a single instant. As we'll see, Blake's painting is constructed according to the idea that our perpetually vanishing present forms a three-dimensional slice of four-dimensional space-time which it divides into two regions, the past (Satan) and the future (Christ). Whereas *The Marriage of Heaven and Hell* title page shows universally Contrary contents of the individual mind according to the doctrine of matter's immateriality, *The Vision of the Last Judgment* extends, and partly inverts, this perspective to show mind's immanence throughout the universe at any given moment in time. More clearly than *The Marriage* design, "The Vision" illustrates Whitehead's observation, noted earlier: "there is a dual aspect to the relationship of an occasion of experience as one relatum and the experienced world as another relatum. The world is included within the occasion in one sense, and the occasion is included in the world in another sense. For example, I am in the room, and the room is an item in my present experience. But my present experience is what I now am."[7] Whitehead's final "But . . ." crucially redirects the Berkeleyan, subjective-idealist thrust of his remarks toward Platonic realism. Similarly, in *The Vision of the Last Judgment* the mutually constitutive relationship between human minute particulars and the universe is seen to locate those particulars as real rather than ideal. Collectively, they make up the universe. Considered singly, however, the subjective contents of each one—its identity here and now as an actual living entity—is defined by the spatiotemporal universe in its totality at every passing moment. To put it differently, the (only locally self-determined) place of each particular within the overall scheme of things constitutes the passing upon it of a Last Judgment.

Blake's painting depicts not the underlying immateriality of humankind's earthly existence so much as the temporal nature of visible space itself, which it portrays as "inspissated" per Henry More. We seem to see a flesh field of forces such as the art historian Richard Shiff describes with reference to the phenomenology of Merleau-Ponty: "The painter's touch makes flesh, as opposed to air or water, a medium of vision. It gives to flesh the character of an ambient fluid."[8] Blake dissolves St. Paul's dichotomy between "the flesh" and "the spirit" by suggesting that to "lay aside the old man" and "put on the new man who has been created in God's image" (Eph. 4:22, 24) is a sensuous antinomian putting on of new flesh in the form of a "spiritual body" (1 Cor. 15:44; "To Tirzah," E 30). In *The Vision of the Last Judgment*, the invisible head outlined in negative space on the *Marriage* title page transforms into humanity's collective inspissation throughout all of positive space as defined by the redeemed Albion's toroidal "bound or outward circumference" (*MHH* 4; E 34), as in the inside-out universe of figure 4.2. Thereby, the painting spatializes the recursive "fractal lattice" of Contrary relationship that generated the multiple parallel subnarratives of *The Book of Urizen* and *Milton*. As in *Jerusalem*'s conclusion, the eternally returning Moment here yields a cyclical vision of Albion the Mundane Soul's ongoing fall and redemption. Blake's painting depicts one such Moment, conceived as a fleshly quantum of space-time: a humanized version of Planck-Einstein-Whitehead's "atoms of action" and what Niels Bohr called the character of "wholeness inherent in atomic processes, going far beyond the ancient idea of the limited divisibility of matter."[9]

In other words, in *The Vision of the Last Judgment* the Vortex shape connecting heaven and hell—a shape merely implied on the *Marriage* title page because its twisting movement if extended would undermine all rational perspective—becomes, itself, the central representation. The multitude of human figures forms a typology of the human spirit: the different "States . . . which . . . every one on Earth is liable to enter into . . . always . . . These States Exist now Man Passes on . . . he passes thro them like a traveller" (*VLJ*; E 557, E 556). That is, the painting makes visible the journeys of different "traveller[s] thro Eternity" across four-dimensional space-time. From the traveler's viewpoint, says *Milton*, his trajectory "roll[s] backward behind / His path, into a globe itself infolding" (*M* 15:21–35; E 109) at the intersection of two perpendicular planes, one composed of Earth's continually shifting horizons or "finitude-nows," the other formed by Eternity's now of perpetual duration. If therefore earth appears to the traveler as "one infinite plane" (*M* 15:32; E 109), that is because of Eternity's immanency at right angles to the horizon, where temporal change constantly brings new finitude-nows into view—a visionary version of Wordsworth's "something evermore about to be."[10]

Accordingly, the Trump of Judgment sounded by the Four Angels at the painting's middle announces an ongoing Creation-Fall whose Moment, the specious present, perpetually spawns globes of three-dimensional perspective in the eyes of earth's "travelers." The painting's midpoint slit depicts the ongoing big bang of an infinitely deep negativity that establishes the very conditions for picturing. Even as the trumpeted Judgment breaks upon the earth, it is already splitting apart to produce the up/down, left/right, and forward/back directional planes requisite to three-dimensional representation. Blake's *Vision of the Last Judgment* recalls Barry's *Elysium and Tartarus* no less than the *Marriage* title page does, even as it transcends the structural binaries of those earlier works.[11] We see that Judgment Day is everywhere every day, all day long. Yet it is not constant or uniform but discrete and discontinuous, such that Eternity's infiltration into time is brought to crisis and there is an agonizingly urgent question whether the present moment will pass. Blake renders *Milton*'s eternally recurring Moment as a quantized sliver of four-dimensional space-time that is always-already in process of cleaving into the past and future that together define our experience of the passage of time, respectively the immutability of hell and the indefinite potentiality of Jesus. Meanwhile, the next Judgment awaits its unfolding in the budlike slit of flames behind the Four Angels at the painting's inmost center, described by Blake as "a fiery Gulph" (*VLJ*; E 558): not hell but the space in-between heaven and hell, the Contrary source of the entire vision. This slit in the painting's middle represents the tip of the Vortex through which Eternity descends into time and redeems earthly existence from hell's unremitting sameness.

For Blake, as for Henry More, God's essential humaneness entails that not even he could create a material universe without first creating an immaterial space of spirit to receive it: "For let it be remembered that creation is. God descending according to the weakness of man" (E 599). So, if Christ's head in the painting is imagined as a four-dimensional "sphere of spheres" that "expands inward" beyond time into a dimension of space perpendicular to three-dimensional representation, then the painting's midpoint blooms into view at the point where Eternity "contracts outward" to generate serial time and three-dimensional space. In the painting's bottom half, this contraction then continues downward into hell, an unextended point of pure abstraction, dearth, and lack, as in the traditional Plotinean conception of matter. A contemporaneous painting, "Satan Calling up his Legions," positions the viewer in the middle of hell's collapsed no-place.[12] Its dense, pervasive blackness expresses truly Coleridgean horror at the prospect of life without vision—what Blake perhaps contemplated during his decade-long period of solipsistic silence following *The Book of Urizen* in late 1794 while writing and rewriting *The Four Zoas*. Considering the influence of Michelangelo's Sistine Chapel and Coppo di Marcovaldo's dome for the Baptistery of Florence (if only by way of Dante), it seems likely Blake conceived the representational scheme of *The Vision of the Last Judgment* according to the analogies of Renaissance and baroque temple wall paintings. His Christ inhabits a space beyond earthly existence in the same way as a figurine suspended by a pendant from the center of a dome possesses greater depth and volume than figures painted on the dome, whose curvature in turn gives those figures a greater dimensionality than images painted on the flat walls below.

Figure C.4 on pages 230–31 attempts to diagram the unified multidimensionality of Blake's Last Judgment vision. It indicates how his Creation-Fall is not a linear process. In *The Vision of the Last Judgment*, Eternity or "Jesus mind" occupies an infinite hyperspace at the back of the human figures whose solidification into three dimensions Blake portrays in the painting's top half. As they arise from Christ's head, these figures progressively skew 90 degrees sideways. Since their fall occurs not just along the painting's vertical plane but across all 360 degrees, the three-dimensional space they make is spherical and "conglobed": the Mundane Egg. So, the cyclical rotation of souls about the painting's circumference, from heaven to hell and back again, occurs also from front to back. We are looking at a densely inspissated sphere in cross-section, similarly to the angelic "globe" in Blake's Nativity Ode illustration (figure 4.5).

Topologically, Blake's vision of the material world as a Mundane Shell—as the thin three-dimensional outside of a four-dimensional sphere with finite volume but no sharp boundaries—corresponds to the shape of a torus, the geometrical form of the inside-out universe of figure 4.2. A ring torus is the surface of revolutions that results when a circle is rotated along a coplanar axis in three-dimensional space. It is the donut-shaped coil of epicycles generated by the forward motion of a person standing on the surface of Earth as it orbits the sun while spinning on its tilted axis. A striking feature of the ring torus is that, as the distance from the circle's center to the axis of revolution decreases to zero, the shape of the ring degenerates to a horn (when the axis is tangent to the circle), then a spindle (when the axis is a chord of the circle), until finally it contracts to a sphere. As we saw in Chapter 4, the torus "conglobes," "into a globe itself infolding" (*M* 15:24; E 109). The four-dimensional toroidal donut—conceived by Blake as the inside-out cosmos of Giant Albion, "One Man" who contains "Multitudes" (*VLJ*; E 556–57)—collapses into the Urizenic space of its hole. The torus loses its temporal dimension and becomes a solid. The resulting globe then appears to "[roll] thro Voidness" (*M* 29:16; E 127)—namely, Newton's immobile absolute space. In this way, the deformation of the ring torus into a globe lends support for Rond d'Alembert's early suggestion, in Diderot's *Encyclopédie*, that duration might be considered as "une quatriéme dimension," and serial time as "produit . . . par la solidité."[13]

However, the flattening of the Vortex in *The Vision of the Last Judgment* does not stop at three dimensions but continues downward to the painting's bottom. As Albion the Mundane Soul's toroidal form reduces to three and then two dimensions, as shown in figure 4.4, he resembles the text placard seen earlier in *Jerusalem,* plate 62 (figure 1.2). Oppositely to the expanding four-dimensional Eternity at *The Last Judgment*'s top, in hell human forms all but disappear into the picture plane, the very ground of representation. The rocky cutaway at the bottom indicates that the falling souls on the left are skewing a further 90 degrees into one-dimensional lines. The innermost circle of hell containing Satan, who appears recessed inside the cutaway, is evidently a nondimensional pit of solid bedrock: pure abstraction and "Perdition" (E 553). Logically, the souls stretched lengthwise along the bottom are not human forms at all. Above them on the left, aspiring souls emerge from this rock, comparably to Urizen as he simultaneously thrusts himself into and strives to exit solid rock in *The Book of Urizen* plate 9 (figure 3.2). Below the pit and on the right, several heads of the damned can be seen, their bodies swallowed by a lake of fire, which Blake conceives as positive

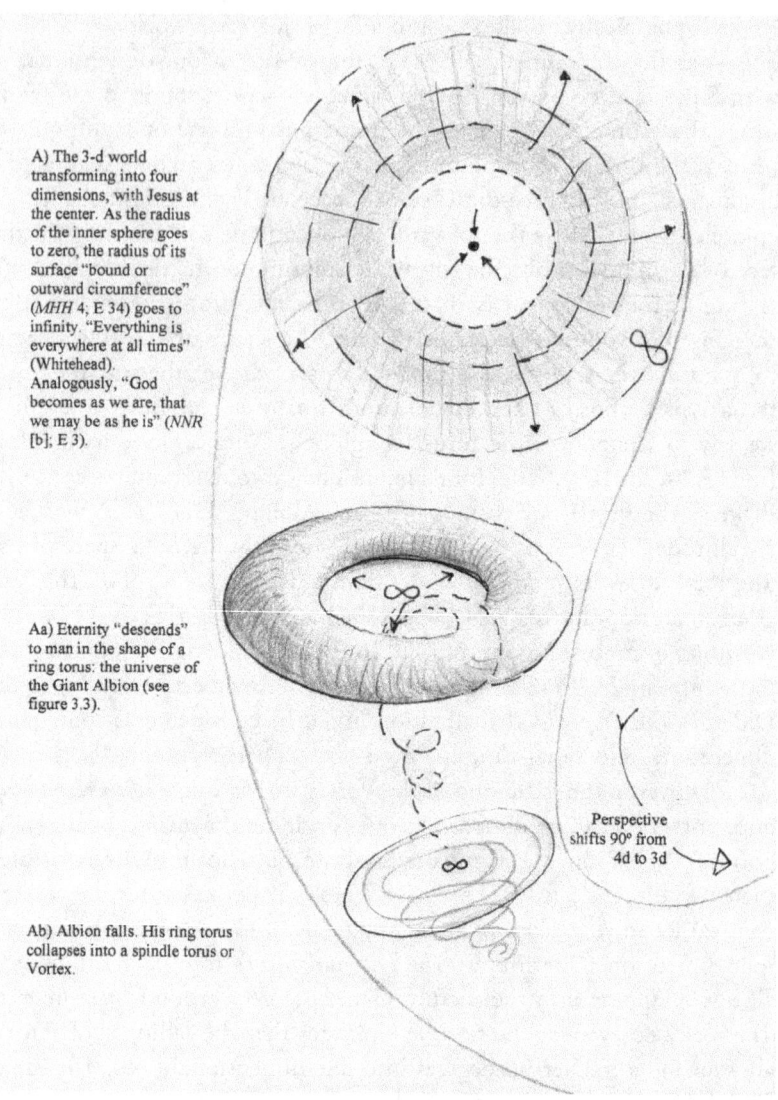

Figure C.4. Different dimensions of Blake's *The Vision of the Last Judgment*. Illustration, Rob DuToit.

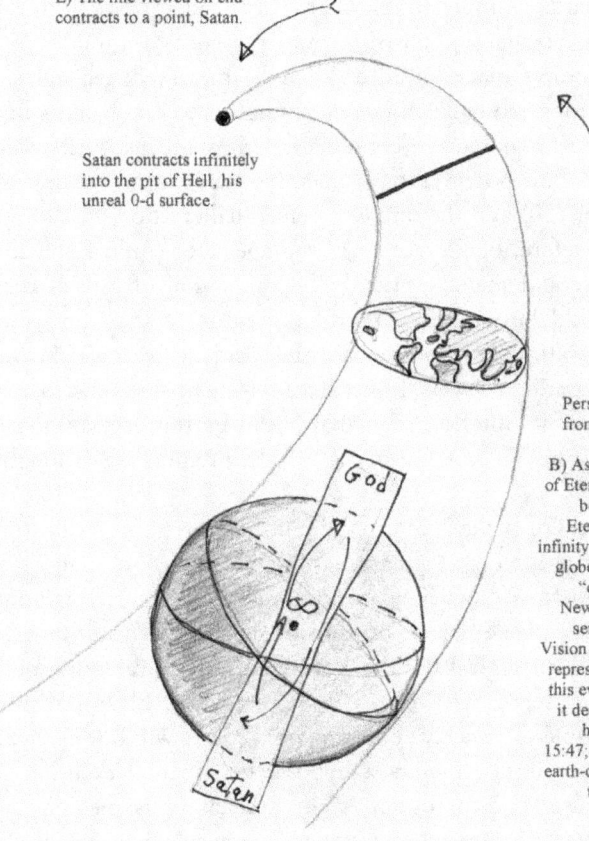

E) The line viewed on end contracts to a point, Satan.

Perspective shifts 90° from 1d to 0d

Satan contracts infinitely into the pit of Hell, his unreal 0-d surface.

D) Viewed on end, the 2-d surface reduces to a line.

Perspective shifts 90° from 2d to 1d

C) Hell begins as Earth's globe is reduced to its 2-d surface, as in a Mercator map or Cartesian coordinate grid.

Perspective shifts 90° from 3d to 2d

B) As it collapses, the space of Eternity everts, its surface becoming a solid globe. Eternity is relegated to an infinity of points all along the globe's surface, i.e., vacant "outer space" viewed in Newtonian terms as God's sensorium. Blake's "The Vision of the Last Judgment" represents a cross-section of this eversion process, which it depicts as an apocalyptic hole "in the Zenith" (*M* 15:47; E 110) through which earth-dwellers can look back to Eternity through the Vortex of its collapse.

darkness visible, a corporeal density beyond all vision, the very essence of Locke's insensible "material substratum."

So, Christ and Satan form opposite vanishing points of the three-dimensional world made visible by the apocalyptic slit in the painting's middle. Whereas Satan flails against formless sensations inside an idealization of the corpuscularians' impenetrable matter, the radiant Son seated in the spacious cranium at the top portrays pure presence-to-mind beyond objecthood. He epitomizes Blake's ideal of a wholly mediated, self-reflexive translucency of perception: not a Coleridgean Symbol based on objects whose God-given biblical meanings require conservation through institutions like the church, but an endless potentiality of meaning made available through "Sublime Allegory" (E 730) and the ongoing perceptual creation of "symbols *about* symbols."[14] Therefore, Blake writes, "If the Spectator could Enter into these Images . . . or could make a Friend & Companion of one of these Images . . . then would he meet the Lord in the Air & then he would be happy" (*VLJ*; E 560). This rapture becomes possible because the real physical distance separating viewers from the painting is coextensive with the imaginary region of its interior, such that it can be traversed by means of "Imagination . . . the Fiery Chariot of . . . Contemplative Thought" (*VLJ*; E 560).

The various spaces of Blake's painting might therefore seem to illustrate Berkeley's claim that the words "situation," "high" and "low," "up" and "down," "right" and "left," by which man interprets "the quantity, relation, and order of the proper visible objects or ideas of sight" in terms of tangible extension, are merely "analogical."[15] The spaces are analogical but not in Berkeley's immaterialist sense. As I've tried to demonstrate, Blake remained a reluctant dualist unable to conceive of immaterial spirits without extended physical bodies. Matter is not merely a hypostasized mental delusion, as Berkeley insisted. Matter is error but it is also real, as is its metaphysical correlative, Satan-Urizen the "mistaken Demon of heaven" (*VLJ* 5:3; E 49). Otherwise, error wouldn't need to be confronted and cast out. Consider how Blake's account of his painting ends, accordingly, with a most un-Berkeley-like call for resistance and change: "Mental Things are alone Real what is Calld Corporeal Nobody Knows of its Dwelling Place it is in Fallacy & its Existence an Imposture Where is the Existence Out of Mind or Thought Where is it but in the Mind of a Fool" (*VLJ*; E 565). Blake's position is a simple syllogism. 1) "Mental Things are alone Real." 2) "[W]hat is Calld Corporeal"—"the Existence Out of Mind or Thought"—exists nowhere "but in the Mind of a Fool." 3) Therefore, error in the minds of fools is real, and truth must be rescued from this destructive contradiction without temporizing attempts to have "Error . . . make a Part of Truth" (E 565). Berkeley's suggestion (examined in chapter 6) that the wise philosopher "ought to *think with the learned, and speak with the vulgar*" here stands refuted.[16] Unlike Berkeley but like Whitehead and Plato himself on Whitehead's "corrected" interpretation (see chapter 5), Blake was a subjective *realist*.

To recapitulate, it is Eternity's eversion into an earthly globe that supplies the basis for the Vortex of *The Last Judgment*. The painting represents a momentary three-dimensional cross-section of that process. The damned falling on the right approach toward the picture plane, ultimately curving sideways along its surface toward hell's abstract point of solid rock at the bottom. Opposite them, the souls rising on the left *gain* in dimensionality as they approach the heavenly presence-room, depicted in recession to indicate its greater volume inversely to the recessed zero-dimensional hell at bottom. The spirits emanating from Christ's head are in the process of forming "new Expanses, . . . / Creating Space, Creating Time according to the wonders Divine / Of Human Imagination" (J 98:30–32; E 258), while the ones drawing toward him on the left are, geometrically, shooting straight up all across the surface of the globe toward the immanency of a Jesus whose hour "is coming, and now is" (John 5:25). In short, Christ's head is the center of a 3-sphere of human consciousness, as seen in figure 4.4.

Relatively to the viewer, then, humanity rotates between two equally unvisualizable dimensions: contracting downward and forward into a mathematical point while also expanding upward and back toward four-dimensional space, each movement forming a "horn" vortex with its apex at Satan and its mouth in Christ. Figure C.4 diagrams the descent portion of this cyclical process, but it implies another, complementary horn toroid showing the ascent from hell to heaven. These movements across right angles explain why the figures seen at the painting's cutaways of heaven and hell appear to be traversing diagonal pleats between different dimensions. Despite its swirly appearance and two opposed spheres at top and bottom, Blake's design discloses at the corners a rectilinear structure based on a series of perpendiculars. By smoothing out right angles, the (still squarish) vortex of the *Last Judgment* invites visionary travel between four perpendicular dimensions occupying the same space. (In contrast, the prototype vortex of the *Visions of the Daughters of Albion* frontispiece merely superimposes different dimensions, producing a confused impression of montage.)

So, the painting's Fall and Redemption occur not just along a median as we take in the artist's personal perspective on the Last Judgment: "I have represented it as I saw it . . . according to . . . a certain order suited to my Imaginative Eye . . . the Last Judgment . . . is seen by . . . Every one according to the situation he holds . . . to different People it appears differently as every thing else does . . . on Earth" (*VLJ*; E 554–55). In addition, Fall and Redemption develop parabolically as Eternity intersects the three-dimensional earthly sphere at right angles in all directions. Hence, the surface of the earth is covered all over with innumerable different visible horizons or finitude-nows, each able to be transcended through movement, as we saw in chapter 3. Blake's painting illustrates Whitehead's observation that in a relativistic universe, judgment, though "categorical" in Kant's sense, "always concerns a proposition true or false in its application to the actual occasion which is the subject making the judgment . . . Judgment concerns the universe as objectified [as data or evidence] from the standpoint of the judging subject."[17] It is in this sense that Blake can assert that "whenever any Individual Rejects Error & Embraces Truth a Last Judgment passes upon that Individual" (*VLJ*; E 562). In a relativistic universe, to judge is to *be* judged by that entire universe. This is one way to understand Jesus's warning, "Judge not, that ye be not judged" (Matt. 7:1).[18]

At the same time, Blake's point is that we *must* judge, because in an empirical world of Bayesian probabilities and less-than-certain outcomes where "We are all subject to Error" (E 769) and "a Prophet" is simply an "honest man" who "utters his opinion" (E 617), the refusal to judge is mental suicide. To abnegate the necessarily partial nature of one's own point of view—to deny that one even has a point of view—is nihilism and insanity. Satan struggling on the floor of Blake's painting images the self-contradictory nature of such subjective relativism—the position, as Thomas Nagel puts it, that all views are relative except for the claim that they *are* all relative. Blake's Last Judgment, whose rightness is right now, makes it a forerunner of Nagel's "last word": that moral judgment which every individual must ultimately "think straight" from their own limited perspective as best they can, without buffering from the Spectrous "qualifying impulse" and its attempts to relativize responsibility by situating it in some wider framework of meta-values, that is, excuses.[19]

For half a century, conventional wisdom has held that Blake retreated into quietism following the collapse of the French Revolution when he published no new Prophetic Books for fifteen years, then he resumed with a cantankerous and inaccessible mythology. This consensus was reinforced by the *engagé* new historicism of the 1980s and 90s which tended to follow E. P. Thompson and Terry Eagleton in making early Blake a Marxist mascot.

But the opposite is true. *The Vision of the Last Judgment* attests the artist's return to social revolution after long silence, exactly as his hero insists in the fierce peroration to *Milton* (*M* 40:29–41:28; E 142–43). Not to minimize the irony—Blake could hardly have ignored it—that his renewed resistance came by means of a difficult, even obfuscated poetry available only in expensive handmade art books. But Blake had always understood it is because prophets refuse compromise that they often perish without renown. In *Visions of the Daughters of Albion* Blake's persona, Ololon, is unmistakably a Cassandra. Fiery youthful works like *America* and *The Marriage of Heaven and Hell*, printed by the same expensive method as *Milton* and *Jerusalem*, prove on inspection no less complexly hedged, ambivalent, and pacifistic.

That is because prophecy, for Blake, was nothing but the transmission of Vision through techniques of mediation, most especially visualization, by which the divine becomes adapted to human sense. The geometry of *The Vision of the Last Judgment* supplies a cosmological perspective to back up the same attack on mealy-mouthed liberal piety and cost-benefit utilitarianism as Blake delivered twenty years earlier in *The Marriage* and Songs of Innocence like "Holy Thursday" and "The Chimney Sweeper," even as the painting's metaphysical space-time also implicitly shows how the claim of purity behind critiques of compromise shares room with dangerously absolutist historical energies.[20] Blake's vision of a four-dimensional temporal universe was indeed "coherent," to use Donald Ault's Whiteheadian term. Nevertheless, it took him the better part of a lifetime to arrive at a fully mapped-out visualization of the relativistic "concept space" he had been seeing by glimpses all along.

Notes

Introduction

1. Donald D. Ault, *Visionary Physics: Blake's Response to Newton* (Chicago: University of Chicago Press, 1974), 4.

2. See Alice Jenkins, *Space and the "March of Mind": Literature and the Physical Sciences in Britain 1815–50* (Oxford: Oxford University Press), introduction.

3. See, for example, John Brenkman's analyses of "London" and "A Poison Tree" in *Culture and Domination* (Ithaca: Cornell University Press, 1987), 102–38.

4. Steven Goldsmith, *Unbuilding Jerusalem: Apocalypse and Romantic Representation* (Ithaca: Cornell University Press, 1993), 189–90.

5. Angela Esterhammer, *Creating States: Studies in the Performative Language of John Milton and William Blake* (Toronto: University of Toronto Press, 1994), 12, 166–67; her italics.

6. Robert N. Essick, "William Blake, William Hamilton, and the Materials of Graphic Meaning," *ELH* 52:4 (1985), 852.

7. Sarah Haggarty, "Blake's *Newton*, Line-Drawing, and Geometry," *Studies in Romanticism* 60:2 (2021), 133, 146.

8. See, for example, Stuart Peterfreund, *William Blake in a Newtonian World* (Norman: University of Oklahoma Press, 1998), chap. 1.

9. Coleridge endowed this "third kind" with Anglican *via media* connotations Blake did not share. Yet Blake did request a Church of England service at his burial in 1827, as Jerome J. McGann points out in *Social Value and Poetic Acts: The Historical Judgment of Literary Work* (Cambridge: Harvard University Press, 1988), 233. This was perhaps a reflection of how political polarization had driven traditional Dissent closer to the center following Napoleon's defeat, as twenty-plus years of revanchist nationalism suddenly faced renewed calls for reform. A trend toward the center is already evident in Blake's *Milton* (c. 1804–11), which grafts a Christian pacifist reaction to the war onto the Latitudinarian metaphysics of Henry More (see chapters 4 and 5).

10. More precisely, the variable-density aether model Newton proposed in the *Opticks* allows the gravitational field to be represented as a series of light-distance differentials based on variation in light speed or refractive index. Eric S. Baird compares it to Einstein's refractive approach to gravitational light-bending in 1911. Hence, he says, "it seems reasonable to interpret Newton's 'absolute space' as an absolute Euclidean embedding-space that acts as a container for non-Euclidean geometry, rather than as an indication that Newton believed that gravity had no effect on measured or perceived distances, times, or 'effective' geometrical relationships." Newton accepted that "Light is propagated in time, spending in its passage from the Sun to us about seven Minutes of time." *Opticks: or, A Treatise of the Reflections, Refractions, Inflections and Colours of Light*, 4th ed. (London: William Innys, 1730), 2. Henry Cavendish went on to calculate the sun's light-bending effect in 1784 and discussed it with Priestley's friend, John Michell; but he never published his results. Curved-space models were not taken seriously until Einstein, despite that *Opticks* Quest. 30 suggested "gross Bodies [i.e., matter] and Light [are] convertible into one another," and 31 broached the intriguing idea that chemically reactive "small Particles" exert short-range forces stronger ("more attractive Powers") than those of gravity, magnetism, and electricity (*Opticks* 374, 376). Baird concludes, "Newton's aether model arguably represents one of the most serious missed opportunities in the history of gravitational physics." Eric S. Baird, "Newton's Aether Model," *arXiv: General Physics* (1 Nov., 2000), n.p.; arXiv:physics/0011003. Along these lines, Wolfgang Rindler demonstrates that the mathematical mechanics for general relativity had already been worked out in the eighteenth century with mounting awareness that the laws of nature are not "discovered" but "invented." For how else could two completely different mathematical models, one mechanistic and the other Einsteinian, both describe nearly the same observations? "General Relativity before Relativity: An Unconventional Overview of Relativity Theory," *American Journal of Physics* 62:10 (1994): 887–93. As we'll see, Blake's Newton print accordingly insinuates Newton, too, was of the Devil's party without knowing it.

11. Isaac Newton, General Scholium, *Principia Mathematica*, 3rd ed., trans. Andrew Motte (Amherst: Prometheus Books, 1995), 443.

12. Letter of 7 December 1675 to Henry Oldenburg, in *The Correspondence of Isaac Newton*, ed. H. W. Turnbull, J. F. Scott, A. Rupert Hall, and Laura Tilling, 7 vols. (Cambridge: Cambridge University Press, 1959–77), 1.362–83; rptd. in *Newton*, ed. I. Bernard Cohen and Richard Westfall (New York: Norton, 1995), 19.

13. Joseph Spence, *Anecdotes, Observations, and Characters, of Books and Men*, ed. Samuel Weller Singer (London: W. H. Carpenter, 1820), 54. Newton's remark was widely distributed long before its publication. Spence's *Observations*, written between 1726 and 1759, are referenced extensively by Warburton, Warton, Samuel Johnson, and Malone.

14. William Wordsworth, "Ode ('There was a time')," ll. 165–70, in *Oxford Authors*, ed. Stephen Gill (New York: Oxford University Press, 1984), 301. Unlike

Newton's simile, however, Wordsworth's metaphor is metaleptic and four dimensional. The sportive "Children" are the immortal souls of the same adults who look back at them from the margin where the ocean of eternity meets the world of time. At this limit where contact occurs, the eternal children offer sustaining hope and comfort to the dejected selves of the temporalized adults they have become. Thus, Wordsworth, like Blake, occasionally imagined a collective Divine Humanity beyond society's conservative Burkean solidarity in the face of history and change. Presumably, the Ode was one instance when "Wordsworth the Natural Man" was *not* "rising up against the Spiritual Man" (E 665). *The Prelude*'s description of the Trinity College Newton statue as "the marble index of a mind for ever / Voyaging through strange seas of Thought, alone" (3:34–35) not only echoes Newton's remark but emphasizes his solitude, like Blake's print.

The Immortality Ode's heightened dimensionality beyond Euclid and Newton is further evident when we consider its echoes of Newton's much-heralded heir, Humphry Davy, whose poem, "The Sons of Genius"—written in 1795, age seventeen—was published in Southey's *Annual Anthology* in 1799. Roger Sharrock 60 years ago pointed out extensive echoes of Davy's *Introductory Discourse* of early 1802 in Wordsworth's famous paragraphs on the exalted character of the poet, which he added to the preface to the third edition of *Lyrical Ballads* in April. "The Chemist and the Poet: Sir Humphry Davy and the Preface to *Lyrical Ballads*," *Notes and Records of the Royal Society of London* 17:1 (1962): 57–76, esp. 69–72. The Ode's image of children sporting upon the shore while revisited by their adult selves develops the central paradox of the antepenultimate stanza 30 of Davy's poem:

> Like yon proud rocks amidst the sea of time,
> Superior, scorning all the billows' rage,
> The living sons of Genius stand sublime,
> Th' immortal children of another age.

"The Sons of Genius," in *Annual Anthology*, ed. R. Southey, 2 vols. (T. N. Longman and O. Rees: Bristol, 1799–1800), 1.99. The Ode's epigraph stating, "The Child Is Father of the Man," is here prefigured. Davy's Romantic-materialist perspective reflects a Baconian view of the so-called ancients as having lived during the "infancy" of scientific reason; hence, their very success empowers successors to go beyond and revise them. The Ode embraces this paradox to conservative effect, replacing scientific discovery with inward truth, heroic advance with elegiac reflection, collective progress with solitary endurance.

15. Newton, *Opticks*, 2; the phrase also appears at pp. 14, 28, 31, 40, and passim. John Gage suggests the print's "encircling darkness may be related to the gloomy bottom of the Cave of the Neoplatonic material world, which had recently been expounded by Thomas Taylor in his translation of [Proclus's] *The Hymns of Orpheus*. . . ." "Blake's 'Newton,'" *Journal of the Warburg and Courtauld Institutes* 34 (1971), 373.

16. Andrew M. Cooper, *William Blake and the Productions of Time* (Aldershot: Ashgate, 2013), chap. 3, esp. 86–87.

17. Wordsworth, *The Prelude* 6:166–67. Letter to James Gillman of 24 October 1826, in *Collected Letters of Samuel Taylor Coleridge*, ed. E. L. Griggs, 6 vols. (Oxford: Clarendon, 1956–71), 6.636; his italics.

18. David Hume, *Enquiries concerning Human Understanding and concerning the Principles of Morals*, ed. L. A. Selby-Bigge, 3rd ed., rev. P. H. Nidditch (Oxford: Oxford University Press, 1975), 25.

19. David Hume, *A Treatise of Human Nature*, ed. L. A. Selby-Bigge, 2nd ed., rev. P. H. Nidditch (1740; Oxford: Clarendon, 1978), 51–53.

20. Matthew Wickman, *Literature after Euclid: The Geometric Imagination in the Long Scottish Enlightenment* (Philadelphia: University of Pennsylvania Press, 2016), 45. Thomas Reid's "Geometry of Visibles" appears in his *Inquiry into the Human Mind, on the Principles of Common Sense* (1764).

21. *De gravitatione et aequipondio fluidorum*, in *Isaac Newton: Philosophical Writings*, ed. Andrew Janiak (Cambridge: Cambridge University Press, 2004), 64–65.

22. *Paradise Lost* 5:622–24, in *The Complete Poetry and Essential Prose of John Milton*, ed. William Kerrigan, John Rumrich, and Stephen M. Fallon (New York: Modern Library, 2007). In turn, the *Paradise Lost* and *Four Zoas* passages both stand in contrast to *Milton*'s celebration of the transient, sense-rich lives of the Children of Los: "gorgeous clothed Flies that dance . . . / . . . / in . . . intricate mazes of delight artful to weave: / . . . / To touch each other & recede; to cross & change & return" (*M* 26:2–7; E 123). Considering the cosmic setting of Milton's angels, it seems possible Blake's flies are dancing on the head of a pin: universal motion made visible at a physical point.

23. Letter to J. H. Reynolds of 3 May 1818, in *Keats: Poems and Selected Letters*, ed. Carlos Baker (New York: Scribner's, 1962), 434.

24. Francis Cornford, *Plato's Cosmology: The "Timaeus" of Plato*, trans. with commentary (London: Routledge, 1935), 179. A. E. Taylor, *A Commentary on Plato's "Timaeus"* (Oxford: Clarendon, 1928), 456–57.

25. As Cornford reminds us, "Plato's Space is not a void which remains completely distinct from particles moving in it; it is a Recipient which affords a basis for images reflected in it, as in a mirror—a comparison that could not be applied to atoms and void. Space is to him the 'room' or place where things are, not intervals or stretches of vacancy where things are not" (*Plato's Cosmology* 200).

26. Blake's view of Newton as inaugurating a Creation-Fall from heaven to earth was surely reinforced by the outpouring of encomia such as Pope's epigraph intended for his monument in Westminster Abbey: "Nature, and Nature's Laws lay hid in Night. / God said, *Let Newton be!* and all was Light." Alexander Pope, *Complete Poetical Works*, ed. Bliss Perry (Boston: Houghton Mifflin, 1903), 135; his italics. At least one editor in Blake's day surmised the epitaph was rejected because it sounds "a little profane." *Works of Alexander Pope*, ed. William Roscoe, 10 vols.

(London: C. and J. Rivington et al., 1824), 3.378. It was well known that Newton was born on Christmas Day: another link to anti-Christ, from a satirical perspective.

27. *Timaeus* 52b, in *Plato IX: Timaeus, Critias, Cleitophon, Menexenus, Epistles*, trans. R. G. Bury (Cambridge: Harvard University Press, 1981), 123.

28. Roger Joseph Boscovich, *A Theory of Natural Philosophy* [*Theoria Philosophiae Naturalis*], trans. J. M. Child, Latin/English, rev. ed. (1763; Chicago: Open Court, 1922), 113; his italics.

29. See John Mulligan, "Blake's Use of Geometry in 'Newton' (1805)," *Notes and Queries* 63:2 (2016): 224–28. John Gage, however, sees "the relationship of the chord of the circle to the triangle [as] closer to Figure 2 of Newton's *Opticks*, bk., i pt. I (Pl. 65c), which illustrates the passage of a ray of light through a prism" (373).

30. Bohr and Heisenberg both spoke at the bicentennial anniversary of Boscovich's *Theory of Natural Philosophy*. *Actes du Symposium International R. J. Bošković 1958* (Beograd: Academie Serbe des Sciences, 1959), 27–28 and 29–30, respectively.

31. That said, Blake remained quite uninterested in waves as such, whether of water, sound, or light. As we'll see, his preferred analogies for the propagation of matter were radiation (Neoplatonism) and vibration (David Hartley). Maybe he intuited how wave-mechanical theories like Erwin Schrödinger's, who in a 1926 paper identified particles as epiphenomenal "whitecaps" (*Schaumkamm*) arising from an underlying wave field, merely reconciled the discontinuities of Boscovichean microphysical atomism with the visualizations of classical physics. By smoothing out the incommensurability of electron "jumps" through appeal to Maxwell and nineteenth-century electromagnetic field theory, Schrödinger ignored the new epistemological need for a principle like Bohr's "complementarity" based in the limited nature of human perception and communication. His famous half-dead, half-alive cat completely misses Bohr's point, since a sizable creature containing a nearly infinite number of quantum interactions cannot itself be observed experimentally at the quantum level. As I argue in chapter 2, a main appeal for Blake of Boscovich's proto-modern quantum theory of nested subatomic spheres of attraction and repulsion (see figures 2.2 and 3.3) was that his force-curve becomes infinitely repulsive as two "atom-points" approach each other. This is because the particles, the indivisible centers of atoms, are immaterial. As probability densities, they are not actually smeared across space but remain delocalized and unobservable except through a kind of complementarity principle beyond sense-based three-dimensional visualization. Call it, "intellectual vision" (E 757).

32. Joseph Priestley, *History and Present State of Electricity*, 4th ed. (London: J. Johnson: 1775), 443–44.

33. Donald D. Ault, *Visionary Physics: Blake's Response to Newton* (Chicago: University of Chicago Press, 1974), 128.

34. Like the man said, "Good fences make good neighbors." Robert Frost, "Mending Wall," ll. 27, 45, in *North of Boston*, 2nd ed. (New York: Henry Holt,

1915), 11, 13. The phrase was a proverb mentioned, for example, in *Blum's Farmer's and Planter's Almanac* for 1850 and again for 1861. See Wolfgang Mieder, "Good Fences Make Good Neighbours [*sic*]: History and Significance of an Ambiguous Proverb," *Folklore* 114:2 (2003), 159. Blake's lifelong idea of the boundary as connecting what it divides underpins his empirical view of the truth/error relation and his corollary critique of Berkeley's idealism, as seen in chapter 6.

35. I allude to W. J. T. Mitchell, "Blake's Divine Comedy: Dramatic Meaning in *Milton*," in *Blake's Sublime Allegory: "The Four Zoas," "Milton," "Jerusalem*," ed. Stuart Curran and Joseph Wittreich, Jr. (Madison: University of Wisconsin Press, 1973), 281–307.

36. Joseph Priestley, *Disquisitions Relating to Matter and Spirit* (London: J. Johnson, 1777), 54; all italics his.

37. Andrea Henderson, "The Physics and Poetry of Analogy," *Victorian Studies* 56:3 (2014), 389–90.

Chapter One

1. Thomas Reid, *Essays on the Intellectual Powers of Man* (1785), in *Works*, ed. William Hamilton, 8th ed., 2 vols. (Edinburgh: James Thin, 1895), 1.263.

2. Paul Youngquist, *Madness and Blake's Myth* (University Park: University of Pennsylvania Press, 1989), 66.

3. Steven M. Rosen, *Topologies of the Flesh* (Athens: Ohio University Press, 2006), 34; his italics. Rosen links Merleau-Ponty's flesh phenomenology with Klein bottle topology on pp. 24–39.

4. Maurice Merleau-Ponty, *The Visible and the Invisible*, trans. Alphonso Lingis (Evanston: Northwestern University Press, 1968), 147, 140, 143.

5. Henri Bergson, *Creative Evolution*, trans. Arthur Mitchell (New York: Henry Holt, 1911), 272.

6. Compare *The Marriage of Heaven and Hell*, where the Blake devil and an angel swap horrific metaphysical visions of each other's "eternal lot" (*MHH* 17; E 41).

7. Morris Eaves, The Counter-Arts Conspiracy: Art and Industry in the Age of Blake (Ithaca: Cornell University Press, 1992), 145.

8. A. N. Whitehead, *Science and the Modern World* (New York: Macmillan, 1925), 91.

9. *Paradise Lost* 4:269–71, in *The Complete Poetry and Essential Prose of John Milton*, ed. William Kerrigan, John Rumrich, and Stephen M. Fallon (New York: Modern Library, 2007).

10. See Alan Bewell, " 'Jacobin Plants': Botany as Social Theory in the 1790s," *Wordsworth Circle* 20:3 (1989): 132–39.

11. David Worrall links Blake's Argument design to Erasmus Darwin's *Botanic Garden* and its discussion of the light discharged by certain flowers. "William Blake

and Erasmus Darwin's Botanic Garden," *Bulletin of the New York Public Library* 78:4 (1974–75), 402–04.

12. Similarly, *Milton* invites readers to identify with "the little winged fly, smaller than a grain of sand" (*M* 15 20:27). The passage turns the traditional image of death, decay, and impermanence into a symbol of delicacy, movement, and receptiveness toward heaven within. In eighteenth-century usage, "fly" refers to flying insects generally, not just *Muscomorpha*. Damselflies resemble dragonflies but, unlike them, fold the wings along the body when at rest, similarly to the figure at the top of *Jerusalem*'s title page.

13. Abraham Tucker, whose influence on Blake I examine in chapter 5, remarks: "The retina of the eye, whereon all our visible objects are painted, takes its name from a net." *The Light of Nature Pursued*, 6 vols. (London: T. Jones and T. Payne, 1768), 2.22. Phineas Fletcher uses the word in this sense in his long allegorical poem on the human body, *The Purple Island* (1633), an important source of anthropomorphic metaphors for Blake: "The forms caught in this net are brought to sight / And to [man's] eye are lively pourtrayed." *The Purple Island*, 5:34–35, in *Giles and Phineas Fletcher: Poetical Works*, ed. Frederick S. Boas, 2 vols. (Cambridge: Cambridge University Press, 1908–09), 60. Robert N. Essick pointed out some time ago that Blake regarded crosshatching, eighteenth-century engraving's visual code for three-dimensional perspective, as a "net" for capturing representations. "Blake and the Traditions of Reproductive Engraving," *Blake Studies* 4 (1972): 59–103. The skintight tunics worn by the figures in Blake's art make visible the perceptual union between the form of these depicted spirits and the netlike *tunicae* in the bulb of the viewer's eye. David V. Erdman notes that Oothoon's "silken nets and traps of adamant" resemble Darwin's description of the Venus flytrap as spreading a "viscous snare" of "magic nets, unseen." *Blake: Prophet Against Empire*, 3rd ed. (Princeton: Princeton University Press, 1977), 511.

14. David Worrall plausibly ascribes Blakean vision to "bright, shimmering or glittering points of light . . . likely to have been migraine aura." "'Seen in my visions': Klüver Form-Constant Visual Hallucinations in William Blake's Paintings and Illuminated Books," *Blake: An Illustrated Quarterly* 55:4 (2022), para. 3, n.p. I suppose the question here concerns *how* Blake's belief in a humanizing, Jesus-centric imagination shaped his perception of the relatively straightforward, roughly geometrical brain-based illusions Worrall refers to, much as Blake himself asserted the apprehension of a miracle depends on belief (see chapter 2). Gleams of light are also Neoplatonism's staple image of pre-existence from Plotinus's *Enneads* to Wordsworth's Immortality Ode and *The Prelude* (1805), notably 1:614–17.

15. Plato, *Theatatus*, 155d. Aristotle, *Metaphysics* A, 982b11–12. Albert Einstein, "What I Believe," *Forum and Century* 84:4 (1930): 192–93. A. N. Whitehead, *Modes of Thought* (New York: Free Press, 1938), 168.

16. Samuel Taylor Coleridge, "Essay on Method," *The Friend*, ed. Barbara E. Rooke, 3 vols. (Princeton: Princeton University Press, 1969), 3.507–14.

17. See Joseph Viscomi, *Blake and the Idea of the Book* (Princeton: Princeton University Press, 1993), 30.

18. Note that *Daughters of Albion*'s copy G frontispiece, printed in 1795 and reproduced in the Princeton edition because its "impressively detailed and sombre" hand coloring is "in keeping with the poem's dark events and brooding mood," makes a very different impression from copy I printed in 1793 (figure 1.1), whose bright semi-transparent washes reflect a still-hopeful political climate before the government crackdown of November 1794. William Blake: *The Early Illuminated Books*, ed. Morris Eaves, Robert N. Essick, and Joseph Viscomi (Princeton: Princeton University Press, 1993), 226.

19. Fredrick Hoerner, "Prolific Reflections: Blake's Contortion of Surveillance in *Visions of the Daughters of Albion*," *Studies in Romanticism* 35:1 (1996): 119–50. Where Hoerner sees ideological "contortion," I see "contracting" or "expanding" dimensionality.

20. The poem's apparent collapse in solipsistic monologue leads Nelson Hilton to claim: "Oothoon finally cannot connect or develop . . . She ends as she began, a reflection . . . Moreover, in suggesting that 'conversing with shadows dire' represents Theotormon's involvement with his own narcissistic projections, the closing lines put forward the depressing possibility that Oothoon herself is one of those projections: herself, to repeat, 'a solitary shadow wailing on the margin of non-entity.' " "An Original Story," in *Unnam'd Forms: Blake and Textuality*, ed. Nelson Hilton and Thomas A. Vogler (Berkeley: University of California Press, 1986), 102; his italics. Indeed, but why not go on and ask in whose head Theotormon's image exists, in turn?

21. As noted by Nelson Hilton, 91–98, Oothoon sounds "flamboyant" at 6:16–21 and 7:16. Her third-person references to herself at 6:18–22, 7:23, and possibly 8:11, waver at the limit between self-transcendence and subjective self-absorption. See also James A. W. Heffernan, "Blake's Oothoon: The Dilemma of Marginality," *Studies in Romanticism* 30:1 (1991): 3–18.

22. Steven Goldsmith, *Blake's Agitation: Criticism and the Emotions* (Baltimore: Johns Hopkins University Press, 2013), 78.

23. Cf. Robert Palter: "In its weakest sense 'field' refers simply to any theory involving continuous distributions in space of the values of some physical magnitude, such as force or energy (in this sense Newton's theory of gravitation is a field theory); sometimes, at the other extreme, 'field' refers to any theory in which all fundamental laws are expressed as partial differential equations with space and time as independent variables, and in which all corpuscular aspects of the phenomena covered by the theory may be derived from these fundamental laws (in this sense Einstein's theory of gravitation is closest approach to a field theory known at present . . .)." *Whitehead's Philosophy of Science* (Chicago: University of Chicago Press, 1960), 193.

24. See Muireann Maguire and Timothy Langen, "Introduction: Countersense and Interpretation," in *Reading Backwards: An Advance Retrospective on Russian Literature* (Cambridge: Open Book, 2021), xiii–xxvi.

25. Jorge Luis Borges, "Kafka and His Precursors," in *Other Inquisitions* (New York: Simon and Schuster, 1965), 108; his italics. Borges's point is that the later writer brings to light in the precursor a "quality" that would never otherwise have been perceived—"in other words, it would not exist" (108)—that is, not in our world. Borges thus arrives at something like Plato's *anamnesis* and Whitehead's stance of Platonic realism rooted in the "objective immortality" of qualities (see chapter 5). For Whitehead, literature, philosophy, and social science are all "engaged in finding linguistic expressions for meanings as yet unexpressed." *Adventures of Ideas* (New York: Macmillan, 1933), 221. So, Borges says, "the word 'precursor' . . . should be cleansed of all connotation of polemics or rivalry." Compare Blake: "I cannot think that Real Poets have any competition None are greatest in the Kingdom of Heaven it is so in Poetry" (E 665; significantly, an annotation to Wordsworth). See Seamus Heaney, Richard Kearney, and Jorge Luis Borges, "Borges and the World of Fiction: An Interview with Jorge Luis Borges," *The Crane Bag* 6:2 (1982): 71–78. See also Michael L. Thomas, "Resisting the Habit of Tlön: Whitehead, Borges, and the Fictional Nature of Concepts," *Philosophy and Literature* 42:1 (2018): 81–96. What Thomas says of the imaginary world of Tlön in Borges's "Tlön, Uqbar, Orbis Tertius" (1940) holds also for Blake's view of Newtonianism: "The inquiry surrounding Tlön models the instantiation of abstractions in reality, indicating that the reality of concepts doesn't require materiality and, in fact, that concepts become material through their realization in human perspectives" (86). In Whiteheadian terms, it's a case of Newton's "fallacy of misplaced concreteness." More positively, Thomas writes of Borges: "As potential objects, fictions provide the starting point for a search that uncovers their presence in reality, which is found in the relations they form in the 'real world' surrounding us" (87). Likewise, Blake's *Daughters of Albion* frontispiece illustrates visually, what the poem goes on to demonstrate, that what we call "the visible world" constitutes an interface between the imaginary and the material, an interface whose essence is the symbolism of language.

26. T. S. Eliot, "Tradition and the Individual Talent" (1919), in *The Sacred Wood: Essays on Poetry and Criticism* (London: Methuen, 1928), 50. Eliot illustrates his idea by appeal to a recent experiment on catalysis (though he neglects to mention water as necessary to produce sulphuric acid). His concern to stress "poetry *is* a science" (his italics) requiring "training and equipment" reflects relativity's prestige at the time. Borges footnotes Eliot.

27. Andrew Franta, "Shelley and the Poetics of Political Indirection," *Poetics Today* 22:4 (2001), 767.

28. Marjorie Levinson, "The New Historicism: Back to the Future," in *Rethinking Historicism: Critical Readings in Romantic History*, ed. Levinson (New York: Blackwell, 1991), 18–63. See also her introduction.

29. See Taylor Schey, "Limited Analogies: Reading Relations in Wordsworth's *The Borderers*," *Studies in Romanticism* 56:2 (2017): 177–201.

30. Coleridge, 1.488, 1.491–92. No doubt, calling Bacon "the British Plato" was a nod to the mood of fevered nationalism during and just after the Napoleonic Wars.

31. A. N. Whitehead, *Modes of Thought* (New York: Macmillan, 1938), 222–23.

32. As Linda Dalrymple Henderson points out, "during the course of the 1920s, Einstein and mathematician Hermann Minkowski's earlier incorporation of time into the four-dimensional space-time continuum had gradually overshadowed cultural memories of the geometrical, spatial fourth dimension." "The Image and Imagination of the Fourth Dimension in Twentieth-Century Art and Culture," *Configurations* 17:1–2 (2009), 133. For the spatial fourth dimension of science-fiction writers like H. G. Wells and mystics like Ouspensky, see Mark Blacklock, "Higher Spatial Form in Weird Fiction," *Textual Practice* 31:6 (2017): 1101–16; also Elizabeth L. Throesch, *Before Einstein: The Fourth Dimension in Fin-de-Siècle Literature and Culture* (New York: Anthem, 2017). Like Henderson, Throesch notes that the late-Victorian "hyperspace philosophy" popularized by C. E. Hinton and E. A. Abbott's *Flatland* (1884) considered time as an illusion by which human sensibility sliced up a four-dimensional space into three-dimensional "fugitive nows": a forerunner of the Einsteinian "block universe." But this changed with "the translation of Bergson into English in the 1910s as well as the popularization of Einstein's work from 1919 onward" (195).

33. See Robert D. Denham, *Northrop Frye and Others: Volume III: Interpenetrating Visions* (Ottawa: University of Ottawa Press, 2018), chap. 6.

34. In 1870, James Joseph Sylvester argued: "Induction and analogy are the special characteristics of modern mathematics, in which theorems have given place to theories, and no truth is regarded otherwise than as a link in an infinite chain." "Inaugural Presidential Address," in *The Laws of Verse or Principles of Versification Exemplified in Metrical Translations: Together with an Annotated Reprint of the Inaugural Presidential Address to the Mathematical and Physical Section of the British Association at Exeter* (London: Longmans, Green, 1870), 118n. Qtd. in Imogen Forbes-Macphail, "Topological Poetics, Gerard Manley Hopkins, Nineteenth-Century Mathematics, and the Principle of Continuity," *ELH* 88:1 (2021), 150–51.

35. Erasmus Darwin, *Zoonomia; or, the Laws of Organic Life*, 2 vols. (London: J. Johnson, 1794–96), 1.1; his italics. Compare physicist Max Planck: "The disjointed data of experience can never furnish a veritable science without the intelligent interference of a spirit actuated by faith. . . . We have a right to feel secure in surrendering to our belief in a philosophy of the world based upon a faith in the rational ordering of this world." *The Philosophy of Physics*, trans. W. H. Johnson (New York: Norton, 1936), 122. Qtd. in *Science and Society: The History of Modern Physical Science in the Twentieth Century*, ed. Peter Gallison, Michael Gordin, and David Kaiser (London: Routledge, 2002), 47–48. Einstein concurred with this view: to believe the universe is at all intelligible is also to recognize its design far exceeds any human mind. So, Einstein remained agnostic toward God (except, of course, for knowing God "does not throw dice"). As a preview to my use of Alfred North

Whitehead in later chapters, consider Bertrand Russell's recollection: "[Whitehead] was impressed by the aspect of unity in the universe, and considered that it is only through this aspect that scientific inferences can be justified." *Portraits from Memory and Other Essays* (1956; New York: Routledge, 2021), 84. Compare with the preceding Darwin quotation the following, almost Orphic passage from Bacon's *Advancement of Learning*:

> Is not the precept of a musician, to fall from a discord or harsh accord upon a concord or sweet accord, alike true in affection? Is not the trope of music, to avoid or slide from the close or cadence, common with the trope of rhetoric of deceiving expectations? Is not the delight upon the quavering upon a stop in music the same with the playing of light upon the water? . . . Are not the organs of the senses of one kind with the organs of reflection, the eye with a glass, the ear with a cave or strait, determined and bounded? Neither are these only similitudes, as men of narrow observation may conceive them to be, but the same footsteps of nature, treading or printing upon several subjects or matters. *Works*, ed. James Spedding, Robert Leslie Ellis, Douglas Denon Heath, 7 vols. (London: Longman, 1857–74), 3.348–49.

Shelley, after quoting that striking final metaphor of nature's footsteps, remarks approvingly that Bacon "considers the faculty which perceives them as the storehouse of axioms common to all knowledge." *A Defence of Poetry*, in *Shelley's Poetry and Prose*, ed. Donald H. Reiman and Neil Fraistadt, 2nd ed. (New York: Norton, 2002), 512.

36. Devin S. Griffiths, "The Intuitions of Analogy in Erasmus Darwin's Poetics," *Studies in English Literature* 51:3 (2011), 654. Griffiths continues: "[Darwin] relies upon the assertion from natural theology that nature is 'stamped' with patterns. Darwin is an analogical realist. But, instead of serving Christian apologetics as proof of God, Darwin's analogy of nature is designed to yield basic insights into the patterns of nature: it argues for a nature that is both coherent and intelligible, hence, accessible to empiricist inquiry." Arguably, this is much the same mythopoeic view of nature as Bacon propounds in *The Wisdom of the Ancients* and *The New Atlantis*—namely, the view that scientific reason originates in a prerational recognition of the unity and coherence of nature's analogies, affinities, and resemblances. See Stephen H. Daniel, "Myth and the Grammar of Discovery in Francis Bacon," *Philosophy and Rhetoric* 15:4 (1982): 219–37.

37. A. N. Whitehead, *Process and Reality* (1929; New York: Free Press, 1978), 205; see also 199–207.

38. A. N. Whitehead, *Science and the Modern World* (New York: Macmillan, 1925), 44.

39. Jonathan Bain, "Whitehead's Theory of Gravity," *Studies in the History and Philosophy of Modern Physics* 29:4 (1998), 557.

40. *The Commentaries of Proclus on the Timaeus of Plato in Five Books; containing a treasury of Pythagoric and Platonic physiology*, trans. Thomas Taylor, 2 vols. in 1 (London: printed for the author, 1820), 404–05; his italics.

41. The chemistry of Humphry Davy, at least, probably was *not* entirely beyond Blake's ken. The "electric flame of Miltons awful precipitate descent" (*M* 20:26; E 114) perhaps owes something to Davy's celebrated second Bakerian Lecture of 19 November 1807, published in 1808, two years before *Milton*. Davy's paper opens by announcing "a more intimate knowledge . . . concerning the true elements of bodies" as revealed by "the electrical powers of decomposition," namely electrolysis. "The Bakerian Lecture: On Some New Phenomena of Chemical Changes Produced by Electricity," *Philosophical Transactions of the Royal Society of London* 98 (Royal Society: London, 1808), 1, 5. Milton's "electric . . . descent" is, similarly, a noninstantaneous decomposition of the Selfhood's "False Body: an Incrustation" (*M* 40:35; E 142). By passing electricity between two carbon sticks to produce the first arc light, Davy demonstrated that flame is a chemical reaction involving not just oxidation but ionization (pace Lavoisier and Priestley both), and that fire is neither a gas nor an elemental substance (phlogiston theory's "combustible earth") but a plasma: a differential state of change in energy occurring between, on one hand, fuel elements such as carbon or charcoal and, on the other hand, spent fumes, i.e., gas molecules and smoke in the form of solid particles. Milton's fiery passage through a series of transitional States within the space of a Moment seems similarly "electric," even plasmatic.

At the same time, the "brilliant phenomena" of "most intense light" ("Bakerian Lecture" 2) that Davy demonstrated, and which his paper describes, exhibit much greater incandescence than the smoky vulcanism by which Blake had mythologized revolution in *America* and used to portray his "melting [of] metals into living fluids" as a correspondingly revolutionary innovation (*MHH* 15; E 40). In one experiment involving flammable liquid sodium, an "active metal" like potassium, Davy writes that "the globules often burnt at the moment of their formation, and sometimes violently exploded and separated into smaller globules, which flew with great velocity through the air in a state of vivid combustion, producing a beautiful effect of continued jets of fire" (6). Clearly, this is a brighter kind of flame than the hellish burning Blake had associated with his method of etching in relief. Is Milton's "precipitate descent" through the sky in *Milton* 20:26 a chemistry pun based on Davy's godlike power to precipitate new metals, a power newly surpassing Blakean engraving as the Contrary to Urizen's search for a primordial "solid without fluctuation" (*BU* 4:11; E 72)? Similarly, Sharon Ruston points out that Coleridge in *Biographia Literaria* (1816) uses "sublime" as a *verb* to describe the poetic imagination, thereby referring to the chemical process of sublimation—not what Burke or Kant had meant by the noun. *Creating Romanticism* (London: Palgrave Macmillan, 2013), 132.

In the years preceding *Milton*'s publication, Davy's experiments became widely known through Jane Marcet's *Conversations on Chemistry* (1806), her best-selling popularization for young adults which presents chemistry as a method of education based in a new form of logic. It seems very plausible Blake knew this work, considering the educational concerns of *Songs of Innocence*. Far more than his predecessor Priestley, Davy epitomized utopian hopes for a unification of the two cultures of art and quantified science, and hence an end to mechanism's "bifurcation of nature" (as Whitehead later termed it) and the redemption of "Bacon & Newton & Locke" alongside "Milton & Shakspear & Chaucer" (*J* 98:9; E 257). As early as April 1802, the visionary fifth and sixth paragraphs which Wordsworth added to the Preface to *Lyrical Ballads* declare "the Poet" as "ready to follow the steps of the Man of science, . . . carrying sensation [via "the grand elementary principle of pleasure"] into the midst of the objects of the science itself." *Oxford Authors*, ed. Stephen Gill (New York: Oxford University Press, 1984), 605–607.

42. Robert Rosenblum, *Transformations in Late Eighteenth Century Art* (Princeton: Princeton University Press, 1967), 169–70. The term "Romantic neo-classicism" appears in Rosenblum's 1956 dissertation, published as *The International Style of 1800: A Study in Linear Abstraction* (New York: Garland, 1976).

43. See Dahlia Porter, *Science, Form, and the Problem of Induction in British Romanticism* (Cambridge: Cambridge University Press, 2018), 73–112.

44. Devin Griffiths, *The Age of Analogy: Science and Literature between the Darwins* (Baltimore: Johns Hopkins University Press, 2016).

45. Joseph Priestley, *History and Present State of Electricity*, 4th ed. (London: J. Johnson: 1775), 14.

46. Erasmus Darwin, "Apology" to *The Botanic Garden* (London: J. Johnson, 1791), vii. Qtd. in Mathew J. A. Green, *Visionary Materialism in the Early Works of William Blake* (New York: Palgrave, 2005), 150.

47. David Hume, *A Treatise of Human Nature*, ed. L. A. Selby-Bigge, 2nd ed., rev. P. H. Nidditch (1740; Oxford: Clarendon, 1978), "Of Unphilosophical Probability," 147.

48. David Hume, *Enquiries concerning Human Understanding and concerning the Principles of Morals*, ed. L. A. Selby-Bigge, 3rd ed., rev. P. H. Nidditch (Oxford: Oxford University Press, 1975), 104.

49. Humphry Davy, "Historical View of the Progress of Chemistry," in *Elements of Chemical Philosophy* (London: J. Johnson, 1812), 2.

50. Humphry Davy, "Introductory Lecture to the Chemistry of Nature," notebook held at the Royal Institution (MS HD/2/C/1, f.1v. RI MS HD/14/E, 166). Qtd. in Sharon Ruston, "Humphry Davy: Analogy, Priority, and the 'true philosopher,'" *Ambix* 66:2–3 (2019), 130.

51. Robert Oppenheimer, "Analogy in Science," *Centennial Review of Arts & Science* 2 (1958): 351–73. See also Mary B. Hesse's important distinction between "positive," "negative," and "neutral" analogy, where the latter often indicates areas for further investigation. *Models and Analogies in Science* (Notre Dame: University of Notre Dame Press, 1966), 8–9.

52. The first two quotations describe Flaxman's style in Albert Boime, *Art in an Age of Revolution: 1750–1800* (Chicago: University of Chicago, 1987), 373; the third describes George Cumberland's style in Rosenblum, *Transformations*, 167.

53. See Vincent A. De Luca, "Blake's Wall of Words," in *Unnam'd Forms: Blake and Textuality*, ed. Nelson Hilton and Thomas A. Vogler (Berkeley: University of California Press, 1986), 218–41.

54. James Clerk Maxwell, "Are There Real Analogies in Nature?" in *The Scientific Letters and Papers*, ed. P. M. Harman, 3 vols. (Cambridge: Cambridge University Press, 1990–2002), 1.377. This essay also appears in Lewis Campbell and William Garnett, *Life of James Clerk Maxwell* (London: Macmillan, 1882), 235–44.

55. *Albert Einstein: Philosopher-Scientist*, ed. Paul Arthur Schilpp, Library of Living Philosophers 7 (Chicago: Open Court, 1949), 669.

56. Stanley Cavell, *In Quest of the Ordinary: Lines of Skepticism and Romanticism* (Chicago: University of Chicago Press, 1988), 31.

57. John Locke, *An Essay concerning Human Understanding*, ed. Alexander Campbell Fraser, 2 vols. (New York: Dover, 1959), 1.230.

58. Planck's term, *weltbild*, is well described by Thomas Ryckman:

> The theoretical physicist is to be understood as engaged in building up a "physical world image," much of which is an admittedly mental construction. Employing symbols for the metaphysically real, the *Weltbild* is not readily expressible in ordinary language, and it may posit elements that outstrip present capacities for observational test. At the same time it must yield consequences that are possible to observationally confirm, and in addition it must be flexible enough to accommodate new phenomena. Though in a sense a creation of mind, its implications purport to refer to a mind-independent real external world. As it is always incomplete, it can never be supposed in satisfactory agreement with mind-independent states of affairs, and in any case such agreement can never be directly ascertained but at most indirectly inferred. Perhaps its most important functions are to serve as a platform within which further thought may develop, as well as to suggest experiments whereby it can be transcended or amended. *Einstein* (London: Routledge, 2017), 295.

59. Leopold Damrosch, Jr., *Symbol and Truth in Blake's Myth* (Princeton: Princeton University Press, 1980), 83; his italics. Damrosch says, "In Blake the symbol only possesses meaning in conjunction with other symbols" because his symbols exist "half way between the real and the symbolic . . . Like the Gnostics, Blake is thus committed to the peculiar richness but also the ambiguity of symbols about symbols, which can only be described in paradoxical terms, as for instance by Gilbert Durand: 'They are symbols of the symbolic function itself, which is—like them!—mediatory between the transcendence of the signified and the manifested world of concrete, incarnated signs that become symbols through it'" (156–57).

60. Wallace Stevens, *The Necessary Angel* (New York: Vintage, 1951), 72.

61. See John Barrell, *The Political Theory of Art from Reynolds to Hazlitt: "The Body of the Public"* (New Haven: Yale University Press, 1986), 225–31.

62. Niels Bohr, "The Unity of Human Knowledge," in *Essays 1958–62: On Atomic Physics and Human Knowledge* (New York: John Wiley, 1963), 10. This is not to ignore what one witness to Bohr's sluggish, often obscure, always grindingly methodical speech in sidebar debates with Einstein during the fifth Solvay Conference of 1927, called the "awful Bohr incantation terminology. Impossible for anyone else to summarize." Paul Ehrenfest, letter to Goudsmit, Ulenbeck, and Dieke, 3 November 1927, in *Foundations of Quantum Physics I (1926–1932)*, ed. Jørgen Kalckar (1985), in *Niels Bohr: Collected Works*, ed. Léon Rosenfeld, Erik Rüdinger, and Finn Aaserud, 13 vols. (Amsterdam: North-Holland/Elsevier, 1972–2008), 6.38. Indeed, some physicists wondered if Bohr had traded in his lab smock for philosopher's robes. (He had.) Nevertheless, Bohr's preoccupation with epistemology, based on his insight that "we are suspended in language, . . . 'reality' is also a word," offered the right scientific approach to describing and communicating phenomena vanishingly small. As Aage Petersen explained, for Bohr there is an analogy between physics and psychology based on their shared concept of observation. "Quantum physics illuminates the dialectics of introspection because the physicist's [experimentally adjustable] partition between system and measuring tool corresponds closely to the epistemologist's [similarly adjustable] partition between object and subject [e.g., the subject can take himself as an object]." "The Philosophy of Niels Bohr," *Bulletin of the Atomic Scientists* 19:7 (1963), 9–10. That is why Bohr told Heisenberg: "When it comes to atoms, language can be used only as in poetry. The poet, too, is not nearly so concerned with describing facts as with creating images and establishing mental connections." Werner Heisenberg, *Physics and Beyond: Encounters and Conversations*, trans. Edward T. Heise (New York: Harper & Row, 1971), 41–42. Accordingly, Heisenberg himself remarked: "The transition from the 'possible' to the 'actual' takes place during the act of observation. If we want to describe what happens in an atomic event, we have to realize that the word 'happens' can apply only to the observation, not to the state of affairs between two observations." *Physics and Philosophy: The Revolution in Modern Science* (New York: Harper & Row, 1958), 54–55.

63. Albert Einstein, letter to J. Stark, 14 December 1908. Qtd. in Abraham Pais, *Subtle Is the Lord: The Science and the Life of Albert Einstein* (Oxford: Oxford University Press, 2005), 13; Einstein's italics. Also qtd. in Schilpp, 683–84.

Not to overstate the conservatism of Einstein's stance in the Einstein-Bohr debate. As David Kaiser points out, Einstein and Bohr both were "realists" insofar as they agreed on the fundamental existence of entities with relations and properties independent of human agents. Their disagreement concerned epistemological access to this reality. Bohr, a great sharer of scientific ideas whose often overlong papers are surprisingly short on scientific equations, restricted "physical reality" to the content of communicable descriptions in common language, a position compatible with Kant. Einstein, the solitary visualizer who once explained, "words or the language, as they are written or spoken, do not seem to play any role in my mechanism of thought," wanted to capture the whole of reality in theories he set forth quite tersely in papers dense with mathematics. Kaiser, "Bringing the Human Actors Back on Stage: The Personal Context of the Einstein-Bohr Debate," *British Journal for the History of Science* 27:2 (1994): 129–52. "A Testimonial from Professor Einstein," in Jacques Hadamard, *The Psychology of Invention in the Mathematical Field* (Princeton: Princeton University Press, 1945), appendix 2, 142.

Chapter Two

1. The most thorough investigation into whether Blake read Priestley is Jeffrey Barclay Mertz, "A Visionary among the Radicals: William Blake and the Circle of Joseph Johnson, 1790–95," D. Phil., University of Oxford 2010, esp. chaps. 1, 5, and the conclusion. Mertz's overall verdict supports the consensus of most Blake scholars: not proved but quite likely. In Mertz's "The Responses of William Blake and Joseph Priestley to Two Swedenborgian Ideas," *Blake: An Illustrated* Quarterly 47:2 (2013), n.p., he finds echoes in *The Marriage of Heaven and Hell* of Priestley's *Letters to the Members of the New Jerusalem Church*. He further remarks that Morton Paley and Graham Pechey both have noticed similarities between Blake's language in *The Marriage* and Priestley's *Disquisitions Relating to Matter and Spirit*. Paley, *Energy and Imagination: A Study of the Development of Blake's Thought* (Oxford: Clarendon, 1970), 8–9. Pechey, "*The Marriage of Heaven and Hell*: A Text and Its Conjuncture," *Oxford Literary Review* 3 (1979), 57. Priestley had been a friend of Johnson since 1764, and Johnson published some 130 of his titles. Thus, the question of Priestley's influence on Blake turns partly on whether Blake met Priestley at Johnson's legendary Tuesday dinners, which Priestley attended. The strongest claim is by Frederick Tatham, Blake's follower from the 1820s, in his less than reliable *Life of Blake* (1832): "He was intimate with a great many of the most learned & eminent men of his time, whom he generally met at Johnsons, the Bookseller of St. Pauls Church Yard." *Blake Records: Documents (1714–1841) concerning the Life of William Blake (1757–1827) and His Family*, ed. G. E. Bentley, Jr., 2nd ed. (New Haven: Yale University Press, 2004), 685. For now, it seems all we can say is that Blake's proximity to the Joseph Johnson circle—his friendship with Fuseli, who was a regular diner and visitor; his numerous commissions from Johnson (82 or 83 plates in 1790–95 alone); and the fact that Johnson sold his Illuminated Books—makes it all but certain that Blake would at least have been familiar with Priestley's ideas about matter, spirit, space, and God. As Jon Mee observes, "Johnson's shop, which Blake attended in the way of business, would itself have been a place of sociability and conversation, sufficiently, anyway, to have given him a sense of the nature

of intellectual debate" at Johnson's dinners. "A Little Less Conversation, a Little More Action: Mutuality, Converse and Mental Fight," in *Blake and Conflict*, ed. Sarah Haggarty and Jon Mee (London: Palgrave, 2009), 133. Things are further complicated by Mertz's evidence in his dissertation that Blake likely read reviews of publications by Johnson's authors, including Priestley, in Johnson's journal, *The Analytical Review*, and had some social and professional contact with the reviewers. The closest anybody has gotten to showing that Blake attended Johnson's dinners is David Erdman, who examined Godwin's two diary references to meetings with "Blake" (as distinct from his dozen references to the radical "A[rthur] Blake") in *Blake: Prophet against Empire: A Poet's Interpretation of the History of His Own Times*, 3rd ed. (Princeton: Princeton University Press, 1977), 157–60. Erdman concluded that he had not found a smoking gun. The question, of course, is what constitutes reasonable doubt. However, if Priestley's "immateriality of matter" reflected broader field-oriented tendencies in late-Enlightenment physics deriving from Newton himself, as I claim, then Blake need not have read Priestley at all.

2. David Hartley, *Observations on Man, his Frame, his Duty, and his Expectations*, 2 vols. (London: S. Richardson, 1749), 1.234.

3. John W. Yolton's term in *Thinking Matter: Materialism in Eighteenth-Century Britain* (Minneapolis: University of Minnesota Press, 1983), 55. Hume, for example, asserts that perceptions are substances. Since "all our perceptions are different from each other, and from every thing else in the universe, they are also distinct and separable, and may be consider'd as separately existent, and may exist separately, and have no need of any thing else to support their existence." *A Treatise of Human Nature*, ed. L. A. Selby-Bigge, 2nd ed., rev. P. H. Nidditch (1740; Oxford: Clarendon, 1978), 233. The non-Cartesian "ontologizing" of perceptions also underlies Mach, Whitehead, and Bohr's respectively "neutral," "objectivist," and descriptivist treatments of the datum of experience as an unprivileged portion of a common, implicitly panexperiential world of real things.

4. Hume, 253. I take it Hume emphasizes here that having perceptions is *not* like watching a camera obscura set up in some definite known location, as we might have supposed from his theater analogy.

5. Blake's position resembles that of Galen Strawson, for example his "Realistic Monism: Why Physicalism Entails Panpsychism," *Journal of Consciousness Studies* 13:10–11 (2006): 3–31.

6. Joseph Priestley, *A Free Discussion of the Doctrines of Materialism, and Philosophical Necessity, in a correspondence between Dr. Price, and Dr. Priestley* (London: J. Johnson 1778), 15–17; all italics Priestley's.

7. Robert E. Schofield, "Joseph Priestley, the Theory of Oxidation and the Nature of Matter," *Journal of the History of Ideas* 25:2 (1964), 293.

8. Roger Joseph Boscovich, *A Theory of Natural Philosophy* [*Theoria Philosophiae Naturalis*], trans. J. M. Child, Latin/English, rev. ed. (1763; Chicago: Open Court, 1922), 51.

9. See, for example, Richard Olson, "The Reception of Boscovich's Ideas in Scotland," *Isis* 60:1 (1969): 91–103. Olson demonstrates that "by the last decade of the eighteenth century almost all of Boscovich's important philosophical ideas were known to the Scots" (96).

10. Dugald Stewart, *Elements of the Philosophy of the Human Mind* (1802), in *Collected Works*, ed. Sir William Hamilton, 11 vols. (Edinburgh: T. Constable, 1854–60), 2.107.

11. Joseph Priestley, *History and Present State of Discoveries Relating to Vision, Light, and Colours*, 2 vols. (J. Johnson: London, 1772), 1.390–91.

12. For example, Ludwik Silberstein noted Boscovich "contains many remarkably clear and radical ideas regarding the relativity of space, time and motion." *The Theory of Relativity*, 2nd ed. (London: Macmillan, 1924), 38. In 1907, J. J. Thomson showed that only certain central orbits are stable for particles under Boscovich's oscillating field of attraction and repulsion, thus leading the way toward Bohr's discrete orbits. J. J. Thomson, *The Corpuscular Theory of Matter* (New York: Scribner's, 1907), 160–61.

13. L. L. Whyte says: "His kinematic method is easy to understand today, thanks to the work of Einstein, Eddington, Milne, and others, but it was so original that until 1922 [the date of J. M. Child's modernized translation from the Latin] the significance of Boscovich's transformation of Newtonian mechanics remained unrecognized." In the previous passage, Whyte alludes to Mach's relativistic reinterpretation of Newton's famous bucket experiment ostensibly demonstrating the reality of absolute space. "Boscovich and Particle Theory," *Nature*, Feb. 9, 1957 (179): 284–85.

14. Letter to Coleridge of 3 October 1832 (unsent), in Robert Perceval Graves, *Life of Sir William Rowan Hamilton* (Dublin: Dublin University Press, 1882), 593.

15. Ibid., 411. Qtd. in Robert Kargon, "William Rowan Hamilton and Boscovichean Atomism," *Journal of the History of Ideas* 26:1 (1965), 137. Quoting from Hamilton's 1832 Introductory Lecture on Astronomy, Brandon C. Yen remarks: "The 'imaginative' nature of physical science consists in its aim to establish an 'analogy' between physical phenomena and 'our own laws and forms of thought'"—a process Hamilton describes in quasi-shamanic terms, via Coleridge's "France: An Ode," as "darting our being through earth, sea, and air." "Poetry and Science: William Wordsworth and his Irish Friends William Rowan Hamilton and Francis Beaufort Edgeworth, *c.* 1829," *Romanticism* 26:1 (2020), 94–95.

16. In *Science and the Modern World* (p.4), Whitehead emphasizes the universality of Francis Bacon's claim in *Sylva Sylvarum*, based on contemporary theories of chemical affinity that anticipate the nineteenth-century kinetic theory of heat: "It is certain that all bodies whatsoever, though they have no sense, yet they have perception: for when one body is applied to another, there is a kind of election to embrace that which is agreeable, and to exclude or expel that which is ingrate; and whether the body be alterant or altered, evermore perception precedeth operation;

for else all bodies would be like one to another." *Works of Francis Bacon*, ed. James Spedding, Robert Leslie Ellis, and Douglas Denon Heath, 7 vols. (London: Longman, 1857–61), 2.602; qtd. slightly differently in Whitehead.

17. Milič Čapek, *Bergson and Modern Physics: A Re-Interpretation and Re-Evaluation* (Dordrecht: D. Reidel, 1971), 267; his italics. Isabelle Stengers makes the same point with respect to Whitehead's philosophy in *Thinking with Whitehead: A Free and Wild Creation of Concepts*, trans. Michel Chase (Cambridge: Harvard University Press, 2002), 172.

18. Milič Čapek, *The Philosophical Impact of Contemporary Physics* (Princeton: Van Nostrand, 1961), 259, also xiv, 224–25, 283, 294, 323. To glance ahead at the corresponding problem of how the macro-level Lockean self persists across lapses of memory or deep unconscious sleep, Blake's answer is that this self is reconstituted upon awakening through a discontinuous quantum event that is a resurrection, as in *Milton*'s conclusion and *Jerusalem*'s beginning (see chapter 5).

19. Sir Isaac Newton, *Opticks: or, A Treatise of the Reflections, Refractions, Inflections and Colours of Light*, 4th ed. (London: William Innys, 1730), 376.

20. Joseph Priestley, *Disquisitions Relating to Matter and Spirit* (London: J. Johnson, 1777), 22; all italics Priestley's.

21. Robert E. Schofield doubts Boscovich's influence on Priestley, but only because he thinks "most of the experimental implications of Boscovich's theory of matter were previously implicit in Newton and Hales," and specifically in "[Priestley's] reading, as a student at Daventry, of John Rowning's *Compendious System of Natural Philosophy*." "Joseph Priestley, Natural Philosopher," *Ambix* 14:1 (1967): 1–15.

22. Boscovich's introduction states: "The law of gravitation, decreasing in the inverse duplicate ratio of the distances, demands that there should be an attraction at very small distances, & that it should increase indefinitely. However, I show that the law is nowhere exactly in conformity with a ratio of this sort . . . ; nor, I assert, can a law of this kind be deduced from astronomy, that is followed with perfect accuracy even at the distances of the planets & the comets, but one merely that is [sic] at most so very nearly correct, that the differences from the law of inverse squares is very slight" (23). See also Arnold Thackray, "'Matter in a Nut Shell': Newton's *Opticks* and Eighteenth-Century Chemistry," *Ambix* 15:1 (1968): 29–53.

23. The *OED* dates the first English use of *incunabulum* to 1861, but it seems reasonable to assume that Blake—well-versed in book history, and a former apprentice to James Basire, member of the Society of Antiquaries—knew the word from Latin and Italian art books, which typically use the form, *incunabulus*. The symbolic book-baby "A Cradle Song" also invites comparison with Gutenberg's original forty-two-line Bible, which was not a wholly printed book; blank spaces were left after printing for rubrication (red-lettering) and the elaboration of initial letter flourishes drawn by hand, a process known as illumination. Blake's touching up of his poem's printed page is enacted within the poem when the mother's empathetic eye redraws and completes the divine image already touching her—an image of the

reader's reciprocation of this dyadic process. The incunabulus of "A Cradle Song" can therefore be seen to contrast with the incubus in Fuseli's sensational painting "The Nightmare" (1781), which chiefly supplied the demon behind the infant "fiend" in "Infant Sorrow." See my "Small Room for Judgment: Geometry and Prolepsis in Blake's 'Infant Sorrow,'" *European Romantic Review* 31:2 (2020): 129–55.

24. Similarly, the goal of Hartley's "truly philosophical Language" is "Progress in pure unmixed Happiness," such that we might "both give to and receive from each other Happiness indefinitely" (1.320). "Mind wandering" names a new field of associationism in contemporary cognitive psychology. Explains Thomas Metzinger, mind wandering occurs when "attentional agency is 'hijacked' by representational processes not focused on the here and now any more [*sic*], . . . leading to stimulus-independent trains of thoughts, for example during episodes of 'daydreaming' or 'zoning out' while reading." "Why are dreams interesting for philosophers? The example of minimal phenomenal selfhood, plus an example for future research," *Frontiers in Psychology* 4 (October 2013), 746.

25. James Barry, *Works*, ed. Edward Fryer, 2 vols. (London: T. Cadell and W. Davies, 1809), 2.585–86.

26. David Francis Taylor, "Picturing Ekphrasis: Image and Text in Shakespeare Painting," *European Romantic Review* 33:4 (2022), 472. Taylor cites the preceding Barry passage.

27. Erwin Panofsky, *Perspective as Symbolic Form*, trans. Christopher S. Wood (New York: Zone Books, 1992), 30, 27.

28. Northrop Frye, *Fearful Symmetry: A Study of William Blake* (Princeton: Princeton University Press, 1947), 350.

29. I've borrowed the thought and phrasing behind the last two sentences from Steven Shaviro's discussion of A. N. Whitehead in *The Universe of Things: On Speculative Realism* (Minneapolis: University of Minnesota Press, 2014), chap. 7, "Aisthesis." "What are we to make of the rampant and unapologetic aestheticism of the late Whitehead?" asks Shaviro. He answers that, for Whitehead, "beauty involves an immediate excess of sensation: something that stimulates thinking but that cannot be contained in, or expressed by, any particular thought."

30. William Porterfield, "An essay concerning the motions of our eyes. Part 1. Of their external motions," *Edinburgh Medical Essays and Observations* 3 (1737), 184–86. Qtd. in Nicholas J. Wade, "The Vision of William Porterfield," in *Brain, Mind and Medicine: Essays in Eighteenth-Century Neuroscience*, ed. Harry A. Whitaker, C. U. M. Smith, and Stanley Finger (New York: Springer, 2007), 168.

31. Richard Allen explains: "Readers of the *Observations* would have been familiar with the words 'decompound' and 'decomposite'—both from the late Latin *decompositus*, a rendering of the Greek *parasynthetos*—in which the 'de-' prefix signifies 'repeatedly' or 'further.' . . . In Hartley's theory, the associations in a complex action or idea are synchronic, while the associations in a decomplex action or idea

are diachronic." "David Hartley," sect. 5, in *The Stanford Encyclopedia of Philosophy*, ed. Edward N. Zalta, https://plato.stanford.edu/archives/sum2020/entries/hartley/.

32. Richard Allen, *David Hartley on Human Nature* (Albany: SUNY Press, 1999), 226.

33. A. N. Whitehead, *Modes of Thought* (New York: Macmillan, 1938), 57.

34. After Bayes' death, his executor Richard Price revised and published his work as "An Essay Towards Solving a Problem in the Doctrine of Chances," *Philosophical Transactions of the Royal Society of London* 53 (1763): 370–418. Hartley summarizes the similar-sounding theory of "an ingenious Friend" in his section "On the Nature of Assent" (1.339). Richard Allen implies the allusion is to Bayes (*David Hartley on Human Nature* 244–46); other scholars have proposed Abraham de Moivre, Daniel Bernoulli, and possibly Nicholas Saunderson. It is perhaps worth pointing out here that relativity quantum theorists came to understand that classical causality never was a strict logical imperative so much as a guide to scientific theory based on rules about sequences of events—not a "law of nature" but only a regular "order of temporal events." As Morris Schlick pointed out, Heisenberg's uncertainty principle, which is an indeterminacy only of prediction, does not overturn causation but renders it probabilistic rather than mechanical. "Causation in Contemporary Physics (I)" (1931), trans. David Rynin, *British Journal for the Philosophy of Science* 12:1 (1961): 177–93. Bayes' theorem led the way. Schlick goes on to observe that Hume (if not Blake) would have concurred with this assessment. "Causation in Contemporary Physics (II)" (1931), trans. David Rynin, *British Journal for the Philosophy of Science* 12:2 (1961), 286–87.

35. Disillusioned with the French Revolution, Wordsworth gave the idea of culture as mankind's "second nature" a conservative, Burkean cast, as James Chandler demonstrates in *Wordsworth's Second Nature: A Study of the Poetry and Politics* (Chicago: University of Chicago Press, 1984). But there is nothing intrinsically conservative about the idea, which was also enlisted to support Revolutionary theories of man's perfectibility through social reform. Blake straddles the difference by regarding man's "second nature" as aboriginal, a God-given inheritance which, by promoting art and vision, can render him a "new man" (Col. 3:10).

36. See, among others, Heather Glen, *Vision and Disenchantment: Blake's Songs and Wordsworth's Lyrical Ballads* (Cambridge: Cambridge University Press, 1983).

37. Stanley Gardner, *Blake's Innocence and Experience Retraced* (New York: St. Martin's, 1986), 11–13.

38. Richard Price, *Four Dissertations* (London: T. Cadell, 1777), esp. 396–401, 413–16.

39. David Hume, *Enquiries concerning Human Understanding and concerning the Principles of Morals*, ed. L. A. Selby-Bigge, 3rd ed., rev. P. H. Nidditch (Oxford: Oxford University Press, 1975), 111, 113.

40. In *Milton*, the Evangelicals "Whitefield" and "Westley" invert Hume's sarcasm by incarnating the true miracle of faith itself: "shew us Miracles! / Can

you have greater Miracles than these? Men who devote / Their lifes whole comfort to intire scorn & injury & death" (*M* 22:62–23:2; E 118).

41. Contrary to what I've suggested elsewhere, Joseph Fletcher argues Blake's early *There Is No Natural Religion* is an attack not on Hume's atheistic *Dialogues concerning Natural Religion* via Priestley's *Letters to a Philosophical Unbeliever*, but on Priestley's *Letters* and Natural Religion via Hume's *Dialogues*. As my preceding remarks would suggest, plainly Fletcher is right. Always for Blake, "Corporeal Friends" are the real enemy (*M* 4:26; E 98; E 728). Fletcher, *William Blake as Natural Philosopher, 1788–1795* (New York: Anthem, 2022), chap. 1. Fletcher's introduction links Blake's "Neoplatonically influenced pantheism" to Green Romanticism and Romantic ecology, speculative realism, and current critiques of "reductive mechanism" in terms that accord well with the present study's twin reference points of Neoplatonism and Whitehead. See also Steven Shaviro's Whitehead-oriented discussion in *The Universe of Things*, esp. introduction, chaps. 1 and 2.

42. Thus, Priestley, Price's friend, criticized Swedenborg on similarly sensational grounds for propounding "a scheme of religion so visionary, and so destitute of all rational *evidence* . . . there is no appeal to *facts*, which any person may examine, such as miracles, obvious to the senses." *Letters to the Members of the New Jerusalem Church, Formed by Baron Swedenborg* (Birmingham and London: J. Johnson, 1791), xii, xiv; his italics. But from Blake's standpoint, Price, Priestley, and Hume are all in the same rationalist boat with respect to miracles.

43. The visionary observation of the early Karl Marx can no longer be resisted: "The forming of the five senses is a labor of the entire history of the world down to the present." "Private Property and Communism," in *Economic and Philosophic Manuscripts of 1844*, trans. Martin Milligan (Moscow: Progress, 1959), 46.

44. Angela Esterhammer, *Creating States: Studies in the Performative Language of John Milton and William Blake* (Toronto: University of Toronto Press, 1994), 12. Qtd. in my introduction.

45. *De utilitate credendi* 16.34, in J. H. S. Burleigh, *Augustine: Earlier Writings* (Philadelphia: Westminster, 1953), "The Usefulness of Belief," 320; his italics.

46. See Hume's early unsent letter to George Cheyne or Dr. John Arbuthnot of March 1734 in *Letters*, ed. J. Y. T. Greig, 3 vols. (Oxford: Clarendon, 1932). His account of how his moral reflections "serve to little other Purpose, than to waste the Spirits, the Force of the Mind meeting with no Resistance, but wasting itself in the Air, like our Arm when it misses its Aim" (1.14), appears to anticipate his later annihilation of causation and personal identity: "I never can catch *myself* at any time without a perception, and never can observe any thing but the perception." *A Treatise of Human Nature*, ed. L. A. Selby-Bigge, 2nd ed., rev. Peter Nidditch (1740; Oxford: Clarendon, 1978), 252; his italics. Without mechanical "Resistance," mind cannot make contact with the self, which must therefore be an illusion or, as today's neuropsychologists say, a "controlled hallucination." John Richetti connects Hume's March 1734 letter with his dramatization of his "forlorn solitude" and "melancholy" in the

conclusion to *A Treatise,* book 1, in "Empiricist Philosophers and Eighteenth-Century Autobiography," in *A History of English Autobiography,* ed. Adam Smyth (Cambridge: Cambridge University Press, 2016), 160.

47. John Locke, *An Essay concerning Human Understanding,* ed. Alexander Campbell Fraser, 2 vols. (New York: Dover, 1959), 123; his italics.

48. Thomas Reid, *Essays on the Intellectual Powers of Man* (1785), in *Works,* ed. William Hamilton, 8th ed., 2 vols. (Edinburgh: James Thin, 1895), 1.93.

49. A. N. Whitehead, *Process and Reality,* ed. David Ray Griffin and Donald W. Sherburne (1929; New York: Free Press, 1978), 283. Qtd. in Shimon Malin, "Whitehead's Philosophy and the Collapse of Quantum States," in *Physics and Whitehead,* ed. Timothy E. Eastman and Hank Keeton (Albany: SUNY Press, 2003), 78.

50. William James, *The Principles of Psychology,* 2 vols. in 1 (New York: Henry Holt, 1890), 2.605–11; James, "A Pluralistic Universe," in *William James: Writings 1902–10,* ed. Bruce Kuklick (New York: Library of America, 1987), 760.

51. Consider the following definition of "picture": "A pictorial representation is a formalism that has an isomorphic relation to the objects it represents such that the visualized structure of the representation corresponds to a similar structure in nature." Jan Faye, "Copenhagen Interpretation of Quantum Mechanics," sect. 6, *Stanford Encyclopedia of Philosophy,* ed. Edward N. Zalta, https://plato.stanford.edu/archives/win2019/entries/qm-copenhagen/. Blake's "Cradle Song" picture functions, similarly, to naturalize the spirituality of the song. At the same time, the reader-viewer sees the picture as, unmistakably, "a formalism."

52. Daniel Heller-Roazen, *Absentees: On Variously Missing Persons* (New York: Zone Books, 2021), 7–12.

53. Compare the last line's "may joy" with the dynamism of "God becomes as we are, that we may be as he is" (*NNR* [b]; E 3).

54. Alfred Lord, *The Singer of Tales* (Cambridge: Harvard University Press, 1960), 124.

55. The paradoxes of Bacon's legacy are well sorted by Joanna Picciotto, *Labors of Innocence in Early Modern England* (Cambridge: Harvard University Press, 2010), esp. 35–40, 129–30.

56. Joseph Priestley, *History and Present State of Electricity,* 4th ed. (London: J. Johnson, 1775), 443–44.

57. Joseph Priestley, *An Essay on the First Principles of Government* (London: J. Dodsley, 1771), 253, 562.

58. Allen, *David Hartley on Human Nature,* 191. In contrast, the idea of matter hit upon by Priestley's friend, the astronomer William Michell, and advocated by Priestley in *Disquisitions* alongside Boscovich's, is pure brickbat (26–27). One is not surprised that Priestley never developed a theory of mind or an epistemology.

59. Qtd. in Barbara J. Shapiro, *John Wilkins 1614–1672: An Intellectual Biography* (Berkeley: University of California Press, 1969), 192.

60. Garrett Stewart, *The Deed of Reading: Literature Writing Language Philosophy* (Ithaca: Cornell University Press, 2015), 41.

61. Giorgio Agamben, *The End of the Poem: Studies in Poetics*, trans. Daniel Heller-Roazen (Stanford: Stanford University Press, 1999), 109.

62. Samuel Palmer, *Letters*, ed. Raymond Lister (Oxford: Oxford University Press, 1974), 322; also qtd. in *Blake Records: Documents (1714–1841) concerning the Life of William Blake (1757–1827) and his Family*, ed. G.E. Bentley, Jr., 2nd ed. (New Haven: Yale University Press, 2004), 435.

63. Giorgio Agamben, *What Is Philosophy?*, trans. Lorenzo Chiaso (Stanford: Stanford University Press, 2017), sect. 19.

64. Thomas Taylor, Preface to *The Elements of the True Arithmetic of Infinites* (London: printed for the author, 1809), vi; his italics.

65. Samuel Taylor Coleridge, *Biographia Literaria*, ed. W. Jackson Bate, 2 vols. in 1 (Princeton: Princeton University Press, 1983), 1.111–12; his italics. The decomposition of perceptual objects into interconnected component parts described here by Coleridge resembles the "reflexive imagery" William Keach analyzes in *Shelley's Style* (New York: Methuen, 1984), 79–117.

66. See, for example, Simon Jarvis, "Prosody as Cognition," *Critical Quarterly* 40:4 (2003): 3–15. See also Jarvis, "Thinking in Verse," in *The Cambridge Companion to British Romantic Poetry*, ed. Maureen N. McLane and James Chandler (Cambridge: Cambridge University Press, 2008), 98–116.

67. See Yolton, *Thinking Matter*, 63–64, and passim. For a summary of how postmodern polysemy establishes "field" conditions that warrant a "topological" approach to reading texts, see Andrew Piper, "Reading's Refrain: From Bibliography to Topology," *ELH* 80:2 (2013): 373–99, esp. 378.

68. *Lycidas*, ll. 190–94, in *The Complete Poetry and Essential Prose of John Milton*, ed. William Kerrigan, John Rumrich, and Stephen M. Fallon (New York: Modern Library, 2007), 109–10. The element of mortality associated with this perhaps pilgrim-like progress is attested by *Lycidas*'s remembrance of the *Et in Arcadia ego* tradition of "Death in person" within earthly paradise. In Erwin Panofsky's translation: "Even in Arcady, there I am." "Et in Arcadia ego: Poussin and the Elegiac Tradition," in *Meaning in the Visual Arts* (New York: Anchor Books, 1955), 307. The fade-away endings of several Songs mentioned below indicate Blake was quite aware of this tradition.

69. I borrow the term "fractal lattice" from Allen, *David Hartley on Human Nature*, 87–88.

70. A. N. Whitehead *Adventures of Ideas* (New York: Macmillan, 1933), 275.

71. "By reasoning that the chemical elements are formed from stable configurations of these fundamental points, whose force curves combine following the usual parallelogram law, one could understand in a qualitative way how different chemical substances might have different force patterns and hence different reactive properties." Boscovich 98.

Chapter Three

1. See Amanda Jo Goldstein, *Sweet Science: Romantic Materialism and the New Logics of Life* (Chicago: University of Chicago Press, 2017), chap. 1. Goldstein's book is an original account of Romantic "neo-Lucretian figurative materialism," but her opening chapter on Blake's Mundane Egg swerves away from the physics of Lucretian epigenesis, atomic "self-organization," and cosmology into more familiar Romantic territory of vitalism, biological imagery, and embryology.

2. A. N. Whitehead, *Process and Reality*, ed. David Ray Griffin and Donald W. Sherburne (1929; New York: Free Press, 1978), 97.

3. Steven Shaviro, "'Striving with Systems': Blake and the Politics of Difference," *boundary 2* 10:3 (1982): 229–50.

4. Angela Esterhammer, "Calling into Existence: *The Book of Urizen*," in *Blake in the Nineties*, ed. Steve Clark and David Worrall (New York: St. Martin's, 1999), 122.

5. Proclus, *Elements of Theology*, Prop. 35, in *The Six Books of Proclus on the Theology of Plato*, trans. Thomas Taylor, 2 vols. (London: printed for the author, 1816), 1.326. I am paraphrasing from Christoph Helmig and Carlos Steel, "Proclus," sect. 3.1, in *The Stanford Encyclopedia of Philosophy*, ed. Edward N. Zalta, https://plato.stanford.edu/archives/fall2020/entries/proclus/.

6. Samuel Taylor Coleridge, *Lay Sermons*, ed. R. J. White (Princeton: Princeton University Press, 1972), 30.

7. As Paul Mann argued some time ago, Urizen is the Bible, the Book, books, bookishness, and literary criticism. "*The Book of Urizen* and the Horizon of the Book," in *Unnam'd Forms: Blake and Textuality*, ed. Nelson Hilton and Thomas A. Vogler (Berkeley: University of California Press, 1986), 49–68.

8. W. J. T. Mitchell, *Blake's Composite Art: A Study of the Illuminated Poetry* (Princeton: Princeton University Press, 1978), 160.

9. The visual paradox of Blake's plate 9 finds a present-day counterpart in the metaphysical mysteriousness of Anish Kapoor's sculptures coated, notoriously, with Vantablack, described by its manufacturer, the technology company Surrey Nanosystems, as "the darkest manmade substance," whose "unrivalled absorption from ultra-violet out beyond the terahertz spectral range" reduces the appearance of three-dimensional objects to 2-d. https://www.surreynanosystems.com/about/vantablack.

10. Joseph Priestley, *Disquisitions Relating to Matter and Spirit* (London: J. Johnson, 1782), 39; all italics Priestley's. I quote from 1782 because the first passage does not appear in the 1777 edition. In both instances, Priestley is quoting from his earlier *History and Present State of Discoveries Relating to Vision, Light, and Colours* (1772).

11. Coste reported Newton's proposal in the second edition of his translation of Locke's *Essay concerning Human Understanding*, in a long footnote to IV.x.18.

At the time of his conversation with Newton, Locke was a permanent guest at the country house of Sir Francis and Lady Masham, the daughter of the Cambridge Platonist Ralph Cudworth, where Newton was an occasional visitor and Coste the resident tutor to their son.

12. Jonathan Bennett and Peter Remnant, "How Matter Might First be Made," *Canadian Journal of Philosophy*, Supplementary Volume 4 (1978), 7.

13. John Locke, *An Essay concerning Human Understanding*, ed. Alexander Campbell Fraser, 2 vols. (Oxford: Clarendon, 1894), 2.321n–22n.

14. Gottfried Wilhelm Leibniz, *New Essays concerning Human Understanding*, ed. and trans. Alfred Gideon Langley (Chicago: Open Court, 1916), bk. 4, chap. 10, 510.

15. See Leibniz's third paper in *The Leibniz-Clarke Correspondence*, ed. H. G. Alexander (Manchester: Manchester University Press, 1956), 25–27. There are close parallels between Leibniz's view of space and Plato's account of the Receptacle in the *Timaeus*, discussed in chapter 5.

16. Robert Palter, *Whitehead's Philosophy of Science* (Chicago: University of Chicago Press, 1960), 213.

17. A. N. Whitehead, *Modes of Thought* (New York: Macmillan, 1938), 136.

18. A. N. Whitehead, *Science and the Modern World* (New York: Macmillan, 1925), 91.

19. *De gravitatione et aequipondio fluidorum*, in *Isaac Newton: Philosophical Writings*, ed. Andrew Janiak (Cambridge: Cambridge University Press, 2004), 69.

20. R. S. Woolhouse, "Reid and Stewart on Lockean Creation," *Journal of the History of Philosophy* 20:1 (1982), 88; his italics.

21. The original editors of Newton's *De gravitatione* note that, despite the unlocalized Deity exalted in the General Scholium to the *Principia*, the analogy Newton makes in *De gravitatione* "is almost tantamount to describing God as the soul of the world and the world as his body." *Unpublished Scientific Papers of Isaac Newton*, ed. A. R. and M. B. Hall (Cambridge: Cambridge University Press, 1962), 82. A prime source of such thought is, of course, the *Timaeus*, where Plato defines physiology as a kind of theology. Proclus says at the outset of his commentary: "It is necessary that true physiology should be suspended from theology, in the same manner as nature is suspended from the Gods." *The Commentaries of Proclus on the Timaeus of Plato in Five Books; containing a treasury of Pythagoric and Platonic physiology*, trans. Thomas Taylor, 2 vols. in 1 (London: printed for the author, 1820), 171. Taylor remarks: "That the design of the Platonic Timaeus embraces the whole of physiology, and that it pertains to the theory of the universe . . . from the beginning to the end, appears to me to be clearly evident" (1).

22. In *Jerusalem*, Blake portrays this collapse in terms of Newton's decomposition of white light into the rainbow, an emblem of materialism as in Plotinus. Eno, a Daughter of Beulah, creates Nature as a bulwark of repose for fallen Albion: "With awful hands she took / A Moment of Time, drawing it out with many tears

& afflictions / . . . / Into a Rainbow of Jewels and gold, a mild Reflection from / Albion's dread Tomb. Eight thousand and five hundred years / Is its extension" (*J* 48:30–37; E 197). Whereas Eternity is an energetic space-time of events, the phenomenal world in all its many-splendored beauty—that is, three-dimensional space and history—arises when the mind forfeits awareness of duration or temporal passage. In reality, that world's appearance, like the rainbow, is a function of the observer's position.

23. Whitehead explains his theory of "symbolic reference" in *Symbolism, Its Meaning and Effect* (New York: Fordham University Press, 1927). Here as throughout his writings he argues, "The truth is that our sense-perceptions are extraordinarily vague and confused modes of experience. Also there is every evidence that their prominent side of external reference is very superficial in its disclosure of the universe" (*Modes of Thought* 210). Whitehead is following William James, who argued for "the re-instatement of the vague to its proper place in our mental life." *The Principles of Psychology*, 2 vols. in 1 (New York: Henry Holt, 1890), 1.254. Whitehead's "causal efficacy" also appears indebted to Leibniz's *petites impressions*: "at every moment there is in us an infinity of perceptions . . . of which we are unaware because they are either too minute and too numerous, or else too unvarying. . . . [B]y virtue of these minute perceptions the present is big with the future and burdened with the past." Gottfried Wilhelm Leibniz, *New Essays on Human Understanding*, ed. Peter Remnant and Jonathan Bennett (New York: Cambridge University Press, 1996), 53, 55. Qtd. in Joseph Fletcher, "Leibniz, the Infinite, and Blake's Early Metaphysics," *Studies in Romanticism* 56:2 (2017), 143.

24. Says Whitehead: "There is a dual aspect to the relationship of an occasion of experience as one relatum and the experienced world as another relatum. The world is included within the occasion in one sense, and the occasion is included in the world in another sense. For example, I am in the room, and the room is an item in my present experience. But my present experience is what I now am" (*Modes of Thought* 224). Blake shares Whitehead's "pre-Kantian," also pre-Cartesian, assumption "that the experience enjoyed by an actual entity is that entity *formaliter* . . . in respect to those forms of its constitution whereby it is that individual entity with its own measure of absolute self-realization" (*Process and* Reality 51; Whitehead's italics). Beyond Berkeley and Chris Townsend, discussed in chapter 6, the term "minute particular" expresses Blake's Thomistic sense that particulars and the universe exist in a mutually constituent relation.

25. Isabelle Stengers, *Thinking with Whitehead: A Free and Wild Creation of Concepts*, trans. Michel Chase (Cambridge: Harvard University Press, 2002), 294–96.

26. Stanley Bates, "The Mind's Horizon," in *Beyond Representation: Philosophy and the Poetic Imagination*, ed. Richard Eldridge (Cambridge: Cambridge University Press, 1996), 169.

27. Not to derogate from Newton, whose successor to the Lucasian Chair at Cambridge remarked: "Sir Isaac, in mathematicks, could sometimes see almost

by intuition, even without demonstration." *Memoirs of the Life and Writings of Mr. William Whiston*, 2nd ed. (London: J. Whiston and B. White, 1753), 35.

28. Joseph Priestley, *History and Present State of Discoveries Relating to Vision, Light, and Colours* (London: J. Johnson, 1772), 711–12. See also Robert Smith, *A Compleat System of Opticks* (Cambridge: printed for the author, 1738), 7.

29. Garrett Stewart, *The Look of Reading: Book, Painting, Text* (Chicago: University of Chicago Press, 2006), 72.

30. David Worrall, "William Blake and Erasmus Darwin's *Botanic Garden*," *Bulletin of the New York Public Library* 78:4 (1974–75), 409–10. Martin Priestman summarizes Darwin's thought in terms that help to explain his considerable influence on Blake:

> At every level, Darwin's science works to integrate the apparent stasis of physical matter with the idea of ceaseless energy and motion. Arguably, all theories of matter have to do this to a certain extent, but Darwin's sense of a continuous evolution from the primal explosion to the constant attraction-repulsion interplay energizing all matter at the atomic level, and thence to the active "volitions" connecting us to the rest of the organic world from our simplest sense perceptions to the most refined sympathies and the "chef d'oeuvre" of sex, is extraordinary in its sweep and range. But if this vision of Nature as endlessly creative—rather than simply created—is the main subject of his poetry, he is not blind to the ways in which this very creativity also entails massive destruction. *The Poetry of Erasmus Darwin: Enlightened Spaces, Romantic Times* (London: Routledge, 2013), 136.

See also Alan Bewell: "For Darwin, nature was inseparably bound up with novelty, fashion, and change. Whereas most of Darwin's contemporaries viewed 'novelty' as a rudimentary and fleeting initial aspect of the appreciation of art, Darwin made it central to his aesthetics." "Erasmus Darwin's Cosmopolitan Nature," *ELH* 76:1 (2009), 21.

31. William Herschel, "Catalogue of a Second Thousand of New Nebulae and Clusters of Stars; With a Few Introductory Remarks on the Construction of the Heavens," *Philosophical Transactions of the Royal Society of London* 79 (1789), 225.

Chapter Four

1. Robert D. Denham, *Northrop Frye and Others: Volume III: Interpenetrating Visions* (Ottawa: University of Ottawa Press, 2018), chap. 6. Denham says Frye eventually read David Bohm's *Wholeness and the Implicate Order*, with its claim that "in the implicate order, the totality of existence is enfolded within each region

of space (and time)." *Wholeness and the Implicate Order* (New York: Routledge, 1980), 219. The present study is hardly the first to investigate Blake in relation to relativity. Twenty-five years ago, Mark Lussier remarked: "Apparently, Blake's visionary thrust leads inward while the endeavors of cosmologists lead outward, yet any careful reading in either field immediately encounters the paradox inherent in visionary poetics and theoretical physics; the inner leads to the outer and vice versa." "Blake's Vortex: The Quantum Tunnel in *Milton*," *Nineteenth Century Contexts* 18:3 (1994), 264. Lussier regards the Vortex as a wormhole in time and space but does not address its topology or dimensions. Elsewhere, Adam Cohen argues that the "perspectival lightness" of *The Marriage of Heaven and Hell* reflects "a theory of metaphysical relativity" that parallels Einstein's "theory of relationships in which no reference frame was given epistemological preference." "Genius in Perspective: Blake, Einstein, and Relativity," *The Wordsworth Circle* 31:3 (2000), 166.

2. Northrop Frye, "The Road of Excess," in *The Stubborn Structure: Essays on Criticism and Society* (London: Routledge, 1970), 163.

3. In contrast to the static synchronism Frye imagines, Paul Younquist argues convincingly: "The stable structure of *Jerusalem* appears to vanish with the activity of reading, which multiplies structures as the poem unfolds over time. . . . In the barely visible gap that opens within 'spatial form' to define its self-identity [via repetition] can be glimpsed its diachronic origins in the *process* of reading the poem." "Reading the Apocalypse: The Narrativity of Blake's *Jerusalem*," *Studies in Romanticism* 34:3 (1993), 604; his italics.

4. Steven Shaviro remarks, "The notion of coherence is so important to Whitehead that he defines it on the first page of *Process and Reality*." *Other Criteria: Kant, Whitehead, Deleuze, and Aesthetics* (Cambridge: MIT Press, 2009), 107. Whitehead explains: " 'Coherence,' as here employed, means that the fundamental ideas, in terms of which the scheme is developed, presuppose each other so that in isolation they are meaningless. This requirement does not mean that they are definable in terms of each other; it means that what is indefinable in one such notion cannot be abstracted from its relevance to the other notions." Against abstraction, coherence affirms that "all actual entities are in the solidarity of one world." *Process and Reality*, ed. David Ray Griffin and Donald W. Sherburne (1929; New York: Free Press, 1978), 3, 67. Whereas Hume destroyed causal relatedness by restricting perception to sense data, Whitehead's *Science and the Modern World* portrays "the Romantic reaction" of Wordsworth, Shelley, and their successors as resisting "the bifurcation of nature" into a world of mathematics apart from basic human intuitions. Like Einstein, Whitehead regarded the chief Humean threat to "coherence" in his own day as the antirealist or instrumentalist (so-called "Copenhagen") interpretation of quantum mechanics that came to be parodied in the 1950s slogan, "Shut up and calculate."

5. See A. N. Whitehead, *Science and the Modern World* (New York: Macmillan, 1925), 60–70. According to Robert D. Denham, *Northrop Frye and Others*, this book was a key influence on Northrop Frye. See also John Herman Randall,

Jr., *The Career of Philosophy: From the Middle Ages to the Enlightenment* (New York: Columbia University Press, 1962), 575–94, esp. 585–89. Aultian Blake's "response to Newton" parallels Whitehead's response, in *Process and Reality*, to Bacon, Descartes, Locke, Spinoza, Berkeley, and Hume when he identifies inconsistencies in their work that prefigure his own "philosophy of organism," but that were ignored by the main empirico-rationalist tradition; see pp. 73, 51–60, and 130–83.

6. Donald D. Ault, *Visionary Physics: Blake's Response to Newton* (Chicago: University of Chicago Press, 1974), 1; his italics.

7. Niels Bohr, *Causality and Complementarity: Supplementary Papers*, in *The Philosophical Writing*, ed. Jan Faye and Henry J. Folse, 4 vols. (Woodbridge: Ox Bow, 1998), 4.152. It is worth noting that the renowned visualizations or thought experiments (*Gedankenexperiment*) that led Einstein to special relativity availed him little in developing the general theory. As early as 1912, Einstein in a much-cited letter remarked: "I have gained great respect for mathematics, whose more subtle parts I considered until now, in my ignorance, as pure luxury! Compared with this problem, the original theory of relativity is childish." Letter to Arnold Sommerfeld, 29 October 1912, in *The Collected Papers of Albert Einstein*, vol. 5, *The Swiss Years: Correspondence, 1902–1914*, ed. Martin J. Klein, A. J. Kox, and Robert Schulmann (Princeton: Princeton University Press, 1993), doc. 421.

8. I take these phrases from Milič Čapek, *The Philosophical Impact of Contemporary Physics* (Princeton: Van Nostrand, 1961), 259, also xiv, 224–25, 283, 294, 323, and passim; he footnotes Reichenbach and Poincaré in support. Elsewhere, Čapek speaks of "our Euclidian kinetic-corpuscular subconscious which is the depositary of our daily individual, as well as ancestral, experience." *Bergson and Modern Physics* (New York: Humanities Press, 1971), 58; here he footnotes Bachelard. The empiricism of Poincaré's argument that the intuitions and experiences underpinning traditional geometry as a science have been determined by natural selection loosely parallels Blake's historicist critique of perspective in *The Book of Urizen*, based on the German higher criticism of the Bible. See Jerome J. McGann, "The Idea of an Indeterminate Text: Blake's Bible of Hell and Dr. Alexander Geddes," *Studies in Romanticism* 25:3 (1986): 303–24.

9. Maurice Merleau-Ponty, *The Visible and the Invisible*, trans. Alphonso Lingis (Evanston: Northwestern University Press, 1968), 141, 142, 146. Whitehead was an important influence on Merleau-Ponty's late work, including his lectures on nature at the Collège de France beginning in 1956. See William S. Hamrick and Jan Van Der Veken, *Nature and Logos: A Whiteheadian Key to Merleau-Ponty's Fundamental Thought* (Albany: SUNY Press, 2011), 1–8.

10. C. S. Lewis, *The Discarded Image: An Introduction to Medieval and Renaissance Literature* (Cambridge: Cambridge University Press, 1964), 119, 116; his italics. Qtd. in William Egginton, "On Dante, Hyperspheres, and the Curvature of the Medieval Cosmos," *Journal of the History of Ideas* 60:2 (1999), 202.

11. Northrop Frye expresses this notion when he writes, "Perception . . . is a mental act. Yet it is equally true that the legs do not walk, but that the mind walks the legs." *Fearful Symmetry: A Study of William Blake* (Princeton: Princeton University Press, 1947), 19.

12. Northrop Frye was obviously thinking of Gamow's humanized, inside-out four-dimensional universe when, working off "the medieval scheme, or at any rate . . . Dante's version of it," he wrote: "In the anagogic phase, literature imitates . . . the thought of a human mind which is at the circumference and not the center of its reality. . . . When we pass into anagogy, nature becomes, not the container, but the thing contained. . . . Nature is now inside the mind of an infinite man who builds his cities out of the Milky Way." *Anatomy of Criticism* (Princeton: Princeton University Press, 1957), 116, 119. See also Frye, *Fearful Symmetry*, 384–85. Dante lays the basis for this inversion in *Inferno* canto 34 when the pilgrim and Virgil cross over Satan, the center of gravity, and must reverse themselves in order to continue their journey beyond Hell. Blake's notes to his diagram of the Circles of Hell (Butlin 812.101) show he perfectly understood this reversal (E 690).

13. Geoffrey Hartman analyzed Wordsworth's "spot syndrome . . . [and] obsession with specific places" in terms of Jungian "centroversion" and "the process of self-discovery or individuation." *Wordsworth's Poetry 1787–1814* (New Haven: Yale University Press, 1971), 84–86, and passim. Thereby, he set Wordsworth's ostensible fear of apocalypse against Blake's embracing of it, laying the way for much invidious comparison of the two poets. But Blake, too, had use for spots in the form of minute particulars, localities, and neighborhoods.

14. *Essay on Man* 1:289–92, in *Complete Poetical Works of Alexander Pope*, ed. Bliss Perry (Boston: Houghton Mifflin, 1903), 141.

15. C. S. Lewis, *The Allegory of Love: A Study in Medieval Tradition* (New York: Oxford University Press, 1958), 45. Or, as Erich Auerbach puts it in his essay, "Figura," "It is we who are the shadows." *Scenes from the Drama of European Literature* (New York: Meridian Books, 1959), 2. Lewis's allegory/symbolism dichotomy is an instance of Romantic Ideology that harkens back to Coleridge; I recall it purely for heuristic purposes. See A. D. Nuttall's criticism: "In tying allegory so closely to the task of representing the inner life, Lewis was, perhaps, bringing it closer to sacramentalism [and symbolism] than he knew." Since the passions are mental actions whose reality remains elusive and indefinite—Lewis simply places them as "the given"—their expression by means of allegory is therefore only relatively more literal and less metaphorical or intellectual than symbolism is. *Two Concepts of Allegory: A Study of Shakespeare's "The Tempest" and the Logic of Allegorical Expression* (New York: Barnes & Noble, 1967), 16–20.

16. "On the Morning of Christ's Nativity," ll. 109–14, in *The Complete Poetry and Essential Prose of John Milton*, ed. William Kerrigan, John Rumrich, and Stephen M. Fallon (New York: Modern Library, 2007), 23.

17. Thomas Banchoff, introduction to Edwin Abbott, *Flatland: A Romance of Many Dimensions*, 2nd rev. ed. (Princeton: Princeton University Press, 1991), xx.

18. Mother M. Christopher Pecheux, "The Image of the Sun in Milton's 'Nativity Ode,'" *Huntington Library Quarterly* 38:4 (1975): 315–33.

19. *St. Augustine: Sermons for Christmas and Epiphany*, Ancient Christian Writers 15, trans. Thomas Lawler (London: Paulist, 1952), 104. Qtd. in Pecheux, 320–21.

20. Giles Fletcher, "Christs Triumph after Death," ll. 97–104, in *Christs victorie, and triumph in Heaven, and earth, over, and after death* (Cambridge: C. Legge, 1610), 70–71.

21. "Swift as thought" describes simulacra in *Titus Lucretius Carus, His Six Books of Epicurean Philosophy*, trans. Thomas Creech, 4th ed. (London: T. Braddyll, 1699), bk. 4, p. 107, also bk. 1, p. 14 as follows:

> Let two broad bodies meet and part again,
> The Air must fill the space that's left between;
> Yet tho suppos'd it [the Air] flies as swift as thought,
> E'en common sense denies it can be brought
> O're all at once; the nearest first possesst,
> And thence 'tis hurried on, and fills the rest.

When Blake punningly short-circuits his description of Milton's descent by saying it occurs "swift as the swallow or swift" (*M* 15:8; E 110), he demonstrates, similarly to Fletcher the Younger, that its sheer physical movement defies visualization. The homology, "swift as the . . . swift," instantiates Lucretius's observation of the fall of atoms and the trajectories of simulacra: "Their race is finish'd when begun" (bk.1, p. 39; also bk. 4, p. 108). Contrast with this the absolute space wherein *Paradise Lost*'s supreme deity creates heaven and earth instantaneously by decree, independent of sensory perception: "Immediate are the Acts of God, more swift / Than time or motion" (7:176–77). Thomas Nail examines Lucretius's figurative language in relation to movement, change, and growth in *Lucretius I: An Ontology of Motion* (Edinburgh: Edinburgh University Press, 2019).

22. Charles Howard Hinton, "What Is the Fourth Dimension?" Qtd. in *The Annotated "Flatland: A Romance of Many Dimensions,"* by Edwin A. Abbott, ed. Ian Stewart (New York: Perseus, 2002), xxii.

23. Andrew Marvell, *The Complete Poems*, ed. Elizabeth Story Donno (New York: Penguin, 1972), 51. Blake quotes Marvell's poem in "The Ecchoing Green," line 27.

24. A similar sphere within a sphere arises when Marvell's Drop of Dew, "recollecting its own light, / Does, in its pure and circling thoughts, express / The greater heaven in an heaven less" (ll. 24–26).

25. Cantos 12:268, 8:55–60, in Edward Benlowes, *Theophilia, or Loves Sacrifice, a Divine Poem* (London: R. N., 1652), 230, 111; his italics. The first-cited

passage proclaims God's incomprehensibility in similar terms: "Search Empires Dawn, unwinde Times Ball again, / . . . / See—*All's vain* (12:268–70). Compare the "golden string" and "ball" of *Jerusalem*'s address "To the Christians" (*J* 77; E231).

26. William Hogarth, *The Analysis of Beauty* (London: J. Reeves, 1753), 7–8.

27. Hannes Ole Matthiessen, "Who Placed the Eye in the Centre of a Sphere? Speculations about the Origins of Thomas Reid's Geometry of Visibles," *Journal of Scottish Philosophy* 14:3 (2016), 240.

28. In his paper of 1838, Charles Wheatstone argued that "the most vivid belief of the solidity of an object of three dimensions arises from two different perspective projections of it being simultaneously presented to the mind." He notes Robert Smith had already described at length a spontaneous stereoscopic experience in his *Compleat System of Opticks* (1738), which I earlier suggested Blake read. Nor had Smith's account escaped Priestley's notice sixty years earlier in his *History of Discoveries Relating to Vision*, along with another by James Jurin (669–72). Wheatstone further remarks that in parallel-eyed gazing there is "no difference between the visual appearance of an object in relief and its perspective projection on a plane surface" (371); thus, a pictorial representation of distant objects sometimes may be mistaken for reality, so long as any surrounding circumstances that might disturb the illusion are excluded. He mentions "the Diorama" as an example of this. "Contributions to the Physiology of Vision.–Part the First. On some remarkable, and hitherto unobserved, Phenomena of Binocular Vision," *Philosophical Transactions of the Royal Society of London* 128 (1838), 380, 371. Might Blake's many noticeably rectilinear, diorama-like illustrations anticipate Wheatstone by goading viewers to recognize the mind's power to construct the illusion of a three-dimensional world out of two-dimensional materials—and thence, by analogy, a visionary four-dimensional world out of the materials of ordinary three-dimensional seeing?

29. George Berkeley, *Works on Vision*, ed. Colin Murray Turbayne (New York: Bobbs-Merrill, 1963), sect. 151.

30. That said, the contrast between Catholic Dante's corporeal Virgil and Blake's inward "vehicular" Milton is no less striking.

31. For example, Milton's proclamation of his intended self-annihilation remains deeply contradictory: "I in my Selfhood am that Satan: I am that Evil One! / He is my Spectre! in my obedience to loose him from my Hells / To claim the Hells, my Furnaces, I go to Eternal Death (*M* 14:30.–31; E 108). Absent awareness of the wider human potentiality represented by Ololon, Milton's classical-heroic "I . . . I . . . I" remains initially contaminated by rigid Calvinistic "obedience" and the egotism of the Miltonic Satan's "Which way I fly is Hell; myself am Hell" (4:75).

32. Commentators note that at *Paradiso* 13:101–02 and 17:15–17 Dante seems to go out of his way to put Euclid down.

33. Mark Peterson, "Dante and the 3-Sphere," *American Journal of Physics* 47:12 (1979): 1031–34.

270 / Notes to Chapter Four

34. Dante Alighieri, *Paradiso* 27:112–14, in *Purgatory and Paradise*, trans. Henry Cary, ed. Henry C. Walsh (Philadelphia: H. Altemus, 1889), 297.

35. Regarded in four-dimensional terms, Blake's Dante design resolves Edward Young's Euclidean bafflement on confronting a similar diagrammatic image of the universe, in lines Blake knew:

> Orb above orb ascending without end!
> Circle in circle, without end, enclosed!
> Wheel, within wheel; Ezekiel! like to thine!
> Like thine, it seems a vision or a dream;
> . . .
> What then the wondrous space through which they roll?
> At once it quite ingulfs all human thought;
> 'T is Comprehension's absolute defeat.

The Complaint: or, Night-Thoughts on Life, Death & Immortality, 3rd ed. (London: William Tegg, 1859), 9:1095–1105.

36. Needless to say, Blake's association of temporal change with Jesus and the freedom of mercy and forgiveness runs counter to Dante's retributive justice, which he condemned (E 690).

37. Irene Samuel remarks that Nativity Ode ll. 173–80 alludes to *Paradiso* 17:31–33, in *Dante and Milton: The "Commedia" and "Paradise Lost"* (Ithaca: Cornell University Press, 1969), appendix A.

38. I simplify. It is Milton's studied refusal to choose between Ptolemaic and Copernican astronomies that leads him to avoid describing concretely many of the poem's different places and spaces. Blake, however, seems to think Milton misses the point that "space" and "place" alike reside in the body and soul of Albion (see chapter 5). For a phenomenology-oriented analysis of body as the source of place in *Paradise Lost*, see John Gillies, "Space and Place in *Paradise Lost*," *ELH* 74:1 (2007): 27–57.

39. Jacob Bronowski, *William Blake, 1757–1827: A Man without a Mask* (New York: Haskell House, 1944), 98. Ault's *Visionary Physics* skirts this Bronowski passage on page 1, where Ault insists his focus will be on "structure," not the shape shiftings of topology. In a note, Bronowski goes on to quote a letter in which the mathematician M. H. A. Newman applies the method of "slicing" to demonstrate that Blake's cosmos is "the solid analogue of the surface of a sphere" (148).

40. Egginton, 199.

41. These hills feature prominently in the verses Blake enclosed in his second letter to Butts of 22 November 1802, "With happiness stretchd across the hills" (E 720–22), which describe an apocalyptic union with Los that looks like a forerunner of the one in *Milton*.

42. Alexander Gilchrist, *Life of William Blake*, 2 vols. (London: Macmillan, 1863), 1.371.

43. Qtd. in Carlo Rovelli, *Reality Is Not What It Seems: The Journey to Quantum Gravity*, trans. Simon Carnell and Erica Segre (New York: Riverhead, 2017), 121.

44. George Berkeley, *Alciphron*, Fourth Dialogue, in *Works on Vision*, 108.

45. Erwin Panofsky, *Perspective as Symbolic Form*, trans. Christopher S. Wood (New York: Zone Books, 1992), 53.

46. Mentioned by Francis Seaman, "Whitehead and Relativity," *Philosophy of Science* 22:3 (1955), 225.

47. Čapek, *Philosophical Impact*, 19; his italics.

48. Hermann Minkowski, "Space and Time" (1908), in *The Principle of Relativity: A Collection of Original Memoirs on the Special and General Theory of Relativity*, trans. W. Perrett and G. Jeffrey (London: Methuen 1923), 76. Qtd. in Florian Cajori, "Origins of Fourth Dimension Concepts," *American Mathematical Monthly* 33:8 (1926), 405.

49. Blake calls it "a Female Space" because he bases it on Plato's Receptacle of becoming, which *Timaeus* 50d likens to the "mother" or "wet nurse" of the cosmos whose father is the Demiurge; see chapter 5.

50. A. A. Robb, *Geometry of Time and Space* (Cambridge: Cambridge University Press, 1936), 5.

51. Frye, 377. Qtd. in Harold Bloom, *Blake's Apocalypse* (Ithaca: Cornell University Press, 1963), 310. Paul Weiss, "Contemporary World," *Review of Metaphysics* 6:3 (1952): 526–39. Qtd. in Čapek, *Philosophical Impact*, 218.

52. David V. Erdman, *Blake: Prophet against Empire*, 3rd ed. (Princeton: Princeton University Press, 1977), 423.

53. *Enchiridion Metaphysicum* (1671) chaps. 27 and 28, trans. into English in Joseph Glanvill, *Saducismus Triumphatus* (1681), in *Philosophical Writings of Henry More*, ed. Flora Isabel MacKinnon (London: Kessinger, 1925), 213.

54. Henry More, *The Immortality of the Soul* (London: J. Flesher, 1659), 13.

55. See Jasper Reid, *The Metaphysics of Henry More* (London: Springer, 2012), 203–05; also, his "The Evolution of Henry More's Divine Absolute Space," *Journal of the History of Philosophy* 45:1 (2007): 79–102.

56. Yeats's illustrations appear in *A Critical Edition of Yeats's "A Vision,"* ed. George Mills Harper and Walter Kelly Hood (1925; London: Macmillan, 1978), 130, 131. Kathleen Raine was not far off when she claimed, "Yeats's natural aptitude" for the "diagrammatic type of thought" about contrary forces "qualified him to understand Blake better than any commentator since." *Yeats the Initiate: Certain Essays on Certain Themes in the Work of W. B. Yeats* (Portlaoise, IRE: Dolmen, 1986), 110. As Yeats well knew, the geometry of the vortex has held spiritual implications since ancient times. The lemniscate by which the Greeks represented infinity as an ouroboros, symbol of reality as "Continually building, continually destroying" (*J*

88:40; E 247), is isomorphic with the Möbius strip, a two-dimensional surface whose twisted loop joins inside and outside. Stepping up in dimensions, the visually similar yin-yang symbol of the Dao, which depicts the interrelationship of light and dark, can be seen as a representation of two spheres trying to occupy the same space. If you imagine each area as a curved three-dimensional tornado or spindle torus being put through its opposite, then the resultant double vortex forms the outer surface of a four-dimensional sphere. In Western philosophy, one could say that the lemniscate and yin-yang symbol figure the underlying paradoxical and Contrary nature of monism (pantheism or panentheism) and dualism (Platonic Christianity). The two have always tended to define themselves in relation to each other—monism, by subsuming dualism into a higher unity founded on relationality, discrimination, and the liminal as expressed in the connective power of "the bounding line" (*DC*; E 550); dualism, by emphasizing the irreducible prior difference necessary for two things even to be in relation. Blake's pun on "bounding" shows his monistic inclination. The line not only establishes an object's boundaries as a body but bounds in doing so, thereby transcending static three-dimensional perspective and disclosing movement as an event-experience in real time.

57. *Psychozoia, or The life of the Soul*, 2.15, in *The Complete Poems of Dr. Henry More (1614–1687)*, ed. Alexander B. Grosart (New York: AMS, 1967), 20.

58. Reid, 320.

59. "The Particular Interpretation" of the 1646 ed. of *Democritus Platonissans*, rptd. in the 1647 ed. of the *Philosophical Poems*, in Grosart 160.

60. *Two Choice and Useful Treatises*, second part, 124 (*Annotations upon Lux Orientalis*, Chap. 14, 121). Qtd. in Reid, 359.

61. See Marjorie H. Nicolson, "The Spirit World of Milton and More," *Studies in Philology* 22:4 (1925): 433–52. Blake echoes *Paradise Lost*'s affirmation of angelic spissitude at 8:622–29 when he relates that in Eternity "Embraces are Cominglings: from the Head even to the Feet" (*J* 69:43; E 223).

63. See Christopher Heppner, *Reading Blake's Designs* (Cambridge: Cambridge University Press, 1995), 220–22.

64. David Hume, *A Treatise of Human Nature*, ed. L. A. Selby-Bigge, 2nd ed., rev. P. H. Nidditch (1740; Oxford: Clarendon, 1978), 319–20; his italics.

65. Blake, who invokes fairies throughout his work, probably recognized the vehicular hypothesis as folklore for intellectuals, a popular form of the anti-Cartesians' *tertium quid*. According to Robert Kirk, fairies "are said to be of a midle Nature betuixt Man and Angel"; they are visible in "light changeable best seen in Twilight"; and their "Bodies be so plyable thorough the Subtilty of the Spirits that agitate them, that they can make them appear or disappear att Pleasure." *The secret commonwealth of elves, fauns & fairies: a study in folk-lore & psychical research / the text by Robert Kirk . . . 1691; the comment by Andrew Lang* (London: D. Nutt, 1893), 5–6. Coleridge appears to have read Kirk's pamphlet, including the appended "Succint Accompt of My Lord Tarbott's Relations, in a Letter to the Honourable

Robert Boyle, Esquire, of the Predictions Made by Seers, Whereof Himself was Ear and Eye-witness." Traces of both writings are evident in *Christabel.*

66. A. N. Whitehead, *The Concept of Nature* (Cambridge: Cambridge University Press, 1920), 53. Following this quotation, I am paraphrasing a comment by Isabelle Stengers, *Thinking with Whitehead: A Free and Wild Creation of Concepts*, trans. Michel Chase (Cambridge: Harvard University Press, 2002), 72.

67. "On the Morning of Christ's Nativity," l. 94, in *The Complete Poetry and Essential Prose of John Milton*, ed. William Kerrigan, John Rumrich, and Stephen M. Fallon (New York: Modern Library, 2007), 22.

68. Compare Blake's pun in the letter to Hayley of 12 March 1804, "Engraving is Eternal Work" (E 743): at once the heroic labor of Eternity and building up Jerusalem, and the endless tedium of unremitting attention to detail.

69. William James, *The Principles of Psychology*, 2 vols. in 1 (New York: Henry Holt, 1890), 2.605–11. James, "A Pluralistic Universe," in *Writings 1902–10*, ed. Bruce Kuklick (New York: Library of America, 1987), 760. Whitehead adopts James's phrase in *Process and Reality* 18; elsewhere, he speaks of "pulses" or "throbs" of experience.

70. A. N. Whitehead, "Process and Reality," in *Essays in Science and Philosophy* (London: Rider, 1948), 127.

71. The Blakean Vortex—open at one end but irreversibly closed at the other—also recalls one of the oldest Greek myths about oracular prophecy. Hesiod relates that Zeus took two wives, Metis and Themis, both omniscient but in opposite ways that correspond to being and becoming. The classicists Marcel Detienne and Jean-Paul Vernant explain:

> The omniscience of Themis relates to an order conceived as already inaugurated and henceforth definitively fixed and stable . . . She spells out the future as if it were already written and since she expresses what will be as if it were what is, she gives no advice but pronounces sentence . . . Metis, by contrast, relates to the future seen from the point of view of its uncertainties; her pronouncements are hypothetical or problematic statements. She advises what should be done so that things may turn out one way rather than another; she tells of the future not as something already fixed but as holding possible good or evil fortunes. Marcel Detienne and Jean-Paul Vernant, *Cunning Intelligence in Greek Culture and Society*, trans. Janet Lloyd (Chicago: University of Chicago Press, 1991), 107–08.

Michael Wood, who quotes this passage, observes that in the end, "we see that Themis was right all along, but only because we and Metis played our parts, [and] took the chances that led to what now looks like necessity." *The Road to Delphi: The Life and Afterlife of Oracles* (New York: Farrar, Strauss and Giroux, 2003), 158.

The Blakean Vortex-Moment can be seen to "glue" the two perspectives together to create a cognitively dissonant crossroads or *kairos* where divine fate meets free will, and the already-written becomes subject to human interpretation and understanding.

72. Abraham Tucker [Edward Search], *The Light of Nature Pursued*, 6 vols. (London: T. Jones and T. Payne, 1768), 2.195.

Chapter Five

1. When Priestley writes that Cartesian dualists who speak of "the omnipresence of the Deity, . . . mean his power of *acting every where*, though he exists *no where*," he echoes More's criticism. Joseph Priestley, *Disquisitions concerning Matter and Spirit* (London: J. Johnson, 1777), 56. Qtd. in my introduction.

2. Henry More, *The Immortality of the Soul* (London: J. Flesher, 1659), 332; More's italics.

3. Kathleen Raine, *Blake and Tradition*, 2 vols. (London: Routledge & Kegan Paul, 1969). George Mills Harper, *The Neoplatonism of William Blake* (Chapel Hill: University of North Carolina Press, 1961).

4. The present chapter will demonstrate how far Blake's position is from the mystical dualism of Thomas Taylor's "Platonic Philosopher's Creed" (1805), which begins: "I BELIEVE in one first cause of all things, whose nature is so immensely transcendent, that it is even super-essential; and that in consequence of this it cannot properly either be named, or spoken of, or conceived by opinion, or be known, or perceived by any being." *Thomas Taylor, the Platonist*, ed. Kathleen Raine and George Mills Harper (Princeton: Princeton University Press, 1969), 439. What more could Calvin say?

5. So, it is not surprising that Whitehead, focused on the same problem of gravity as Blake, should exhibit Neoplatonic tendencies himself. Shimon Malin interprets the "overflowing" that results in Neoplatonism's emanations as a Whiteheadian dynamic process, in *Nature Loves to Hide: Quantum Physics and the Nature of Reality, a Western Perspective*, rev. ed. (Oxford: Oxford University Press, 2012), 191–243. See also David Rodier, "Alfred North Whitehead: Between Platonism and Neoplatonism," in *Neoplatonism and Contemporary Thought, Part One*, ed. R. Baine Harris (Albany: SUNY Press, 2002), 183–204.

6. Thus, John Rogers notes that Priestley's "immateriality of matter" recalls radical seventeenth-century monistic and vitalist conceptions of change as the product of spiritualized matter in motion, in contrast to the Calvinistic "voluntarist doctrine of God's predestinary manipulation of the course of human history." *The Matter of Revolution: Science, Poetry, and Politics in the Age of Milton* (Ithaca: Cornell University Press, 1996), 53–55. Robert E. Schofield emphasizes that "in Priestley there came together four of the five major streams in which Cambridge Neoplatonism flows into the eighteenth century," namely, Newton, Locke, Hartley, and the rationalist

idealism of the liberal dissenting academies. "Joseph Priestley, Eighteenth-century British Neoplatonism, and S. T. Coleridge," in *Transformation and Tradition in the Sciences: Essays in Honour of I. Bernard Cohen*, ed. Everett Mendelsohn (Cambridge: Cambridge University Press, 2003), 237–54.

    7. Critics tend to overlook how anachronistically they use the word "radical" in a context of severe government censorship where an author's refusal to make allowances could mean career suicide and jail time. See John Bugg, "How Radical was Joseph Johnson and Why Does Radicalism Matter?" *Studies in Romanticism* 57:2 (2018): 173–95. Johnson was imprisoned in November 1798 and finally sentenced in February 1799 to a further six months. Bugg notices that subsequently Johnson's publications tended to favor natural-science authors whose work, though sometimes "radical," the government apparently did not consider threatening. It seems likely Blake's intense involvement in science during the early 1800s was a deliberate redirection of his earlier political prophesizing. See also Leslie F. Chard II, "Joseph Johnson in the 1790s," *The Wordsworth Circle* 33:3 (2002): 95–100.

    8. Philip C. Almond, "Henry More and the Apocalypse," *Journal of the History of Ideas* 54:2 (1993), 190.

    9. Jasper Reid, *The Metaphysics of Henry More* (London: Springer, 2012), 241.

    10. Henry More, *An Explanation of the Grand Mystery of Godliness*, bk. 6, chap. 5, sect. 2, in *Theological Works* (London: Joseph Downing, 1708), 156; his italics.

    11. Henry More, *Two Choice and Useful Treatises* (London: James Collins, 1682), part 2, *Annotations upon Lux Orientalis*, chap. 14, 124. Qtd. in Reid, 359.

    12. A. N. Whitehead, *Process and Reality*, ed. David Ray Griffin and Donald W. Sherburne (1929; New York: Free Press, 1978), 150–51. Thus, Whitehead claimed in his philosophy, "The ancient doctrine that 'no one crosses the same river twice' is extended. No thinker thinks twice; and, to put the matter more generally, no subject experiences twice" (43).

    13. Alexander Koyré, *From the Closed World to the Infinite Universe* (Baltimore: Johns Hopkins University Press, 1957), 132.

    14. John Henry, "A Cambridge Platonist's Materialism: Henry More and the Concept of Soul," *Journal of the Warburg and Courtauld Institutes*, 49:1 (1986): 172–95. Blake would have appreciated Cassandra Gorman's insight that "[More's] poems are . . . the groundwork of his philosophical career." "Allegorical Analogies: Henry More's Poetical Cosmology," *Studies in Philology* 114:1 (2017), 150–51. Gorman claims that More's philosophy "associates allegory with the embodiment of ineffables that are at the heart of physical life, paradoxically both beyond and held open to sensorial perception," much as Psyche's veil "both hides the truth and, in the act of closely shrouding it, indicates its presence" (167). The same can be said of Blake's mediatory "symbols *about* symbols," the transparent skintight bodysuits his figures wear throughout his art, and Ololon's Jesus-enwrapping Garment at *Milton's* climax. Gorman cites Christopher Burlinson's point that Spenserian allegory is not

so much the sense beneath an image as "what enwraps the meaning, the image as well as that (or even rather than that) beyond it." *Allegory, Space and the Material World in the Writings of Edmund Spenser* (Cambridge: D. S. Brewer, 2006), 12, 14. As in Spenser's Bower of Bliss episode where Guyon encounters the lady "All in a vele of silke and siluer thin, / That hid no whit her alabaster skin, / But rather shewd more white" (2:12:77), the allegory extends the image of her skin by means of a Plotinean metaphor that dresses up, substantializes, and instantiates that image as such without quite rendering the skin visible, thereby generating the liminality of the veil (Gorman 165). Gorman's account of the role of "Vital Congruity" in More's Neoplatonic scheme also applies to Blake's emanationist mythology: "When every material substance and action is considered analogous to its perfect divine source, the life of the soul between the terrestrial and celestial realms begins to assume an allegorical form" (148). This is what Keats recognized when he wrote, "A Man's life of any worth is a continual allegory . . . Shakespeare led a life of Allegory; his works are the comments on it." Letter to George and Georgiana Keats, 11 February–3 May 1819, in *Keats: Poems and Selected Letters*, ed. Carlos Baker (New York: Scribner's, 1962), 460–61. As we'll see later in this chapter, Keats's phrase "continued allegory" appears in Tucker's definition of "Hypothesis" as "such a representation of things as may be the real case for anything that can be shown to the contrary." Abraham Tucker [Edward Search], *The Light of Nature Pursued*, 6 vols. (London: T. Jones and T. Payne, 1768), 2.5–6. Clearly, Keats and Blake both read this passage from Tucker.

15. "The Interpretation Generall: 'Body,'" in *The Complete Poems of Dr. Henry More (1614–1687)*, ed. Alexander B. Grosart (New York: AMS, 1967), 160.

16. See Jasper Reid, "The Evolution of Henry More's Theory of Divine Absolute Space," *Journal of the History of Philosophy* 45:1 (2007): 79–102.

17. *Henry More's Manual of Metaphysics* [*Enchiridion Metaphysicum*], trans. Alexander Jacob, 2 vols. (Hildesheim: Georg Olms, 1995), 1.69–70; More's italics.

18. Henry More, *Divine Dialogues* (London: Joseph Downing, 1713), 53–54.

19. In order: Henry More, *An Antidote Against Atheism*, in *A Collection of Several Philosophical Writings*, 4th ed. (London: Joseph Downing, 1712), 44; *Immortality of the Soul*, in *A Collection*, 223; *Antidote Against Atheism*, 46.

20. John Locke, *An Essay concerning Human Understanding*, ed. Alexander Campbell Fraser, 2 vols. (Oxford: Clarendon, 1894), 2.321n.–22n.

21. According to Whitehead's "ontological principle," "the general potentiality of the universe must be somewhere . . . this 'somewhere' is the non-temporal actual entity," God. But this is only God "in his primordial nature, alone with himself" (*Process and Reality* 46). Whitehead's qualifier is critical. Newton collapses the difference between God's actuality in his "consequent nature" as he becomes affected by Creation and God's merely "proximate relevance" to actualities as a "primordial mind," that is, a world of abstract Ideas or concepts.

22. John W. Yolton, *Thinking Matter: Materialism in Eighteenth-Century Britain* (Minneapolis: University of Minnesota Press, 1983), 64.

23. For example, J. M. McGuire identifies Henry More as the source of Newton's phrase while observing that More's identification of God with space, based on his idea that both are incorporeally extended, "is precisely what Newton rejects." "Ideas and Texts: Newton and the Intellectual History of Science," *Sartoniana* 24 (2011), 41, 47. Space may be God's sensorium, but it is not God himself.

24. Martial Gueroult, *Descartes' Philosophy Interpreted According to the Order of Reasons: The Soul and God, The Soul and the Body*, trans. Roger Ariew, 2 vols. (1954; Minneapolis: University of Minnesota Press, 1985).

25. Lisa Landoe Hedrick, *Whitehead and the Pittsburgh School: Pre-empting the Problem of Intentionality* (Lanham MD: Rowman & Littlefield, 2021), 8–11. My account of Whitehead's relation to Taylor in this paragraph is much indebted to Landoe. Relatedly, David Rodier points out that the traditional interpretation of the Forms as world-despising and timelessly ideal, and the concomitant rejection of Neoplatonism as a decadent religious corruption of Plato, originates in the efforts of the Arnolds and Jowett to construct a secularized Platonism and present Plato as "the patron of a liberal democratic state." "Alfred North Whitehead: Between Platonism and Neoplatonism," 184.

26. A. E. Taylor, *A Commentary on Plato's "Timaeus"* (Oxford: Clarendon, 1928), 60.

27. *Timaeus* 52b, in *Plato IX: Timaeus, Critias, Cleitophon, Menexenus, Epistles*, trans. R. G. Bury (Cambridge: Harvard University Press, 1981), 123.

28. As Plotinus puts it, matter "not being itself any form, is not itself seen," and so "becomes similar to an object in a mirror." *Select works of Plotinus, the great restorer of the philosophy of Plato: and extracts from the treatise of Synesius on Providence*, trans. Thomas Taylor (London: printed for the author, 1817), 160.

29. A. N. Whitehead, *Adventures of Ideas* (New York: Macmillan, 1933), 134. As Whitehead said in conversation: "Complete abstraction is impossible. There is a unity, an essential togetherness in the universe—this is the notion of the Receptacle." Joseph Gerard Brennan and Alfred North Whitehead, "Whitehead on Plato's Cosmology," *Journal of the History of Philosophy* 9:1 (1971), 76.

30. A. E. Taylor, *Plato: The Man and His Work* (London: Methuen, 1955), 456; see also 190n.1.

31. Surprisingly, it was Whitehead, with his interest in education, who coined the term "creativity" and gave it currency. See Paul Oskar Kristeller, " 'Creativity' and 'Tradition,' " *Journal of the History of Ideas* 44:1 (1983), 105. Not that Kristeller approved. Also see Steven Meyer, "Introduction" (special issue, "Whitehead Now"), *Configurations* 13:1 (2005), 2–4.

32. Even Francis Cornford, who took Taylor to task for his Whiteheadian bias, observed: "Neither the Form nor Space can act as the ultimate moving cause" of becoming because they are both immutable and eternal. Thus, "the moving cause can only be the World-Soul. It becomes more than ever difficult to resist the inference that the Demiurge is to be identified with the Reason in the World-

Soul." *Plato's Cosmology: The "Timaeus" of Plato*, trans. with commentary (London: Routledge, 1935), 197.

33. Hans Robert Jauss, *Aesthetic Experience and Literary Hermeneutics: Theory and History of Literature*, vol. 3, trans. Michael Shaw (Minnesota: University of Minnesota Press, 1982), 32. Qtd. in Thor Magnus Tangerås, *Literature and Transformation: A Narrative Study of Life-Changing Reading Experiences* (London: Anthem, 2020), chap. 1.

34. Giorgio Agamben, *Potentialities: Collected Essays in Philosophy*, trans. Daniel Heller-Roazen (Stanford: Stanford University Press, 1999), 218.

35. Isaac Newton, Scholium to Book 1, *Principia Mathematica*, 3rd ed., trans. Andrew Motte (Amherst: Prometheus, 1995), 15.

36. Samuel Taylor Coleridge, *Biographia Literaria*, ed. James Engell and W. Jackson Bate, 2 vols. in 1 (Princeton: Princeton University Press, 1983), 2.15.

37. Aristotle, *Metaphysics* IX.1.1046a12–13, in *Basic Works*, ed. Richard McKeon (New York: Modern Library, 2001), 820.

38. *Enneads* 2.4.5, in *Select Works of Plotinus, Thomas Taylor's Translation*, ed. G. R. S. Mead (London: G. Bell, 1914), 32–33.

39. Giorgio Agamben, *What Is Philosophy?*, trans. Lorenzo Chiaso (Stanford: Stanford University Press, 2017), sect. 19; partially quoted in my chapter 1. Ian Hacking examines the broader historical context within which Blake could have arrived at an Agambenian view of Plato's khora. Beginning in the late seventeenth-century, Hacking writes, "probability" and "possibility" began to possess an "essential duality . . . both epistemic and aleatory. Aleatory probabilities have to do with the physical state of coins or mortal humans. Epistemic probabilities concern our knowledge. . . . [I]n virtue of its ambiguities 'possibility' could usefully define an unclear concept of probability . . . as something jointly physical and epistemological." Hacking goes on to observe that "aleatory probability is *de re*, having to do with the physical characteristics of things, while epistemic probability is *de dicto*, for it concerns what we know and hence what can be expressed in propositions." *The Emergence of Probability*, 2nd ed. (Cambridge: Cambridge University Press, 1996), 122–24.

40. Noting that imagery of construction appears frequently in Descartes, a recent commentator emphasizes: "l'*absolu* ou le *fondement* qui permet la déduction de connaissances nouvelles n'est pas, pour Descartes, ce qui compose *réellement* les choses; il est plutôt ce dont dépend la *connaissance* de toutes les choses." Benoît Timmermans, "L'analyse cartésienne et la construction de l'ordre des raisons," *Revue Philosophique de Louvain* 94:2 (1996), 205; his italics. Contra Descartes, for Blake and Whitehead alike "the foundation" is a real order of *things*. As Jules Vuillemin pointed out some time ago, Descartes claimed in *Règles pour la direction de l'esprit* 12 ("quatrième lieu") that whereas I can conclude my own existence from the existence of God, I cannot conclude God's existence from mine. Thus, even though Descartes models the foundation of his method on ancient geometry, he

tacitly concedes that philosophical reasoning lacks the reversibility of mathematical reasoning, because the former is "asymétrique," "réflexive," and "transitive." "Propriétés formelles et matérielles de l'ordre cartésen des raisons," in *Études sur l'histoire de la philosophie en hommage à Martial Gueroult* (Paris: Fischbacher, 1964), 43–58; qtd. in Timmermans 210–11. This contradiction can be seen to originate in the Master argument of Diodorus Cronus regarding possibles, contingent propositions, and modal logic—an argument Plato, Aristotle, and Proclus, among others, saw as leading to an aporia. Timmermans observes that repeatedly in the *Règles*, "Descartes décrit cette rencontre, ce face-à-face de l'analyse avec un ordre absent, 'troublé,' interrompu, ou indirect" (211).

41. Donald Ault, *Narrative Unbound: Re-Visioning William Blake's "The Four Zoas"* (Barrytown, NY: Station Hill, 1987), 4, xi, xiii, xi.

42. Ault published his book just as debate about the poem's intents and purposes, success or failure, was coming to a head. See Paul Mann, "The Final State of *The Four Zoas*," *Blake: An Illustrated Quarterly* 18 (1985), 208. See also Peter Otto, "Final States, Finished Forms, and *The Four Zoas*," *Blake: An Illustrated Quarterly* 20 (1987): 144–47; the reply by Mann, "Finishing Blake," *Blake: An Illustrated Quarterly* 22 (1989): 139–42; and Otto's further reply, "Is There a Poem in This Manuscript?" *Blake: An Illustrated Quarterly* 22 (1989): 142–44.

43. Donald D. Ault, *Visionary Physics: Blake's Response to Newton* (Chicago: University of Chicago Press, 1974), 157–58.

44. Urizen externalizes and arrests vortexes traveled originally by the human senses. Before he falls, "The eyelids [were] expansive as morning & the Ears / As a golden ascent winding round to the heavens of heavens." Afterwards, "the wing like tent of the Universe beautiful surrounding all / Or drawn up or let down at the will of the immortal man / Vibrated in such anguish the eyelids quiverd / Weak & Weaker their expansive orbs began shrinking" (*FZ* 73:37–74:6; E 350–51). *Milton*'s Vortex reverses this fall. The eyelids' "wing like tent" reappears in the punning conflation of the Blake pilgrim's Saul of Tarsus-like "left foot" with his opening eyes (*M* 15:49; E 110). OED *tarsus*, sense 1a: "the first or posterior part of the foot"; sense 2: "the thin plate of condensed connective tissue found in each eyelid." In *Milton*, this "tent of the Universe . . . surrounding all" becomes visible when data from the delicate "eyelids" achieve concrete spatiotemporal location via the poet's Felpham-grounded "foot."

45. Blake's Notebook describes "the little blasts of fear / That the hireling blows into my ear" (473). Commentators observe that this poem looks like a response to the Royal Proclamation against seditious writings of 21 May 1792. Yet, note that "the hireling" is no mere government spy but also the poet's inner voice of caution warning him not to speak out.

46. Blake's revisions to *The Four Zoas* can be seen to exemplify the kind of driven pseudo-productivity W. J. T. Mitchell has discerned: "Blake occupies an often ambiguous borderline between the divine madness of inspiration, and the demonic

madness of incapacity and false or fruitless labor, a madness of irrationality, slavery, and compulsive repetition." "Dangerous Blake," *Studies in Romanticism* 21:3 (1982), 413. The conversionary Truchsessian letter of 1804—a symbolic date Blake carved on the title pages of *Milton* and *Jerusalem* both—speaks of a period of misguided "industry," "Incessantly labouring and incessantly spoiling what I had done well" (E 756–57).

47. John Herman Randall, Jr., "The Intelligible Universe of Plotinos [*sic*]," *Journal of the History of Ideas* 30:1 (1969), 11; his capitals. Randall says, Plotinus "reads back the existent Order of things into a logically prior Structure, then generalizes the principle, and pushes it to apply to the very fact of Structure, of Intelligible Order, itself" (12). Randall then goes so far as to compare Plotinus's approach with the "logical realism" of Whitehead: "the *Enneads* can be called the *Principia Mathematica*, in the Whitehead and Russell sense, of Athenian science" (14). For our purposes, the point to notice here is that, as practically every biography or summary of his life has stressed, Whitehead's career in philosophy led *away* from the *Principia* and Russell to the agnostic religious stance he eventually termed *Platonic* realism. Randall emphasizes how Plotinus denies "process" to uphold "structure." Accordingly, if Blake's *Four Zoas* sometimes appears oversaturated with compositional process, one concludes it is partly because the poem sacrifices the artificial structure of deliberate fictional narrative in favor of the momentary processual reality of the poet's private Plotinean *visions* of structure.

48. Leopold Damrosch, Jr., *Symbol and Truth in Blake's Myth* (Princeton: Princeton University Press, 1980), 83, also 280–301.

49. Gordon Tesky, "Bent Abstraction," *ELH* 88:2 (2021), 322. In Spenser, Tesky says, "Such a place can extend into vast distances while remaining inside something smaller than itself, such as the cranium, or a cave by the sea" (333). This seems an apt description of Albion the Mundane Soul as manifested in the *Visions of the Daughters of Albion* frontispiece (figure 1.1).

50. Compare Robert N. Essick: "We all hear a voice in our heads. Most of us believe this to be the stream of our own thoughts; the schizophrenic grants the voice independent identity. Blake told Thomas Butts that he had 'written' his long 'Poem from immediate Dictation twelve or sometimes twenty or thirty lines at a time without Premeditation & even against my Will' (letter of 25 April 1803, E 729)—a comment that should dissuade us from normalizing the reference to dictation in *Jerusalem* as only a traditional bow to the muse." "*Jerusalem* and Blake's Final Works," in *The Cambridge Companion to William Blake*, ed. Morris Eaves (Cambridge: Cambridge University Press, 2003), 257.

51. Gerald Oster, "Phosphenes," *Scientific American* 222:2 (1970), 83.

52. Philip Gerans, "Pathologies of hyperfamiliarity in dreams, delusions and déjà vu," *Frontiers in Psychology* (20 February 2014): 5, "Dreams." According to G. W. Domhoff, dreams represent an extreme case of hyper-associative default processing, "a unique and more fully developed form of mind wandering, and therefore . . . the quintessential cognitive simulation." "The Neural Substrate for

Dreaming: Is It a Subsystem of the Default Network?" *Consciousness and Cognition* 20:4 (2011), 1172. *The Four Zoas* is modeled as a series of dreams after Edward Young's *Night Thoughts* (1742–45).

53. Blake's model for using language in this way goes back through Ossian and Milton to Homer. In Homer, Gregory Nagy explains, "To interpret is really to formalize the speech-act that is radiating from the dream or omen." Likewise, "the 'quoting' of words in Homeric poetry is equivalent to performing the 'quotation.'" *Homeric Responses* (Austin: University of Texas Press, 2003), 24, 22. This is just the opposite of an Austinian speech act that "does what it says" according to social convention. Non-Homeric examples of Nagy's point include *Paradise Lost* 3:372–415, *Milton* 22:15–25, and "Nurse's Song" (*SI*), discussed in chapter 1. See Käte Hamburger's argument that the lyric subject's utterances are real, direct statements about the world and not, as in novels or dramas, framed "as semblance, as fiction or illusion." "The Lyrical Genre," in *The Logic of Literature*, trans. Marilynn J. Rose, 2nd rev. ed. (Bloomington: Indiana University Press, 1993), 271.

54. Blake's mentor in such reader manipulations is surely the Milton of Stanley Fish's *Surprised by Sin: The Reader in "Paradise Lost"* (Cambridge: Harvard University Press, 1967).

55. Henri Bergson, *Duration and Simultaneity*, trans. Leon Jacobson (1922; New York: Bobbs-Merrill, 1965), 34. Bergson's example invites comparison with Augustine in the well-known section of *Confessions* entitled "Time in the human mind, which expects, considers, and remembers," where he instances the co-presence of present, past, and future: "I am about to repeat a psalm that I know. Before beginning, the poem exists in my expectation; when I have just finished, in my memory; but as I am reciting it, it is extended in my memory, on account of what I have already said; and in my expectation, on account of what I have yet to say. What takes place with the entirety of the poem takes place also in each verse and each syllable. This also holds true of the larger action of which the poem is part, and of the individual destiny of a man, which is composed of a series of actions, and of humanity, which is a series of individual destinies." *Confessions* 11.28.38. Qtd. in Jorge Luis Borges, "A History of Eternity," trans. Esther Allen, in *Selected Non-Fictions*, ed. Eliot Weinberger (New York: Penguin, 1999), 136. I quote from this simplified translation for clarity's sake. Borges omits the opening sentence of Augustine's paragraph, which I have added.

Bergson's account is closer still to Victor Zuckerkandl's classic phenomenological description: "The hearing of a melody is a hearing with the melody . . . It is even a condition of hearing melody that the tone present at the moment should fill consciousness entirely, that nothing should be remembered, nothing except it or beside it be present in consciousness . . . Hearing a melody is hearing, having heard, and being about to hear, all at once . . . Every melody declares to us that the past can be there without being remembered, the future without being foreknown." *Sound and Symbol* (Princeton: Princeton University Press, 1956), 71; his italics.

56. A. N. Whitehead, *The Concept of Nature* (Cambridge: Cambridge University Press, 1920), 55. Discussed in Stengers, 56.

57. See Ronald Desmet, "Did Whitehead and Einstein Actually Meet?" in *Researching with Whitehead: System and Adventure*, ed. Franz Riffert and Hans-Joachim Sander (Freiburg: Verlag Karl Alber, 2008), 154.

58. To Vero and Bice Besso, 21 March 1955, in *Albert Einstein. Michele Besso. Correspondance 1903–1955*, ed. and trans. into French, Pierre Speziali (Paris: Hermann, 1972), no. 211. English translations of the passage vary; I've given my own from the German in Speziali.

59. "There can be no Good-Will. Will is always Evil It is pernicious to others or selfish" (E 602). "Good-Will" is evil because it reflects moral Christianity's contradictory insistence that the will is depraved yet can be exerted for good. "Will is always Evil" because, like Urizen, it asserts itself against the general relatedness of things.

60. Compare Tucker, who postulates "various states of life, various forms of Being" through which divine particles of the Mundane Soul migrate constantly. Hence, he writes, "Individuals change," but "the spirits inhabiting [these states]" are eternal. Therefore "life is a journey through the vale of mortality, but the deliverance from Hyle [matter] a return home and resurrection to immortality again" (2.164). Einstein's block universe of permanently existing time slices is sometimes termed "eternalism," but it is very different from Blake's dynamic Eternity.

61. Accordingly, Blake's Notebook poem, "You don't believe I wont attempt to make ye," recommends listening to the meaning of birdsong as a truly strong *Gedankenexperiment* in believing:

> Reason says Miracle. Newton says Doubt
> Doubt Doubt & don't believe without experiment
> That is the very thing that Jesus meant
> When he said Only Believe Believe & try
> Try Try & never mind the Reason why. (E 501)

And if at first you don't succeed, "Try Try" again. Unlike Newton's self-doubting attempts to exempt the observer from the scene of observation, trying to believe without controls inductively broadens the test to include the observer. Scientific trial and error thereby acquire the metaphysical character of a Last Judgment. This also explains why, per the preceding footnote, "Will is always Evil" (602): vision can only be first-personal. At the same time, Blake's "On the Virginity of the Virgin Mary & Joanna Southcott," set down directly above "You don't believe I wont attempt to make ye," also recognizes how wishful thinking can become self-delusion. Southcott's belief in her messianic purpose created its own solipsistic reality, a hysterical pregnancy, therefore the belief itself lies beyond public judgment: "Whether tis good or evil none's to blame" (E 501). Self-deception may be honest. Compare

Los's dilemma in *Milton*: "What could Los do? How could he judge, when Satans self, believ'd / He had not oppress'd anyone" (*M* 7:39–40; E 101).

62. A. N. Whitehead, "Science," in *Essays in Science and Philosophy* (Rider: London, 1948), 176. Compare Arkady Plotnitsky: "According to Blake, through the workings of the Poetic Genius human imagination has the capacity to break the envelope (or, successively, envelopes) of individual minute particulars and reach the ultimate (infinite) vision of order of the world as a continuous assembly of minute particulars. It is as if we could actually see how a given phenomenal field, even a line or a point, is continuously, one by one, *constituted* by its ultimate constituents ("minute particulars"), to *see* the intensive infinity of the continuum, rather than only being able to see certain partial configurations continuous or discontinuous, or mixed, of such constituent points." "Minute Particulars and Quantum Atoms: The Invisible, the Indivisible, and the Visualizable in William Blake and Niels Bohr," *ImageTexT: Interdisciplinary Comics Studies* 3:2 (2006–07), para. 34; his italics. http://imagetext.english.ufl.edu/archives/v3_2/plotnitsky/. This seeing is, I suggest, a vision of Whitehead's extensive continuum—a fleeting glimpse that, for all its cosmic "thickness," immediately collapses back into the continuum of event-objects that supply the content of the vision. Plotnitsky compares Blake's minute particulars with Bohr's atomistic but non-Democritean quanta, while conceding "Blake . . . would have been . . . disturbed by the uncircumventable limits to our imagination that Bohr's vision implies" (para. 33). That is why Whitehead's visualizable idea of "actual entities" spread in an immanent "extensive continuum" offers a better guide to Blake than Bohr's unimaginable mathematic formalisms.

63. Says Whitehead, "An electron within a living body is different from an electron outside it, by reason of the plan of the body." *Science and the Modern World* (Cambridge: Cambridge University Press, 1926), 79.

64. A. N. Whitehead, *The Concept of Nature* (Cambridge: Cambridge University Press, 1920), 148.

65. Isabelle Stengers, *Thinking with Whitehead: A Free and Wild Creation of Concepts*, trans. Michel Chase (Cambridge: Harvard University Press, 2002), 168.

66. *The Commentaries of Proclus on the Timaeus of Plato in Five Books; containing a treasury of Pythagoric and Platonic physiology*, trans. Thomas Taylor, 2 vols. in 1 (London: printed for the author, 1820), 1.470; his italics.

67. Christoph Helmig and Carlos Steel, "Proclus," sect. 3.5, *The Stanford Encyclopedia of Philosophy*, ed. Edward N. Zalta, https://plato.stanford.edu/archives/fall2020/entries/proclus/.

68. Gregory Nagy, "Performing Homer," *New York Review of Books*, 9 April 1998, 81. To be sure, Nagy's view is controversial: this letter to the editor is a response to a highly critical review of one of his books.

69. Did relativity supply a basis for Milman Parry's contemporaneous insight into a collective Homer whose every oral performance is an improvised new composition, even as it reprises and advances a long tradition of such performances?

70. See Robert F. Gleckner, "Blake's 'Dark Visions of Torment' Unfolded: *Innocence* to *Jerusalem*," *SAQ* 95:4 (1996), 708–10.

71. Considering Hazlitt's influence on Keats, this must be the edition he also read; see note 14.

72. Hazlitt argues that the "deadening . . . spell" of "custom" has permitted "a mere play of words" based on the "word, *self*—a "nominal abstraction" of "extended consciousness" connecting past and present experiences—to become illicitly stretched by Lockean empiricists to include anticipations of our *future* self. Rather, he says, "The imagination, by means of which alone I can anticipate future objects, or be interested in them, must carry me out of myself into the feelings of others by one and the same process by which I am thrown forward as it were into my future being, and interested in it." William Hazlitt, *An Essay on the Principles of Human Action*, ed. John R. Nabholz (Gainesville: Scholar's Facsimiles, 1969), 3–11, 117–19; his italics.

73. I borrow here from Niall O'Flaherty, *Utilitarianism in the Age of Enlightenment: The Moral and Political Thought of William Paley* (Cambridge: Cambridge University Press, 2019), 66. Christopher Z. Hobson explores Blake's view of Wesleyan "holiness and perfection" as "illusory," even Satanic, in "Blake, Methodism, and Christian Perfection," *Blake: An Illustrated Quarterly* 55:2 (2021), para. 22, n.p.

74. Thomas Birch, *History of the Royal Society of London* (London: L. Davis and C. Reymers, 1760), 3.250–51.

75. Isaac Newton, letter to Henry Oldenburg of 25 January 1675/76, in *The Works of the Honourable Robert Boyle*, ed. Thomas Birch, 5 vols. (London: A. Millar, 1744), 1.74; italics original. Rptd. in *The Correspondence of Isaac Newton*, ed. H. W. Turnbull, J.F. Scott, A. Rupert Hall, and Laura Tilling, 7 vols. (Cambridge: Cambridge University Press, 1959–771), 1.364–65.

76. P. M. Heimann, "'Nature is a Perpetual Worker'": Newton's Aether and Eighteenth-Century Natural Philosophy," *Ambix* 20:1 (1973), 8. That Newton's letters were not published until 1744 and 1756, respectively, likely reinforced later eighteenth-century enthusiasm for linking aether theory with the new sciences of chemistry and electricity.

77. This is not to deny my chapter 4 suggestion that the "friend with whom [the 'traveller thro eternity'] livd benevolent" (*M* 15:27; E 109) might be Hayley. The evangelical brotherliness of Blake's 1804–05 letters to Hayley indicates he came to regard the healing of their rupture as a local instance of his restored connection with Albion and Jesus.

78. Others hypothesized a multiplicity of different aethers, each targeted to explain a single phenomenon.

79. The occasionalist doctrine, which Berkeley among many others upheld, originated with Thomas Aquinas: "God does not maintain things in existence by

any new action, but by the continuation of the act whereby he bestows *esse*." *Summa Theologiae* (New York: McGraw-Hill, 1964), Ia.104.1.

80. David Hartley, *Observations on Man, His Frame, His Duty, and His Expectations*, 2 vols. (S. Richardson: London, 1749), 1.83, 2.287; his italics. Joseph Priestley's more millenarian and Baconian rhetoric of progress expresses the same vision: "Whatever was the beginning of this world, the end will be glorious and paradisiacal, beyond what our imaginations can now conceive." *An Essay on the First Principles of Government* (London: J. Dodsley, 1771), 4–5.

81. Compare Coleridge's Christian panentheist consolation in his 6 April 1799 letter to Poole from Germany following news of the death of his nine-month-old baby, Berkeley:

> My Baby has not lived in vain—this life has been to him what it is to all of us, education and development! Fling yourself forward into your immortality only a few thousand years, & how small will not the difference between one year old & sixty years appear! Consciousness—! it is not otherwise necessary to our conceptions of future Continuance than as connecting the *present link* of our Being with the one *immediately* preceding it; & *that* degree of Consciousness, *that* small portion of *memory*, it would not only be arrogant, but in the highest degree absurd, to deny even to a much younger Infant. . . . What if the vital force which I sent from my arm into the stone, as I flung it in the air & skimm'd it upon the water—what if even that did not perish! It was *life*—! it was a particle of *Being*—! it was *Power*!—and *how could* it perish—?" *Collected Letters*, ed. E. L. Griggs, 6 vols. (Oxford: Clarendon, 1956–71), 1.379; Coleridge's italics.

Notice these thoughts have a Hartleyan basis: "Life . . . is . . . education and development." In his next letter of 8 April to his wife, Sara, Coleridge explains his viewpoint by means of a Symbol: "What an unknown Being one's own Infant is to one!—a fit of sound—a flash of light—a summer gust that is as it were *created* in the bosom of the calm Air, that rises up we know not how, and goes we know not whither! But we say well; it goes! it is gone!—and only . . . when we sport and juggle with abstract phrases, instead of representing our feelings and ideas . . . , only then we say it *ceases!*" (1.381; his italics). Interestingly, the Coleridgean *anamnesis* of these letters is not especially redemptionist. Nor, by the same token, is Whitehead's *anamnesis* as redemptive as the theodicies of process-theologians like Charles Hartshorne make it out. If Coleridge's claim that nothing is ever lost recalls the truth of his friend Wordsworth's "We are Seven," published the year before, it also recalls Ma-tsu's more violent exemplification of a similar point in the eighth-century Zen koan that stands as one of this book's epigraphs: "How could the wild geese ever have flown away?" Comments Hans Peter Duerr, in words that reiterate Coleridge's distinction between saying, "It goes! it is gone!" and falsely saying, "It *ceases*": "The wild geese flying by *are* wild geese flying by. The wild geese that flew by *are* wild geese that flew by." *Dreamtime: concerning the Boundary between Wilderness and Civilization*, trans. Felicitas Goodman (Oxford: Basil Blackwell, 1985), 119; his italics. Whitehead's abandonment of agnosticism and development of a new kind of theology is sometimes linked to the death of his son, Eric, in World War I.

82. Mark Currie, *About Time: Narrative, Fiction and the Philosophy of Time* (Edinburgh: Edinburgh University Press, 2007), 51–73, esp. 60–61. For Priestley, see *History and Present State of Electricity*, 4th ed. (London: J. Johnson: 1775), 443–44; qtd. in my introduction.

83. Like Tucker, Whitehead concedes that loss of first-person perspective and "subjective immediacy" is necessary in order for the individual to become translated into the wider life of the world process and its complex order of "everlastingness" (*Process and Reality* 21, 346, 348). Nancy Frankenberry points out that Whitehead's "everlastingness" "signifies not timelessness but coincidence with all possible times." "The Logic of Whitehead's Intuition of Everlastingness," *Southern Journal of Philosophy* 21:1 (1983), 33. The same goes for the later Blake's use of "evermore" and the biblical "alway."

84. See James Chandler, "The Languages of Sentiment," *Textual Practice* 22:1 (2008): 21–39; also Kyoko Takanashi, "Mediation, Reading, and Yorick's Sentimental Vehicle," *Novel* 39:3 (2016): 486–503. More's contemporary Ralph Cudworth examines "the . . . Vehicle or Chariot of the Soul" and explicitly compares it to Paul's "Spiritual Body." *An Abridgment of Dr. Cudworth's True Intellectual System of the Universe*, 2 vols. (London: John Oswald, 1782), 2.573.

85. John Locke, *An Essay concerning Human Understanding*, ed. Alexander Campbell Fraser, 2 vols. (1689; New York: Dover, 1959), 1.212.

86. "Chinks," a word Tucker uses again at 2.253–54, derives from Edward Young's *Night Thoughts*: "Thro' chinks, styl'd organs, dim life peeps at night" (3:450). Yet, Tucker's linking the word to expanded sense perception runs quite opposite to Young, a Calvinistic Deist for whom the body is a prison. This makes the Tucker passage at 2.100 a sign of his influence on Blake's *Marriage* almost twenty years before *Milton*.

87. George Berkeley, *A Treatise concerning the Principles of Human Knowledge*, sect. 14, in *Works of George Berkeley*, ed. Alexander Campbell Fraser, 4 vols. (Oxford: Clarendon, 1901), 1.163. See the analysis by Barry Stroud, "Berkeley v. Locke on Primary Qualities," *Philosophy* 55:2 (1980): 149–66, esp. 154–56.

88. That light had velocity was already widely accepted, including by Newton and Huygens. See, e.g., Richard Collins: "A Ray of Light, compleats its nimble Race / From Sun to Earth, in near ten Minutes Space." *Nature Display'd. A Poem* (London: J. Crokatt, 1727), 49.

89. "Autobiographical Notes," in *Albert Einstein—Philosopher Scientist*, ed. Paul Arthur Schilpp, Library of Living Philosophers 7 (Chicago: Open Court, 1949), 52–53.

90. Proclus, *Elements of Physics* 142a. Qtd. in Erwin Panofsky, *Perspective as Symbolic Form*, trans. Christopher S. Wood (New York: Zone Books, 1992), 49. Angus Fletcher makes a similar point with respect to *Paradise Lost*: "Since the effulgent light of the Divine Creator traverses the universe with the greatest possible velocity, . . . it would not be necessary to wait for Maxwell's equations, the Michelson-Morley experiment, or Einstein's relativity theory to see that the boundary of the universe was defined by the speed of light." *The Topological Imagination: Spheres, Edges and Islands* (Cambridge: Harvard University Press, 2016), 146.

91. For all their differences, A. E. Taylor and Francis Cornford largely agree on this. Taylor, *A Commentary on Plato's "Timaeus"* (Oxford: Clarendon, 1928), 410; Cornford, *Plato's Cosmology: The "Timaeus" of Plato* (London: Routledge, 1937), 247. Cornford adds, such light "is similar to the visual current of 'pure fire' which is so fine that it alone can filter through the close texture of the eyeball. We may infer that it consists of particles of smaller grades than flame or glowing heat."

92. Isidoros Katsos, "Chasing the Light: What Happened to the Ancient Theories?" *Isis* 110:2 (2019), 279.

93. Gillian Beer remarks, "If allegory is narrative metaphor, analogy is predictive metaphor. . . . As in hypothesis, the arc of desire seeks to transform the conditional into the actual. And again, as in hypothesis, such a transformation is seen as changing fiction into a truth." *Darwin's Plots: Evolutionary Narrative in Darwin, George Eliot and Nineteenth-Century Fiction*, 3rd ed. (Cambridge: Cambridge University Press, 2009), 80.

94. Since his vision remains loosely tethered to human physiology, Tucker only partly illustrates Anne Janowitz's claim that early modern vehicular journeys commonly traveled to the moon or outer space. "Response: Chandler's 'Vehicular Hypothesis' at Work," *Textual Practice* 22:1 (2008): 41–46. Tucker travels *inward* to the Mundane Soul, within which "the stars with their several systems of planets . . . , light, ether, and . . . [a] variety of bodies" are already "floating about" (2.71). Similarly, Blake's vehicular Milton seeks a more inward return home to Jesus and Eternity, much as the very writing of *Milton* finally manifests as, itself, a vehicle by which the lost Odyssean poet can get back to himself and his wife.

95. Compare, for example, the old huntsman's disproportionate gratitude at the narrator's simple act of kindness in Wordsworth's "Simon Lee" in *Lyrical Ballads* (1798) and that poem's prophetic, if Burkean, witnessing to a human solidarity beyond death.

96. Northrop Frye, *Fearful Symmetry* (Princeton: Princeton University Press, 1947), 18–19. Frye comments, "Insofar as a man is perceived by others (or, in fact, by himself), he is a form or image, and his reality consists in the perceived thing

which we call a 'body.' 'Body' in Blake means the whole man as an object of perception." One sees, then, how "a man" might include multiple "vehicles" according to how he is perceived by others. Lucretius's theory of *simulacra* is drawing near. In Amanda Jo Goldstein's Derridean interpretation, "*De rerum natura* presumes figuration to be central to the reality of things (*res*) in their being and appearing alike: the figural improprieties of perceptual and linguistic representation are taken to originate in and index this general reality, rather than to betray an insuperable gap between human knowledge and other things." *Sweet Science: Romantic Materialism and the New Logics of Life* (Chicago: University of Chicago Press, 2017), 4.

97. Adam Phillips, *Missing Out: In Praise of the Unlived Life* (London: Penguin, 2013), xii.

Chapter Six

1. Niels Bohr, "The Unity of Human Knowledge" (1960), in *Essays 1958–62: On Atomic Physics and Human Knowledge* (New York: John Wiley, 1963), 10.

2. *A Treatise concerning the Principles of Human Knowledge*, sect. 14, in *Works of George Berkeley*, ed. Alexander Campbell Fraser, 4 vols. (Oxford: Clarendon, 1901), 1.105; all italics Berkeley's.

3. K. R. Popper, "A Note on Berkeley as Precursor of Mach," *British Journal for the Philosophy of Science* 4:13 (1953): 26–36.

4. Mach has recently undergone rehabilitation by Eric C. Banks, *The Realistic Empiricism of Mach, James, and Russell: Neutral Monism Reconceived* (Cambridge: Cambridge University Press, 2014).

5. Cf. Stephen Weinberg's famous conclusion, which Blake would have considered to be, indeed, mechanistic materialism's logical dead end: "The more the universe seems comprehensible, the more it also seems pointless." *The First Three Minutes: A Modern View of the Origin of the Universe*, 2nd ed. (New York: Basic Books, 1993), 149. Whitehead makes a similar point within a very different perspective: "A way of life is something more than the shifting relations of bits of matter in space and in time. . . . But, in abstraction from the atmosphere of feeling, one behaviour pattern is as good as another; and they are all equally uninteresting." "Memories," in *Essays in Science and Philosophy* (London: Rider, 1948), 22.

6. Angela Esterhammer, *Creating States: Studies in the Performative Language of John Milton and William Blake* (Toronto: University of Toronto Press, 1994), 12, 166–67; her italics. Qtd. in my introduction.

7. F. E. L. Priestley, "Berkeley and Newtonianism: The *Principles* (1710) and the *Dialogues* (1713)," in *The Practical Vision: Essays in English Literature in Honour of Flora Roy*, ed. Jane Campbell and James Doyle (Waterloo, ONT: Wilfrid Laurier University Press, 1978), 49.

8. *Henry More's Manual of Metaphysics* [*Enchiridion Metaphysicum*], trans. Alexander Jacob, 2 vols. (Hildesheim: Georg Olms, 1995), 1.69–70. Qtd. in chapter 5.

9. George Berkeley, *The Analyst; or, a Discourse Addressed to an Infidel Mathematician* (London: J. Tonson, 1734), 59. Blake's letter to George Cumberland of 12 April 1827 links "The Indefinite" with "Newtons Doctrine of the Fluxions of an Atom. A Thing that does not Exist" (E 783).

Worth mentioning here is my namesake L. J. Cooper's argument that Blake rejected Berkeley's critique in *De Motu* of Newton's absolute space. Whereas Berkeley was influenced (Cooper claims) by popular Newtonianism's representation of absolute space as purely nonsensible, Blake as an empiricist recognized the difference between Newtonianism's "intransigent materialism" and Newton's own, more nuanced position. In contrast to Berkeley's "binary of the sensual and the 'spiritual,' " Blake's appreciation "that the indiscernible can be accessed by empirical, bodily means" led him to identify absolute space with Eternity, which *Milton*'s conclusion shows to be accessible through bodily Vision. "William Blake's Aesthetic Reclamation: Newton, Newtonianism, and Absolute Space in *The Book of Urizen* and *Milton*," *European Romantic Review* 29:2 (2018), 260–61. That Blakean Vision arises out of complexities in Newton's idea of matter, I quite agree. But L. J. Cooper's Blake not only ignores Berkeley's correctly relativist interpretation in *De Motu* of Newton's prime experimental evidence for absolute space, the famous bucket experiment described in the *Principia*; his Blake also embraces the mathematic formalism of Newton's own interpretation of that experiment. Such a position doesn't seem very "empirical." Berkeley pointed out that if Newton's absolute space is "infinite, immovable, indivisible, and insensible," and so cannot be "perceived by sense or pictured by the imagination," then "it must necessarily be quite useless for the distinguishing of motions" and so amounts to "mere nothing." *De Motu*, in *Space from Zeno to Einstein: Classic Readings with a Contemporary Commentary*, ed. Nick Huggett (Cambridge: MIT Press, 1999), 169–72. This is the basis of Whitehead's own (Blakean) critique of Newton. Berkeley says that the "truly circular [i.e., absolute] motion" of the water in Newton's bucket "is strangely compounded of the motions, not alone of the bucket or sling, but also of the daily motion of the earth round her own axis"—and ultimately of the rotational frame of the fixed stars and all the matter in the universe, as Mach and, later, Einstein saw.

None of this impugns Cooper's cardinal insight that "Blake's body, insofar as it is a sensorial signification of 'Eternity' not unlike Newton's bucket, functions as the point of convergence between the indiscernible and a material reality that makes this process [viz., Vision] possible" (262). But Cooper's body-bucket comparison is susceptible to an interpretation nearly opposite to his own. For Newton, the centrifugal force that pushes the water up against the bucket's side is inertia, the resistance of a body to any change in its speed or direction of motion. What Mach and, implicitly, Berkeley saw was that inertia, rather than having a mechanical cause, might simply be defined as the angular momentum of the universe in its entirety. If we posit an analogy between the concave swirl of water in Newton's bucket and Blake's Vortex connecting Eternity at the open end and Earth at the apex where *Milton*'s Blake poet stands, then the water pushing up at right angles to the bucket's axis of rotation can be seen as spreading toward "a universe of starry majesty" (*M* 15:25; E 109)—namely, Eternity, which extends everywhere at right angles to three-dimensional earthly existence. That is why the poet sees Milton at the "perpendicular" (15:48; E 110) when he looks up from the bottom of the vortex of Milton's descent. The cone-shaped spinning mass of water in the bucket is analogous to a spindle torus boring into the earthly sphere and opening it up. When the torus finally collapses, the poet's vision passes, and earth everts back into a solid globe. It seems the same experiment Newton deemed to demonstrate the ulterior reality of absolute space illustrated, for Blake, an apocalyptic experience of cosmic space-time.

    10. *Blake Records: Documents (1714–1841) concerning the Life of William Blake (1757–1827) and his Family*, ed. G. E. Bentley, Jr., 2nd ed. (New Haven: Yale University Press, 2004), 500. Compare Proclus's less than fully idealist view of Euclidean geometry, as summarized by Glenn R. Morrow: "The objects of mathematical inquiry—numbers, points, lines, planes, and all their derivatives—are neither empirical things nor pure forms. 'Mathematical being,' as Proclus puts it in his opening sentence, occupies an intermediate position between the simple immaterial realities of the highest realm and the extended and confusedly complex objects of the sense world. . . . '[M]athematicals' . . . do possess a kind of extendedness. The series of numbers consists of discrete numbers, and geometrical figures are divisible into parts." "Introduction" to Proclus, *A Commentary on the First Book of Euclid's Elements*, trans. Glenn R. Morrow (Princeton: Princeton University Press, 1992), lvii.

    11. David Hume, *A Treatise of Human Nature*, ed. L. A. Selby-Bigge, 2nd ed., rev. P. H. Nidditch (1740; Oxford: Clarendon, 1978), 51–53. Qtd. in my introduction.

    12. A. N. Whitehead, *The Principle of Relativity with Applications to Physical Science* (Cambridge: Cambridge University Press, 1922), 73. Qtd. in Robert Palter, *Whitehead's Philosophy of Science* (Chicago: University of Chicago Press, 1960), 194.

    13. A. N. Whitehead, *Science and the Modern World* (New York: Macmillan, 1925), 69.

    14. George Berkeley, *Three Dialogues between Hylas and Philonous*, in *Works*, 1.336–37; all italics Berkeley's.

    15. A dangerous metaphor, since Hume went on to argue that without Berkeley's God and his truth guarantees, sense experience and personal identity amount to "nothing but a bundle or collection of different impressions" (*Treatise of Human Nature* 252). The physical question here seems to concern the relative tightness of the bundles: are they hard molecules or do they form a spread-out field?

16. Thomas Reid, *Philosophical Works*, ed. Sir William Hamilton, 8th ed., 2 vols. (Edinburgh: James Thin, 1895), 1.286–87.

17. Alexander Jacob, *De Nature Natura: A Study of Idealistic Conceptions of Nature and the Unconscious* (Stuttgart: Franz Steiner Verlag, 1991), 55; his italics.

18. Abraham Tucker [Edward Search], *The Light of Nature Pursued*, 6 vols. (London: T. Jones and T. Payne, 1768), 2.5–6.

19. Chris Townsend, "Visionary Immaterialism: Berkeleian Empiricism in Blake's Poetry," *Studies in Romanticism* 58:3 (2019), 360–61.

20. Townsend here quotes Simon Jarvis, "Blake's Spiritual Body," in *The Meaning of "Life" in Romantic Poetry and Poetics*, ed. Ross Wilson (New York: Routledge, 2009), 24. As Townsend notes, Jarvis follows the phenomenologist Michel Henry in emphasizing how the Cartesian self includes, as ideas, experiences such as pain, delight, fear, and hunger. Townsend says, "If we . . . identify the act of thinking as the essence of self or soul, as posited in the statement 'I think, therefore, I am,' then these ideas that require a body mean that we must include some notion of embodiment in our understanding of the soul" (368). This is the same point we've already seen Whitehead make more comprehensively in relation to physical space-time when he stresses that Descartes "in his own philosophy conceives the thinker as creating the occasional thought," inasmuch as "Descartes insists in *Meditations 2* that the pronouncement, 'I think, therefore I am,' 'is necessarily true each time that I pronounce it, or that I mentally conceive it'" (*Process and Reality* 150–51). Why settle for context-thin French phenomenology when you can have history-rich English metaphysics? Whitehead says that Berkeley's intellectualistic "haste to have recourse to an idealism with its objectivity grounded in the mind of God" shows that he remains wedded to "the notion of simple [purely spatial] location" (*Science and the Modern World* 67ff.). The deliquescent imagery of *The Book of Thel* seems to associate Berkeleyan phenomenalism with Christian-immaterialist fear of the body.

21. Chris Townsend, "Nature and the Language of the Sense: Berkeley's Thought in Coleridge and Wordsworth," *Romanticism* 25:2 (2019): 129–42.

22. George Berkeley, *Alciphron: or, the Minute Philosopher*, in *Works*, 2.155.

23. George Berkeley, *Siris: a Chain of Philosophical Reflexions and Inquiries concerning the virtues of Tar-water*, in *Works*, 2.476. Qtd. in Townsend, 376.

24. Donald Viney, "Process Theism," sect. 4.1, *The Stanford Encyclopedia of Philosophy*, ed. Edward N. Zalta, https://plato.stanford.edu/archives/sum2020/entries/process-theism/.

25. Alexander Campbell Fraser, "Prologomena," in Locke, *Essay concerning Human Understanding*, 1.cxxxi.

26. Whitehead, "Uniformity and Contingency," in *Essays on Science and Philosophy*, 102.

27. Baxter's book was widely known despite its stunning mediocrity. See John Beer, "Coleridge and Andrew Baxter on Dreaming," *Dreaming* 7:2 (1997): 157–69.

28. See, for example, Christopher Fox, *Locke and the Scriblerians: Identity and Consciousness in Early Eighteenth-Century Britain* (Berkeley: University of California Press, 1988).

29. Andrew Baxter, *An Enquiry into the Nature of the Soul*, 3rd ed., 2 vols. (London: A. Millar, 1745), 2.238–39; all italics his.

30. For example, the design to "The Little Boy Lost" (*Innocence*) shows the boy sleepwalking. Priestley, though he opposed Cartesian dualism, condemns as unphilosophical both vehicles and the idea "that there is something *intermediate* between the soul and the gross body." For, he asks, "if there be a soul distinct from the body, and [yet] it be sensible of all the changes that take place in the corporeal system to which it is attached, why does it not perceive that state of the body which is termed *sleep*; and why does it not *contemplate* the state of the body and brain during sleep?" (*Disquisitions* 108–09; his italics). Nevertheless, such perceiving and contemplating are exactly what Blake's Milton does on his Death Couch. Does Milton thereby parody the Demiurge in *Timaeus* 30c contemplating the form of the Cosmic Animal or, more generally, Neoplatonism's hypostasis of the Soul through the Intellect's contemplation of the world of forms and ideas?

31. See Robert E. Schofield, *The Enlightenment of Joseph Priestley: A Study of His Life and Work from 1733–1773* (University Park: Pennsylvania State University Press, 1997), 47–52.

32. Joseph Priestley, *Disquisitions Relating to Matter and Spirit* (London: J. Johnson, 1777), 39. Here and elsewhere, all italics Priestley's.

33. Dugald Stewart, *Philosophical Essays* (1810), in *Collected Works*, ed. Sir William Hamilton, 11 vols. (Edinburgh: T. Constable, 1854–60), 5.94n.1.

34. John W. Yolton, *Perceptual Acquaintance from Descartes to Reid* (Minneapolis: University of Minnesota Press, 1984), 84.

35. Coleridge's usage of this proverb is summarized in Kathleen Wheeler, *Sources, Processes and Methods in Coleridge's "Biographia Literaria"* (Cambridge: Cambridge University Press, 1980), 56–57.

36. Thus, G. J. Warnock reasonably asked why Berkeley did not simply remain a materialist by discarding the other, nonmental half of Locke's double universe of "primary" and "secondary" qualities based on his correspondence theory of perception. Introduction to Berkeley, *"The Principles of Human Knowledge" and "Three Dialogues of Hylas and Philonous"* (London: Fontana, 1962), 31–32.

37. David Hume, *Enquiries concerning Human Understanding and concerning the Principles of Morals*, ed. L. A. Selby-Bigge, 3rd ed., rev. P. H. Nidditch (Oxford: Oxford University Press, 1975), 155n.; his italics.

38. Samuel Taylor Coleridge, *Collected Letters*, ed. E. L. Griggs, 6 vols. (Oxford: Clarendon, 1956–71), 1.192–3; his italics.

39. Priestley's doctrine, which led the later Blake to regard Creation as a merciful setting of limits to the Fall, accords with Richard Mohr's argument that Plato's Demiurge is "chiefly bent on improving the world's intelligibility," first by

crafting the world-soul as rational, and then by creating time as "a giant clock" that "makes science possible." "What the Demiurge does is to introduce standards or measures into the phenomenal realm by imaging as best he can the nature of Forms where Forms are construed as standards or measures." "Plato's Theology Reconsidered: What the Demiurge Does," *History of Philosophy Quarterly* 2:2 (1985), 131. This greatly complicates, even refutes, the nonmetric Eternity Donald Ault attributes to Blake in *Visionary Physics: Blake's Response to Newton* (Chicago: University of Chicago Press, 1974).

40. Wayne Glausser, "Atomistic Simulacra in the Enlightenment and in Bake's Post-Enlightenment," *The Eighteenth Century* 32:1 (1991), 80.

41. Cudworth offers a possible source for Blake's *Marriage of Heaven and Hell* myth of the Giants "who formed this world into its sensual existence and now seem to live in it in chains" (*MHH*; E 40): "*Plato* tells us, that there had been always, . . . as he calls it, a kind of Gigantomachy betwixt these two Parties or Sects of men; The one that held there was no other Substance in the World besides Body; The Other that asserted Incorporeal Substance." Ralph Cudworth, *The True Intellectual System of the Universe, The first part* (London: Richard Royston, 1678), 18; his italics.

42. Joseph Priestley, *Institutes of Natural and Revealed Religion*, 3rd ed., 2 vols. (London: Priestley, 1794), 1.26.

43. See Andrew M. Cooper, *William Blake and the Productions of Time* (Aldershot: Ashgate, 2013), chap. 7. Conceivably, Blake's interest in neurology was also inspired directly by Michelangelo, whose God-cloud in *Creation of Adam* is an anatomically accurate picture of the human brain, as argued by Ian Suk and Rafel J. Tamargo, "Concealed Neuroanatomy in Michelangelo's *Separation of Light from Darkness* in the Sistine Chapel," *Neurosurgery* 66:5 (2010): 851–61.

44. Robert E. Schofield, "Joseph Priestley, Eighteenth-Century British Neoplatonism, and S. T. Coleridge," in *Transformation and Tradition in the Sciences: Essays in Honour of I. Bernard Cohen*, ed. Everett Mendelsohn (Cambridge: Cambridge University Press, 1984), 237–54. Coleridge even titled an early sonnet "To Priestley" (1794).

45. "The Eolian Harp," l. 48, in *Complete Poetical Works of Samuel Taylor Coleridge*, ed. Ernest Hartley Coleridge, 2 vols. (Oxford: Clarendon, 1912), 1.102. Indeed, the delicate dualism of Hartley's mechanically vibrating nerve fibers is better illustrated by Coleridge's passive harp than by the windy vehicular bagpipes of Tucker's Morean "Vision," whose movements can be controlled by voluntary effort and practice—as if bodily motion, including spontaneous associations of ideas, always entailed conscious cognition as in Descartes. Tucker was a better fabulist and storyteller than metaphysician. Wordsworth, too, mentions how in childhood "A plastic power / Abode with me, a forming hand, at times / Rebellious, acting in a devious mood, / A local spirit of its own, at war / With general tendency." This well describes the opposition, for example in *Songs of Innocence*, between

the local quantal particularity of the child's emerging Whiteheadian extensive continuum and "adult" empiricist generalizations founded on Newton's weak force of gravitation. *Prelude* (1805), 2:381–85, in *William Wordsworth*, Oxford Authors, ed. Stephen Gill (New York: Oxford University Press, 1984).

46. On the other hand, Priestley himself, ever the binary rationalist, attacks "the vehicle of the soul," including Hartley's *"infinitesimal elementary body,"* for creating a vicious regress. He entirely misses the vehicle's hypothetical role as an intercommunicating "third kind" or *tertium quid* between mind and matter (*Disquisitions* 74–81). Indeed, Priestley overlooks the third-kind significance of his own clever Neoplatonic-Newtonian "immateriality of matter" based in matter's supposed porosity and tenuity. His blindness is attributable to the polemical thrust of the *Disquisitions*, which deliberately shoehorns physics (along with Boscovich's mathematical force-points) into a politico-religious attack on the atheistic drift of Cartesian dualism.

47. "The Everlasting Gospel" upholds an antinomian Jesus against those who, by "Asking Pardon of his Enemies," abnegate their own identity and finally merge with "Caesar himself / Like dr Priestly & Bacon & Newton" (E 519). Blake is likely recalling Priestley's empiricist emphasis on public miracles over personal vision in his *Letters to the Members of the New Jerusalem Church* (1791), mentioned in chapter 1.

48. I quote Karl Kroeber's phrase for what he deems as Blake's central psychological insight, in *British Romantic Art* (Berkeley: University of California Press, 1986), 17, 22.

49. This, in contrast to the solid-state "grammar of symbolism" rightly ascribed to Northrop Frye's *Fearful Symmetry* in "William Blake," *The English Romantic Poets and Essayists: A Review of Research and Criticism*, ed. Carolyn Washburn Houtchens and Lawrence Huston Houtchens, 2nd ed. (New York: New York University Press, 1966), 21. By contrast, the progressive "grammar of discovery" Stephen H. Daniels discerns in Bacon's *Advancement of Learning*, founded on nature's observed analogies, affinities, and resemblances, seems a likely basis for Blake's project. "Myth and the Grammar of Discovery in Francis Bacon," *Philosophy and Rhetoric* 15 (1982): 219–37.

Conclusion

1. Thomas Paine, *Common Sense* (Philadelphia: W. & T. Bradford, 1776), 86.

2. See *William Blake: The Early Illuminated Books*, ed. Morris Eaves, Robert N. Essick, and Joseph Viscomi (Princeton: Princeton University Press), 236.

3. Peter Otto notes that in Swedenborg's *Heaven and Hell*, hell's "infernal mansions" lie immediately below heaven's "every mountain, hill, rock, plain, and valley" (he quotes Swedenborg), so that the ground of heaven is the roof of hell. Blake "places Earth in the cramped zone (on the line) between these realms, where it is marked by both." "Organizing the Passions: Minds, Bodies, Machines, and

the Sexes in Blake and Swedenborg," *European Romantic Review* 26:3 (2015), 368. Otto quotes from Emmanuel Swedenborg, *A Treatise concerning Heaven and Hell, and of the Wonderful Things therein, as Heard and Seen*, trans. William Cookworthy and Thomas Hartley, 2nd ed. (London: R. Hindmarsh, 1784), 588.

4. Jon Kear sees Tartarus as "open[ing] like a tumultuous wound . . . its skin peeled back across the surface . . . Tartarus is presented as a space that powerfully menaces the picture's thresholds, including the threshold between the picture space and actual space of the viewer." "Staring into the Abyss: James Barry and British History," in *In Elysium: Prints by James Barry*, ed. Ben Thomas (Canterbury: University of Kent, 2010), 17–18. John Barrell explores the influence on Blake of Barry's republican ideology of civic humanism in *The Political Theory of Painting from Reynolds to Hazlitt: "The Body of the Public"* (New Haven: Yale University Press, 1986), 163–221.

5. "Autographic" is how Joseph Viscomi characterizes Blake's relief etching in *Blake and the Idea of the Book* (Princeton: Princeton University, 1993), 30.

6. John Locke, *An Essay concerning Human Understanding*, ed. Alexander Campbell Fraser, 2 vols. (New York: Dover, 1959), 1.149. Abraham Tucker likewise calls this site the "royal presence chamber." *The Light of Nature Pursued*, 6 vols. (London: T. Jones and T. Payne, 1768), 2.59–60.

7. A. N. Whitehead, *Modes of Thought* (New York: Macmillan, 1938), 224. Partially quoted in chapter 3, n.20.

8. Richard Shiff, "Breath of Modernism (Metonymic Drift)," in *In Visible Touch: Modernism and Masculinity*, ed. Terry Smith (Chicago: University of Chicago Press, 1997), 209.

9. Niels Bohr, "Quantum Physics and Philosophy: Causality and Complementarity," in *Essays 1958–62: On Atomic Physics and Human Knowledge* (New York: John Wiley, 1963), 2.

10. *The Prelude* (1805), 6:542, in *William Wordsworth*, Oxford Authors, ed. Stephen Gill (New York: Oxford University Press, 1984), 464. Compare Yeats: "Line and plane are combined in a gyre, and as one tendency or the other must be always the stronger, the gyre is always expanding or contracting. For simplicity of representation the gyre is drawn as a cone. Sometimes this cone represents the individual soul, and that soul's history—these things are inseparable—general life. When general life, we give to its narrow end, to its unexpanded gyre, the name of *Anima Hominis*, and to its broad end, or its expanded gyre, *Anima Mundi* . . ." *A Critical Edition of Yeats's "A Vision,"* ed. George Mills Harper and Walter Kelly Hood (1925; London: Macmillan, 1978), 129; his italics.

11. William Pressly's description of Barry's frieze of paintings as "the first attempt to synthesize in a single canvas through the means of figural art the intellectual history of the whole of humanity" holds even better for the unified space-time of Blake's painting than it does for Barry's six-part narrative sequence. *The Life and Art of James Barry* (New Haven: Yale University Press, 1981), 119. Pressly points

out the influence of Barry's murals on John Martin's "Last Judgement" in *James Barry: The Artist as Hero* (London: Tate Gallery, 1983), 84.

12. Butlin catalog nos. 661 and 662.

13. "J'ai dit plus haut qu'il n'étoit pas possible de concevoir plus de trois dimensions. Un homme d'esprit de ma connoissance croit qu'on pourroit cependant regarder la durée comme une quatriéme dimension, & que le produit du tems [*sic*] par la solidité seroit en quelque maniere un produit de quatre dimensions; cette idée peut être contestée, mais elle a, ce me semble, quelque mérite, quand ce ne seroit que celui de la nouveauté." "Dimension," *Encyclopédie, ou dictionnaire raisonné des sciences, des arts et des métiers*, ed. Denis Diderot and Jean le Rond d'Alembert, 17 vols. (Paris: Briasson, David, Le Breton, and Durand, 1751–72), 4:1010. Carl Friedrich Gauss had been working on the mathematics of non-Euclidean geometry since the early 1790s but deemed the implications too explosive to publish at the time.

14. Leopold Damrosch, Jr., *Symbol and Truth in Blake's Myth* (Princeton: Princeton University Press, 1980), 83; his italics.

15. George Berkeley, *The Theory of Vision or Visual Language*, in *Works on Vision*, ed. Colin Murray Turbayne (New York: Bobbs-Merrill, 1963), 141.

16. George Berkeley, *A Treatise concerning the Principles of Human Knowledge*, in *Works*, ed. Alexander Campbell Fraser, 4 vols. (Oxford: Clarendon, 1901), 1.51; his italics.

17. A. N. Whitehead, *Process and Reality*, ed. David Ray Griffin and Donald W. Sherburne (1929; New York: Free Press, 1978), 200.

18. In Einsteinian terms, *The Vision of the Last Judgment* affirms that while space and time are relative, space-time as a four-dimensional coordinate system is absolute. John Barrell has it backwards when his Marxist critique of liberal classism leads him to detect authoritarianism behind the ostensible universality of Blake's painting: "Different viewing positions . . . are not conceived as different points scattered at various distances from the object of vision, but as occurring at different points along a unilinear approach to eternity." *The Political Theory of Art*, 250.

19. See Thomas Nagel, *The Last Word* (New York: Oxford University Press, 1997), 19, 37.

20. So, for example, the universal doctrine of Contraries set forth on *The Marriage* plate 3 becomes garbled once the Revolutionary Devil expounds it on plate 4, much as the Devil's claim that Milton was "of the Devils party without knowing it" (*MHH* 6; E 35) is only half true. That truth's other half is what Milton learns from *Milton*'s Bard—effectively, that his support of a great leader, Cromwell, in 1648 made him sponsor of the wrong revolution, not the leveling one of 1642 but the monarchist settlement of 1688. Early and late, *The Marriage* and *Milton* alike criticize the debasement of prophecy to politics. What *The Vision of the Last Judgment* makes visually apparent is *why* "the politics of prophecy" is an eternal, if not always avoidable, contradiction.

Works Cited

Abbott, Edwin A. *The Annotated "Flatland: A Romance of Many Dimensions."* Ed. Ian Stewart. New York: Perseus, 2002.
Agamben, Giorgio. *The End of the Poem: Studies in Poetics.* Trans. Daniel Heller-Roazen. Stanford: Stanford University Press, 1999.
Agamben, Giorgio. *Potentialities: Collected Essays in Philosophy.* Trans. Daniel Heller-Roazen. Stanford: Stanford University Press, 1999.
Agamben, Giorgio. *What Is Philosophy?* Trans. Lorenzo Chiaso. Stanford: Stanford University Press, 2017.
Albert Einstein. Michele Besso. Correspondance 1903–1955. Ed. and trans. (into French) Pierre Speziali. Paris: Hermann, 1972.
Albert Einstein: Philosopher-Scientist. Ed. Paul Arthur Schilpp. Chicago: Open Court, 1949.
Allen, Richard. "David Hartley." *The Stanford Encyclopedia of Philosophy.* Ed. Edward N. Zalta. https://plato.stanford.edu/archives/sum2020/entries/hartley/.
Allen, Richard. *David Hartley on Human Nature.* Albany: SUNY Press, 1999.
Almond, Philip C. "Henry More and the Apocalypse." *Journal of the History of Ideas* 54 (1993): 189–200.
Aristotle, *Basic Works.* Ed. Richard McKeon. New York: Modern Library, 2001.
Augustine. *Earlier Writings.* Ed. J. H. S. Burleigh. Philadelphia: Westminster, 1953.
Augustine. *Sermons for Christmas and Epiphany.* Trans. Thomas Lawler. London: Paulist, 1952.
Ault, Donald D. *Narrative Unbound: Re-Visioning William Blake's "The Four Zoas."* Barrytown, NY: Station Hill, 1987.
Ault, Donald D. *Visionary Physics: Blake's Response to Newton.* Chicago: University of Chicago Press, 1974.
Bain, Jonathan. "Whitehead's Theory of Gravity." *Studies in the History and Philosophy of Modern Physics* 29 (1998): 547–74.
Banchoff, Thomas. "Introduction." *Flatland: A Romance of Many Dimensions,* 2nd rev. ed. Princeton: Princeton University Press, 1991.

Barrell, John. *The Political Theory of Art from Reynolds to Hazlitt: "The Body of the Public."* New Haven: Yale University Press, 1986.

Barry, James. *Works.* Ed. Edward Fryer. 2 vols. London: T. Cadell and W. Davies, 1809.

Bates, Stanley. "The Mind's Horizon." *Beyond Representation: Philosophy and the Poetic Imagination.* Ed. Richard Eldridge. Cambridge: Cambridge University Press, 1996. 151–74.

Baxter, Andrew. *An Enquiry into the Nature of the Soul,* 3rd ed. 2 vols. London: A. Millar, 1745.

Beer, John. "Coleridge and Andrew Baxter on Dreaming." *Dreaming* 7 (1997): 157–69.

Benlowes, Edward. *Theophilia, or Loves Sacrifice, a Divine Poem.* London: R. N., 1652.

Bennett, Jonathan, and Peter Remnant. "How Matter Might First be Made." *Canadian Journal of Philosophy,* Supplementary Volume 4 (1978): 1–11.

Bergson, Henri. *Creative Evolution.* Trans. Arthur Mitchell. New York: Henry Holt, 1911.

Bergson, Henri. *Duration and Simultaneity.* Trans. Leon Jacobson. New York: Bobbs-Merrill, 1965.

Berkeley, George. *The Analyst; or, a Discourse Addressed to an Infidel Mathematician.* London: J. Tonson, 1734.

Berkeley, George. *"The Principles of Human Knowledge" and "Three Dialogues of Hylas and Philonous."* Ed. G. J. Warnock. London: Fontana, 1962.

Berkeley, George. *Works of George Berkeley.* Ed. Alexander Campbell Fraser. 4 vols. Oxford: Clarendon, 1901.

Berkeley, George. *Works on Vision.* Ed. Colin Murray Turbayne. New York: Bobbs-Merrill, 1963.

Birch, Thomas. *History of the Royal Society of London.* London: L. Davis and C. Reymers, 1760.

Blacklock, Mark. "Higher Spatial Form in Weird Fiction." *Textual Practice* 31 (2017): 1101–16.

Blake, William. *The Complete Poetry and Prose.* Ed. David V. Erdman. Garden City, NJ: Doubleday, 1982.

Blake Records: Documents (1714–1841) concerning the Life of William Blake (1757–1827) and his Family. Ed. G. E. Bentley, Jr., 2nd ed. New Haven: Yale University Press, 2004.

Bloom, Harold. *Blake's Apocalypse.* Ithaca: Cornell University Press, 1963.

Bohm, David. *Wholeness and the Implicate Order.* New York: Routledge, 1980.

Bohr, Niels. *The Philosophical Writings of Niels Bohr.* Ed. Jan Faye and Henry J. Folse, 4 vols. Woodbridge: Ox Bow, 1998.

Bohr, Niels. "The Unity of Human Knowledge." *Essays 1958–62: On Atomic Physics and Human Knowledge.* New York: John Wiley, 1963.

Boime, Albert. *Art in an Age of Revolution: 1750–1800.* Chicago: University of Chicago, 1987.

Borges, Jorge Luis. "Kafka and His Precursors." *Other Inquisitions*. New York: Simon and Schuster, 1965.

Boscovich, Roger Joseph. *A Theory of Natural Philosophy [Theoria Philosophiae Naturalis]*. Trans. J. M. Child. Chicago: Open Court, 1922.

Boyle, Robert. *Works*. Ed. Thomas Birch. 5 vols. London: A. Millar, 1744.

Brennan, Joseph Gerard, and Alfred North Whitehead, "Whitehead on Plato's Cosmology." *Journal of the History of Philosophy* 9 (1971): 67–78.

Bronowski, Jacob. *William Blake, 1757–1827: A Man without a Mask*. New York: Haskell House, 1944.

Burtt, E. A. *The Metaphysical Foundations of Modern Science*, 2nd rev. ed. Garden City: Anchor Books, 1932.

Cajori, Florian. "Origins of Fourth Dimension Concepts." *American Mathematical Monthly* 33 (1926): 397–406.

Čapek, Milič. *Bergson and Modern Physics: A Re-Interpretation and Re-Evaluation*. Dordrecht: D. Reidel, 1971.

Čapek, Milič. *The Philosophical Impact of Contemporary Physics*. Princeton: Van Nostrand, 1961.

Cavell, Stanley. *In Quest of the Ordinary: Lines of Skepticism and Romanticism*. Chicago: University of Chicago Press, 1988.

Chandler, James K. "The Languages of Sentiment." *Textual Practice* 22 (2008): 21–39.

Chandler, James K. *Wordsworth's Second Nature: A Study of the Poetry and Politics*. Chicago: University of Chicago Press, 1984.

Cohen, Marc S., and C. D. C. Reeve. "Aristotle's Metaphysics." Ed. Edward N. Zalta. *The Stanford Encyclopedia of Philosophy*. https://plato.stanford.edu/archives/win2020/entries/aristotle-metaphysics/.

Coleridge, Samuel Taylor. *Biographia Literaria*. Ed. W. Jackson Bate, 2 vols. Princeton: Princeton University Press, 1983.

Coleridge, Samuel Taylor. *Collected Letters*. Ed. E. L. Griggs, 6 vols. Oxford: Clarendon, 1956–71.

Coleridge, Samuel Taylor. *Complete Poetical Works*. Ed. Ernest Hartley Coleridge, 2 vols. Oxford: Clarendon, 1912.

Coleridge, Samuel Taylor. *The Friend*. Ed. Barbara E. Rooke, 3 vols. Princeton: Princeton University Press, 1969.

Coleridge, Samuel Taylor. *Lay Sermons*. Ed. R. J. White. Princeton: Princeton University Press, 1972.

Cooper, Andrew M. "Small Room for Judgment: Geometry and Prolepsis in Blake's 'Infant Sorrow.'" *European Romantic Review* 31 (2020): 129–55.

Cooper, Andrew M. *William Blake and the Productions of Time*. Aldershot: Ashgate, 2013.

Cornford, Francis. *Plato's Cosmology: The "Timaeus" of Plato*. Trans. with commentary. London: Routledge, 1935.

Cudworth, Ralph. *An Abridgment of Dr. Cudworth's True Intellectual System of the Universe*, 2 vols. London: John Oswald, 1782.
Currie, Mark. *About Time: Narrative, Fiction and the Philosophy of Time*. Edinburgh: Edinburgh University Press, 2007.
Curtis, F. B. "Blake and the 'Moment of Time': An Eighteenth-Century Controversy in Mathematics." *Philological Quarterly* 51 (1972): 460–70.
Damrosch, Jr., Leopold. *Symbol and Truth in Blake's Myth*. Princeton: Princeton University Press, 1980.
Daniel, Stephen H. "Myth and the Grammar of Discovery in Francis Bacon." *Philosophy and Rhetoric* 15 (1982): 219–37.
Dante Alighieri. *Purgatory and Paradise*. Trans. Henry Cary. Ed. Henry C. Walsh. Philadelphia: H. Altemus, 1889.
Darwin, Erasmus. *Zoonomia; or, the Laws of Organic Life*, 2 vols. London: J. Johnson, 1794–96.
Davy, Humphry. "The Bakerian Lecture: On Some New Phenomena of Chemical Changes Produced by Electricity." *Philosophical Transactions of the Royal Society of London* 98 (1808): 1–44.
Davy, Humphry. "Historical View of the Progress of Chemistry." *Elements of Chemical Philosophy*. London: J. Johnson, 1812.
De Luca, Vincent A. "Blake's Wall of Words." *Unnam'd Forms: Blake and Textuality*. Ed. Nelson Hilton and Thomas A. Vogler. Berkeley: University of California Press, 1986. 218–41c
Denham, Robert D. *Northrop Frye and Others: Volume III: Interpenetrating Visions*. Ottawa: University of Ottawa Press, 2018.
Desmet, Ronald. "Did Whitehead and Einstein Actually Meet?" *Researching with Whitehead: System and Adventure*. Ed. Franz Riffert and Hans-Joachim Sander. Freiburg: Verlag Karl Alber, 2008.
Detienne, Marcel, and Jean-Paul Vernant. *Cunning Intelligence*. Trans. Janet Lloyd. Chicago: University of Chicago Press, 1991.
Diderot, Denis, and Jean le Rond d'Alembert, eds. *Encyclopédie, ou dictionnaire raisonné des sciences, des arts et des métiers*, 17 vols. Paris: Briasson, David, Le Breton, and Durand, 1751–72.
Eaves, Morris. *The Counter-Arts Conspiracy: Art and Industry in the Age of Blake*. Ithaca: Cornell University Press, 1992.
Egginton, William. "On Dante, Hyperspheres, and the Curvature of the Medieval Cosmos." *Journal of the History of Ideas* 60 (1999): 195–216.
Einstein, Albert. *Collected Papers: The Swiss Years: Correspondence, 1902–1914*. Ed. Martin J. Klein, A. J. Kox, and Robert Schulmann. Princeton: Princeton University Press, 1993.
Eliot, T. S. *The Sacred Wood: Essays on Poetry and Criticism*. London: Methuen, 1928.
Erdman, David V. *Blake: Prophet against Empire*, 3rd ed. Princeton: Princeton University Press, 1977.

Essick, Robert N. "*The Four Zoas*: Intention and Method." *Blake: An Illustrated Quarterly* 18 (1985): 216–19.

Esterhammer, Angela. "Calling into Existence: *The Book of Urizen*." *Blake in the Nineties*. Ed. Steve Clark and David Worrall. New York: St. Martin's, 1999. 114–32.

Esterhammer, Angela. *Creating States: Studies in the Performative Language of John Milton and William Blake*. Toronto: University of Toronto Press, 1994.

Faye, Jan. "Copenhagen Interpretation of Quantum Mechanics." *Stanford Encyclopedia of Philosophy*. Ed. Edward N. Zalta. https://plato.stanford.edu/archives/win2019/entries/qm-copenhagen/.

Fish, Stanley. *Surprised by Sin: The Reader in "Paradise Lost."* Cambridge: Harvard University Press, 1967.

Fletcher, Giles. *Christs victorie, and triumph in Heaven, and earth, over, and after death*. Cambridge: C. Legge, 1610.

Franta, Andrew. "Shelley and the Poetics of Political Indirection." *Poetics Today* 22 (2001): 765–93.

Frye, Northrop. *Fearful Symmetry: A Study of William Blake*. Princeton: Princeton University Press, 1947.

Frye, Northrop. "The Road of Excess." *The Stubborn Structure: Essays on Criticism and Society*. London: Routledge, 1970. 160–74.

Gardner, Stanley. *Blake's Innocence and Experience Retraced*. New York: St Martin's, 1986.

Gerans, Philip. "Pathologies of Hyperfamiliarity in Dreams, Delusions and Déjà Vu." *Frontiers in Psychology* (20 February 2014): 5.

Gilchrist, Alexander. *Life of William Blake*, 2 vols. London: Macmillan, 1863.

Gillies, John. "Space and Place in *Paradise Lost*." *ELH* 74 (2007): 27–57.

Glausser, Wayne. "Atomistic Simulacra in the Enlightenment and in Blake's Post-Enlightenment." *The Eighteenth Century* 32 (1991): 73–88.

Glen, Heather. *Vision and Disenchantment: Blake's Songs and Wordsworth's Lyrical Ballads*. Cambridge: Cambridge University Press, 1983.

Goldsmith, Steven. *Blake's Agitation: Criticism and the Emotions*. Baltimore: Johns Hopkins University Press, 2013.

Goldsmith, Steven. *Unbuilding Jerusalem: Apocalypse and Romantic Representation*. Ithaca: Cornell University Press, 1993.

Goldstein, Amanda Jo. *Sweet Science: Romantic Materialism and the New Logics of Life*. Chicago: University of Chicago Press, 2017.

Graves, Robert Perceval. *Life of Sir William Rowan Hamilton*. Dublin: Dublin University Press, 1882.

Green, Mathew J. A. *Visionary Materialism in the Early Works of William Blake*. New York: Palgrave, 2005.

Griffiths, Devin S. *The Age of Analogy: Science and Literature between the Darwins*. Baltimore: Johns Hopkins University Press, 2016.

Griffiths, Devin S. "The Intuitions of Analogy in Erasmus Darwin's Poetics." *Studies in English Literature* 51 (2011): 645–55.
Gueroult, Martial. *Descartes' Philosophy Interpreted according to the Order of Reasons: The Soul and God, The Soul and the Body.* Trans. Roger Ariew. 2 vols. Minneapolis: University of Minnesota Press, 1985.
Hamrick, William S., and Jan Van Der Veken. *Nature and Logos: A Whiteheadian Key to Merleau-Ponty's Fundamental Thought.* Albany: SUNY Press, 2011.
Hansen, Niels Viggo. "Spacetime and Becoming: Overcoming the Contradiction between Special Relativity and the Passage of Time." *Physics and Whitehead: Quantum, Process, and Experience.* Ed. Timothy E. Eastman and Hank Keeton. Albany: SUNY Press, 2004. 136–63.
Harper, George Mills. *The Neoplatonism of William Blake.* Chapel Hill: University of North Carolina Press, 1961.
Hartley, David. *Observations on Man, his Frame, his Duty, and his Expectations,* 2 vols. London: S. Richardson, 1749.
Hazlitt, William. *An Essay on the Principles of Human Action.* Ed. John R. Nabholz. Gainesville: Scholar's Facsimiles, 1969.
Hedrick, Lisa Landoe. *Whitehead and the Pittsburgh School: Pre-empting the Problem of Intentionality.* Lanham MD: Rowman & Littlefield, 2021.
Heimann, P. M. "'Nature is a Perpetual Worker'": Newton's Aether and Eighteenth-Century Natural Philosophy." *Ambix* 20 (1973): 1–25.
Helmig, Christoph, and Carlos Steel. "Proclus." *The Stanford Encyclopedia of Philosophy.* Ed. Edward N. Zalta. https://plato.stanford.edu/archives/fall2020/entries/proclus/.
Henderson, Andrea. "The Physics and Poetry of Analogy." *Victorian Studies* 56 (2014): 389–97.
Henderson, Linda Dalrymple. "The Image and Imagination of the Fourth Dimension in Twentieth-Century Art and Culture." *Configurations* 17 (2009): 131–60.
Henry, John. "A Cambridge Platonist's Materialism: Henry More and the Concept of Soul." *Journal of the Warburg and Courtauld Institutes* 49 (1986): 172–95.
Henry More's Manual of Metaphysics [Enchiridion Metaphysicum]. Trans. Alexander Jacob, 2 vols. Hildesheim: Georg Olms, 1995.
Heppner, Christopher. *Reading Blake's Designs.* Cambridge: Cambridge University Press, 1995.
Hesse, Mary B. *Models and Analogies in Science.* Notre Dame: University of Notre Dame Press, 1966.
Hoerner, Fredrick. "Prolific Reflections: Blake's Contortion of Surveillance in *Visions of the Daughters of Albion*." *Studies in Romanticism* 35 (1996): 119–50.
Hogarth, William. *The Analysis of Beauty.* London: J. Reeves, 1753.
Hume, David. *Enquiries concerning Human Understanding and concerning the Principles of Morals.* Ed. L. A. Selby-Bigge, 3rd ed., rev. P. H. Nidditch. Oxford: Oxford University Press, 1975.

Hume, David. *A Treatise of Human Nature*. Ed. L. A. Selby-Bigge, 2nd ed., rev. P. H. Nidditch. Oxford: Clarendon, 1978.
Jacob, Alexander. *De Nature Natura: A Study of Idealistic Conceptions of Nature and the Unconscious*. Stuttgart: Franz Steiner Verlag, 1991.
James, William. *The Principles of Psychology*, 2 vols. New York: Henry Holt, 1890.
James, William. "A Pluralistic Universe." *Writings 1902–10*. Ed. Bruce Kuklick. New York: Library of America, 1987.
Janowitz, Anne. "Response: Chandler's 'Vehicular Hypothesis' at Work." *Textual Practice* 22 (2008): 41–46.
Jarvis, Simon. "Blake's Spiritual Body." *The Meaning of "Life" in Romantic Poetry and Poetics*. Ed. Ross Wilson. New York: Routledge, 2009. 13–32.
Jarvis, Simon. "Prosody as Cognition." *Critical Quarterly* 40 (2003): 3–15.
Jarvis, Simon. "Thinking in Verse." *The Cambridge Companion to British Romantic Poetry*. Ed. Maureen N. McLane and James Chandler. Cambridge: Cambridge University Press, 2008. 98–116.
Jauss, Hans Robert. *Aesthetic Experience and Literary Hermeneutics: Theory and History of Literature*. Trans. Michael Shaw. Minnesota: University of Minnesota Press, 1982.
Kargon, Robert. "William Rowan Hamilton and Boscovichean Atomism." *Journal of the History of Ideas* 26 (1965): 137–40.
Keach, William. *Shelley's Style*. New York: Methuen, 1984.
Kear, Jon. "Staring into the Abyss: James Barry and British History." *In Elysium: Prints by James Barry*. Ed. Ben Thomas. Canterbury: University of Kent, 2010. 16–18.
Keats, John. *Poems and Selected Letters*. Ed. Carlos Baker. New York: Scribner's, 1962.
Kirk, Robert. *The secret commonwealth of elves, fauns & fairies: a study in folk-lore & psychical research / the text by Robert Kirk . . . 1691; the comment by Andrew Lang . . . 1893*. London: D. Nutt, 1893.
Koyré, Alexander. *From the Closed World to the Infinite Universe*. Baltimore: Johns Hopkins University Press, 1957.
Kroeber, Karl. *British Romantic Art*. Berkeley: University of California Press, 1986.
Leibniz, Gottfried Wilhelm. *New Essays concerning Human Understanding*. Ed. and trans. Alfred Gideon Langley. Chicago: Open Court, 1916.
The Leibniz-Clarke Correspondence. Ed. H. G. Alexander. Manchester: Manchester University Press, 1956.
Lewis, C. S. *The Allegory of Love: A Study in Medieval Tradition*. New York: Oxford University Press, 1958.
Lewis, C. S. *The Discarded Image: An Introduction to Medieval and Renaissance Literature*. Cambridge: Cambridge University Press, 1964.
Locke, John. *An Essay concerning Human Understanding*. Ed. Alexander Campbell Fraser, 2 vols. New York: Dover, 1959.
Lord, Alfred. *The Singer of Tales*. Cambridge: Harvard University Press, 1960.

Titus Lucretius Carus, His Six Books of Epicurean Philosophy. Trans. Thomas Creech. London: T. Braddyll, 1699.

Lussier, Mark. "Blake's Vortex: The Quantum Tunnel in *Milton*." *Nineteenth Century Contexts* 18 (1994): 263–91.

Malin, Shimon. *Nature Loves to Hide: Quantum Physics and the Nature of Reality, a Western Perspective.* Oxford: Oxford University Press, 2012.

Malin, Shimon. "Whitehead's Philosophy and the Collapse of Quantum States." *Physics and Whitehead.* Ed. Timothy E. Eastman and Hank Keeton. Albany: SUNY Press, 2003. 74–83.

Mann, Paul. "The Book of Urizen and the Horizon of the Book." *Unnam'd Forms: Blake and Textuality.* Ed. Nelson Hilton and Thomas A. Vogler. Berkeley: University of California Press, 1986. 49–68.

Mann, Paul. "The Final State of *The Four Zoas*." *Blake: An Illustrated Quarterly* 18 (1985): 204–08.

Marvell, Andrew. *The Complete Poems.* Ed. Elizabeth Story Donno. New York: Penguin, 1972.

Matthiessen, Hannes Ole. "Who Placed the Eye in the Centre of a Sphere? Speculations about the Origins of Thomas Reid's Geometry of Visibles." *Journal of Scottish Philosophy* 14 (2016): 231–51.

Maxwell, James Clerk. "Are There Real Analogies in Nature?" *The Scientific Letters and Papers.* Ed. P. M. Harman, 3 vols. Cambridge: Cambridge University Press, 1990–2002.

McGann, Jerome J. "The Idea of an Indeterminate Text: Blake's Bible of Hell and Dr. Alexander Geddes." *Studies in Romanticism* 25 (1986): 303–24.

Merleau-Ponty, Maurice. *Phenomenology of Perception.* Trans. C. Smith. New York: Routledge, 1962.

Merleau-Ponty, Maurice. *The Visible and the Invisible.* Trans. Alphonso Lingis. Evanston: Northwestern University Press, 1968.

Milton, John. *The Complete Poetry and Essential Prose.* Ed. William Kerrigan, John Rumrich, and Stephen M. Fallon. New York: Modern Library, 2007.

Minkowski, Hermann. "Space and Time." *The Principle of Relativity: A Collection of Original Memoirs on the Special and General Theory of Relativity.* Trans. W. Perrett and G. Jeffrey. London: Methuen 1923. 152–53.

Mitchell, W. J. T. "Blake's Comedy: Dramatic Meaning in *Milton*." *Blake's Sublime Allegory: "The Four Zoas," "Milton," "Jerusalem."* Ed. Stuart Curran and Joseph Wittreich, Jr. Madison: University of Wisconsin Press, 1973. 281–307.

Mitchell, W. J. T. *Blake's Composite Art: A Study of the Illuminated Poetry.* Princeton: Princeton University Press, 1978.

Mitchell, W. J. T. "Dangerous Blake," *Studies in Romanticism* 21 (1982): 410–16.

Mohr, Richard. "Plato's Theology Reconsidered: What the Demiurge Does." *History of Philosophy Quarterly* 2 (1985): 131–44.

More, Henry. *A Collection of Several Philosophical Writings*, 4th ed. London: Joseph Downing, 1712.
More, Henry. *The Complete Poems*. Ed. Alexander B. Grosart. New York: AMS, 1967.
More, Henry. *Divine Dialogues*, 2nd ed. London: Joseph Downing, 1713.
More, Henry. *Enchiridion Metaphysicum*. Trans. Joseph Glanvill. In *Philosophical Writings of Henry More*. Ed. Flora Isabel MacKinnon. London: Kessinger, 1925.
More, Henry. *An Explanation of the Grand Mystery of Godliness. Theological Works*. London: Joseph Downing, 1708.
More, Henry. *Immortality of the Soul*. London: J. Flesher, 1659.
More, Henry. *Two Choice and Useful Treatises*. London: James Collins, 1682.
Nagel, Thomas. *The Last Word*. New York: Oxford University Press, 1997.
Nagy, Gregory. *Homeric Responses*. Austin: University of Texas Press, 2003.
Nagy, Gregory. "Performing Homer." *New York Review of Books*. 9 April 1998. 81–82.
Nail, Thomas. *Lucretius I: An Ontology of Motion*. Edinburgh: Edinburgh University Press, 2019.
Newton, Isaac. *Correspondence*. Ed. H. W. Turnbull, J. F. Scott, A. Rupert Hall, and Laura Tilling, 7 vols. Cambridge: Cambridge University Press, 1959–77.
Newton, Isaac. *Opticks: or, A Treatise of the Reflections, Refractions, Inflections and Colours of Light*, 4th ed. London: William Innys, 1730.
Newton, Isaac. *Philosophical Writings*. Ed. Andrew Janiak. Cambridge: Cambridge University Press, 2004.
Newton, Isaac. *Principia Mathematica*, 3rd ed. Trans. Andrew Motte. Amherst: Prometheus Books, 1995.
Newton, Isaac. *Unpublished Scientific Papers*. Ed. A. R. and M. B. Hall. Cambridge: Cambridge University Press, 1962.
Nicolson, Marjorie H. "The Spirit World of Milton and More." *Studies in Philology* 22 (1925): 433–52.
O'Flaherty, Niall. *Utilitarianism in the Age of Enlightenment: The Moral and Political Thought of William Paley*. Cambridge: Cambridge University Press, 2019.
Olson, Richard. "The Reception of Boscovich's Ideas in Scotland." *Isis* 60 (1969): 91–103.
Oppenheimer, Robert. "Analogy in Science." *Centennial Review of Arts & Science* 2 (1958): 351–73.
Oster, Gerald. "Phosphenes." *Scientific American* 222 (1970): 82–87.
Otto, Peter. "Final States, Finished Forms, and *The Four Zoas*." *Blake: An Illustrated Quarterly* 20 (1987): 144–47.
Otto, Peter. "Is There a Poem in This Manuscript?" *Blake: An Illustrated Quarterly* 22 (1989): 142–44.
Pais, Abraham. *Subtle Is the Lord: The Science and the Life of Albert Einstein*. Oxford: Oxford University Press, 2005.

Paley, Morton. *The Continuing City: William Blake's "Jerusalem."* Oxford: Clarendon, 1983.
Paine, Thomas. *Common Sense.* Philadelphia: W. & T. Bradford, 1776.
Palmer, Samuel. *Letters.* Ed. Raymond Lister. Oxford: Oxford University Press, 1974.
Palter, Robert. *Whitehead's Philosophy of Science.* Chicago: University of Chicago Press, 1960.
Panofsky, Erwin. *Perspective as Symbolic Form.* Trans. Christopher S. Wood. New York: Zone Books, 1991.
Pecheux, Mother M. Christopher. "The Image of the Sun in Milton's 'Nativity Ode.'" *Huntington Library Quarterly* 38 (1975): 315–33.
Peterson, Mark. "Dante and the 3-Sphere." *American Journal of Physics* 47 (1979): 1031–34.
Planck, Max. *The Philosophy of Physics.* Trans. W. H. Johnson. New York: Norton, 1936.
Plato, *Timaeus, Critias, Cleitophon, Menexenus, Epistles.* Trans. R. G. Bury. Cambridge: Harvard University Press, 1981.
Plotinus. *Select Works, Thomas Taylor's Translation.* Ed. G. R. S. Mead. London: G. Bell, 1914.
Plotnitsky, Arkady. "Minute Particulars and Quantum Atoms: The Invisible, the Indivisible, and the Visualizable in William Blake and Niels Bohr." *ImageTexT: Interdisciplinary Comics Studies* 3 (2006–07): n.p.
Pope, Alexander. *Complete Poetical* Works. Ed. Bliss Perry. Boston: Houghton Mifflin, 1903.
Popper, K. R. "A Note on Berkeley as Precursor of Mach." *British Journal for the Philosophy of Science* 4 (1953): 26–36.
Porter, Dahlia. *Science, Form, and the Problem of Induction in British Romanticism.* Cambridge: Cambridge University Press, 2018.
Pressly, William. *James Barry: The Artist as Hero.* London: Tate Gallery, 1983.
Pressly, William. *The Life and Art of James Barry.* New Haven: Yale University Press, 1981.
Price, Richard. *Four Dissertations.* London: T. Cadell, 1777.
Priestley, F. E. L. "Berkeley and Newtonianism: The *Principles* (1710) and the *Dialogues* (1713)." *The Practical Vision: Essays in English Literature in Honour of Flora Roy.* Ed. Jane Campbell and James Doyle. Waterloo, ONT: Wilfrid Laurier University Press, 1978. 49–70.
Priestley, Joseph. *Disquisitions Relating to Matter and Spirit.* London: J. Johnson, 1777, 1782.
Priestley, Joseph. *An Essay on the First Principles of Government.* London: J. Dodsley, 1771.
Priestley, Joseph. *A Free Discussion of the Doctrines of Materialism, and Philosophical Necessity, in a correspondence between Dr. Price, and Dr. Priestley.* London: J. Johnson, 1778.

Priestley, Joseph. *History and Present State of Discoveries Relating to Vision, Light, and Colours*, 2 vols. London: J. Johnson, 1772.

Priestley, Joseph. *History and Present State of Electricity*. London: J. Johnson, 1775.

Priestley, Joseph. *Institutes of Natural and Revealed Religion*, 3rd ed. 2 vols. London: J. Johnson, 1794.

Priestman, Martin. *The Poetry of Erasmus Darwin: Enlightened Spaces, Romantic Times*. London: Routledge, 2013.

Proclus. *The Commentaries on the Timaeus of Plato in Five Books; containing a treasury of Pythagoric and Platonic physiology*. Trans. Thomas Taylor, 2 vols. in 1. London: printed for the author, 1820.

Raine, Kathleen. *Blake and Tradition*, 2 vols. London: Routledge & Kegan Paul, 1969.

Randall, Jr., John Herman. *The Career of Philosophy: From the Middle Ages to the Enlightenment*. New York: Columbia University Press, 1962.

Randall, Jr., John Herman. "The Intelligible Universe of Plotinos [*sic*]." *Journal of the History of Ideas* 30 (1969): 3–16.

Reid, Jasper. "The Evolution of Henry More's Divine Absolute Space." *Journal of the History of Philosophy* 45 (2007): 79–102.

Reid, Jasper. *The Metaphysics of Henry More*. London: Springer, 2012.

Reid, Thomas. *Works*. Ed. William Hamilton, 8th ed., 2 vols. Edinburgh: James Thin, 1895.

Robb, A. A. *Geometry of Time and Space*. Cambridge: Cambridge University Press, 1936.

Rodier, David. "Alfred North Whitehead: Between Platonism and Neoplatonism." *Neoplatonism and Contemporary Thought, Part One*. Ed. R. Baine Harris. Albany: SUNY Press, 2002. 183–204.

Rosen, Steven M. *Topologies of the Flesh*. Athens: Ohio University Press, 2006.

Rosenblum, Robert. *The International Style of 1800: A Study in Linear Abstraction*. Garland: New York, 1976.

Rosenblum, Robert. *Transformations in Late Eighteenth Century Art*. Princeton: Princeton University Press, 1967.

Rovelli, Carlo. *Reality Is Not What It Seems: The Journey to Quantum Gravity*. Trans. Simon Carnell and Erica Segre. New York: Riverhead, 2017.

Russell, Bertrand. *Portraits from Memory and Other Essays*. New York: Routledge, 2021.

Ruston, Sharon. "Humphry Davy: Analogy, Priority, and the 'True Philosopher.'" *Ambix* 66 (2019): 121–39.

Ryckman, Thomas. *Einstein*. London: Routledge, 2017.

Samuel, Irene. *Dante and Milton: The "Commedia" and "Paradise Lost."* Ithaca: Cornell University Press, 1969.

Schey, Taylor. "Limited Analogies: Reading Relations in Wordsworth's *The Borderers*." *Studies in Romanticism* 56 (2017): 177–201.

Schofield, Robert E. *The Enlightenment of Joseph Priestley: A Study of His Life and Work from 1733–1773*. University Park: Pennsylvania State University Press, 1997.

Schofield, Robert E. "Joseph Priestley, Eighteenth-century British Neoplatonism, and S. T. Coleridge." *Transformation and Tradition in the Sciences: Essays in Honour of I. Bernard Cohen*. Ed. Everett Mendelsohn. Cambridge: Cambridge University Press, 1984. 237–54.

Schofield, Robert E. "Joseph Priestley, Natural Philosopher." *Ambix* 14 (1967): 1–15.

Schofield, Robert E. "Joseph Priestley, the Theory of Oxidation and the Nature of Matter." *Journal of the History of Ideas* 25 (1964): 285–94.

Shapiro, Barbara. *John Wilkins 1614–1672: An Intellectual Biography*. Berkeley: University of California Press, 1969.

Shaviro, Steven. *Other Criteria: Kant, Whitehead, Deleuze, and Aesthetics*. Cambridge: MIT Press, 2009.

Shaviro, Steven. "'Striving with Systems': Blake and the Politics of Difference." *boundary 2* 10 (1982): 229–50.

Shaviro, Steven. *The Universe of Things: On Speculative Realism*. Minneapolis: University of Minnesota Press, 2014.

Shelley, Percy. *Complete Poetical Works*. Ed. Thomas Hutchinson. London: Oxford University Press, 1914.

Shiff, Richard. "Breath of Modernism (Metonymic Drift)." *In Visible Touch: Modernism and Masculinity*. Ed. Terry Smith. Chicago: Power, 1997.

Silberstein, Ludwik. *The Theory of Relativity*. London: Macmillan 1924.

Smith, Robert. *A Compleat System of Opticks*. Cambridge: printed for the author, 1738.

Spence, Joseph. *Anecdotes, Observations, and Characters, of Books and Men*. Ed. Samuel Weller Singer. London: W. H. Carpenter, 1820.

Stengers, Isabelle. *Thinking with Whitehead: A Free and Wild Creation of Concepts*. Trans. Michel Chase. Cambridge: Harvard University Press, 2002.

Stevens, Wallace. *The Necessary Angel*. New York: Vintage, 1951.

Stewart, Dugald. *Collected Works*. Ed. Sir William Hamilton, 11 vols. Edinburgh: T. Constable, 1854–60.

Stewart, Garrett. *The Deed of Reading: Literature Writing Language Philosophy*. Ithaca: Cornell University Press, 2015.

Stewart, Garrett. *The Look of Reading: Book, Painting, Text* (Chicago: University of Chicago Press, 2006).

Strawson, Galen. "Realistic Monism: Why Physicalism Entails Panpsychism." *Journal of Consciousness Studies* 13 (2006): 3–31.

Stroud, Barry. "Berkeley v. Locke on Primary Qualities." *Philosophy* 55 (1980): 149–66.

Suk, Ian and Rafel J. Tamargo. "Concealed Neuroanatomy in Michelangelo's *Separation of Light from Darkness* in the Sistine Chapel." *Neurosurgery* 66 (2010): 851–61.

Takanashi, Kyoko. "Mediation, Reading, and Yorick's Sentimental Vehicle." *Novel* 39 (2016): 486–503.

Taylor, A. E. *Commentary on Plato's "Timaeus."* Oxford: Clarendon, 1928.

Taylor, A. E. *Plato: The Man and His Work*. London: Methuen, 1955.

Taylor, David Francis. "Picturing Ekphrasis: Image and Text in Shakespeare Painting." *European Romantic Review* 33:4 (2022), 461–78.
Taylor, Thomas. *The Elements of the True Arithmetic of Infinites*. London: printed for the author, 1809.
Tesky, Gordon. "Bent Abstraction." *ELH* 88 (2021): 315–41.
Thackray, Arnold. "'Matter in a Nut Shell': Newton's *Opticks* and Eighteenth-Century Chemistry." *Ambix* 15 (1968): 29–53.
Thomas Taylor, the Platonist. Ed. Kathleen Raine and George Mills Harper. Princeton: Princeton University Press, 1969.
Thomson, J. J. *The Corpuscular Theory of Matter*. New York: Scribner's, 1907.
Throesch, Elizabeth L. *Before Einstein: The Fourth Dimension in Fin-de-Siècle Literature and Culture*. New York: Anthem, 2017.
Townsend, Chris. "Nature and the Language of the Sense: Berkeley's Thought in Coleridge and Wordsworth." *Romanticism* 25 (2019): 129–42.
Townsend, Chris. "Visionary Immaterialism: Berkeleian Empiricism in Blake's Poetry." *Studies in Romanticism* 58 (2019): 357–82.
Tucker, Abraham [Edward Search]. *The Light of Nature Pursued*, 6 vols. London: T. Jones and T. Payne, 1768.
Viney, Donald. "Process Theism." *The Stanford Encyclopedia of Philosophy*. Ed. Edward N. Zalta. https://plato.stanford.edu/archives/sum2020/entries/process-theism/.
Viscomi, Joseph. *Blake and the Idea of the Book*. Princeton: Princeton University, 1993.
Wade, Nicholas J. "The Vision of William Porterfield." *Brain, Mind and Medicine: Essays in Eighteenth-Century Neuroscience*. Ed. Harry A. Whitaker, C. U. M. Smith, and Stanley Finger. New York: Springer, 2007.
Weiss, Paul. "Contemporary World." *Review of Metaphysics* 6 (1953): 525–38.
Whitehead, Alfred North. *Adventures of Ideas*. New York: Macmillan, 1933.
Whitehead, Alfred North. *The Concept of Nature*. Cambridge: Cambridge University Press, 1920.
Whitehead, Alfred North. *Essays in Science and Philosophy*. Rider: London, 1948.
Whitehead, Alfred North. *Modes of Thought*. New York: Macmillan, 1938.
Whitehead, Alfred North. *The Principle of Relativity with Applications to Physical Science*. Cambridge: Cambridge University Press, 1922.
Whitehead, Alfred North. *Process and Reality*. New York: Free Press, 1978.
Whitehead, Alfred North. *Science and the Modern World*. New York: Macmillan, 1925.
Whitehead, Alfred North. *Symbolism: Its Meaning and Effect*. New York: Fordham University Press, 1927.
Whyte, L. L. "Boscovich and Particle Theory." *Nature* 179 (1957): 284–85.
Wickman, Matthew. *Literature after Euclid: The Geometric Imagination in the Long Scottish Enlightenment*. Philadelphia: University of Pennsylvania Press, 2016.
Wood, Michael. *The Road to Delphi: The Life and Afterlife of Oracles*. Farrar, Strauss and Giroux: New York, 2003.

Woolhouse, R. S. "Reid and Stewart on Lockean Creation." *Journal of the History of Philosophy* 20 (1982): 84–90.
Wordsworth, William. *Oxford Authors*. Ed. Stephen Gill. New York: Oxford University Press, 1984.
Worrall, David. "William Blake and Erasmus Darwin's *Botanic Garden*." *Bulletin of the New York Public Library* 78 (1974–75): 397–417.
Yeats, William Butler. *A Critical Edition of Yeats's "A Vision."* Ed. George Mills Harper and Walter Kelly Hood. London: Macmillan, 1978.
Yoder, Paul. "Unlocking Language: Self-Similarity in Blake's *Jerusalem*." *Romantic Circles: Praxis Series: Romantic Complexity*: March 2001, n.p.
Yolton, John W. *Perceptual Acquaintance from Descartes to Reid*. Minneapolis: University of Minnesota Press, 1984.
Yolton, John W. *Thinking Matter: Materialism in Eighteenth-Century Britain*. Minneapolis: University of Minnesota Press, 1983.
Youngquist, Paul. *Madness and Blake's Myth*. University Park: University of Pennsylvania Press, 1989.
Youngquist, Paul. "Reading the Apocalypse: The Narrativity of Blake's *Jerusalem*." *Studies in Romanticism* 34 (1993): 601–25.
Zuckerkandl, Victor. *Sound and Symbol*. Princeton: Princeton University Press, 1956.

Index

Abbott, Edwin A., *Flatland*, 113–18, 246n32
abstraction, 13–15, 29–30, 35–36, 40–41, 96–97, 102, 130–32, 166–67, 173–75, 197–99, 202–4, 215–16, 228–33
action, 1–4, 24, 28, 47–49, 64–66, 69–70, 95–96, 142–43, 150–51, 160–62, 199; at a distance, 13–14, 53, 79–80, 141, 152–53; and ideal causes, 179–80; and imagination, 180; voluntary, 180
aesthetics: immanence of, 59; and nature, 128–30; and politics, 28, 158–59
aether, 3–6, 9–11, 28–29, 49–51, 53–54, 66–67, 137–38, 151–52, 181–84, 204, 238n10. *See also* Newton, Isaac
affect, 81, 141. *See also* feelings; vehicles
Agamben, Giorgio, 73–74, 158–62, 167–70; "The End of the Poem," 79–80
allegory, 27–28, 30–32, 41, 55–56, 68, 88, 163–65, 191–92, 275n14; moral, 37–38
Allen, Richard, 59–60, 72–73
analogy, 14–15, 20–21, 26, 28–29, 32–42, 92–93, 96–99, 111–13, 119–20, 175–76; and disanalogizing, 37–38, 43; and hypothesis, 191–92, 201–2. *See also* symbol
anamnesis, 208–9, 245n25, 285n81. *See also* Coleridge, Samuel Taylor; Plato
Anglicanism, 149–50, 237n9
Apocalypse, apocalypticism, 19–21, 28, 66–68, 124–26, 145, 148–50, 168, 232
appearances, 9–10, 40, 63–64, 91–92, 97–98, 119, 162, 195–96, 206. *See also* geometry
Archytas, 127–28
Aristotle, 24–26, 127–28, 159–60; *De anima* ("On the Soul"), 204
asceticism, 5, 27–28, 154–55
Association for Preserving Liberty and Property against Republicans and Levellers, 27–28
associationism, 45–48, 65–68, 72–82, 88, 92–93. *See also* Hartley, David
atoms, atomic theory, 4–6, 40–43, 45–54, 92–93, 107, 137–41, 151, 176–77, 188–89, 199, 210–14, 226, 241n31. *See also* Boscovich, Roger Joseph
Augustine, Saint, 113–14
Ault, Donald: *Narrative Unbound*, 162–65, 167–72; *Visionary Physics*,

Ault, Donald *(continued)*
1–4, 8–9, 12–14, 106, 124, 139–40, 147, 215–16
Avatamsaka Sutra, 105–6

Bacon, Francis, 8, 32–33, 36, 70–71, 191–92, 246n35, 247n36
Bain, Jonathan, 34
Banchoff, Thomas, 113
Barry, James, *Account of a Series of Pictures, in the Great Room of the Society of Arts*, 56–57; *Elysium and Tartarus or the State of Final Retribution*, 221–22, 227
Bates, Stanley, 98
Baxter, Andrew, 213–14; *Enquiry into the Nature of the Human Soul*, 208–9
Bayes, Thomas, 60–63, 257n34
Benlowe, Edward, *Theophilia*, 116–17
Bennett, Jonathan, 94
Bentley, Richard, 64
Bergson, Henri, 13, 19–20, 107, 186–87, 224–25, 281n55
Berkeley, George, 40–41, 52–55, 72–73, 119, 127–28, 147, 189–90, 192, 195–97, 202–7, 210–14; *Alciphron*, 202–4, 208; *Principles of Human Knowledge*, 197–98, 202, 212–13, 216–17; *De Motu*, 289n9; *Siris*, 204, 206; *Three Dialogues*, 199–202, 205–6, 212–13
Blake, William, 252n1; and analogy, 33, 37–38, 42–43; and anticipation of relativity, 107–8, 264n1; and associationism, 45–48, 74–75; cosmology of, 1–3, 14–15, 20, 29, 34–35, 43, 54–55, 61–67, 105–6, 147–56, 158–62, 165–66, 175, 179–80, 196–97, 206, 210–12, 214–17, 232–33, 235, 241n31, 275n14; and dreams, 207–12, 292n30; and empiricism, 198–208, 215–16; and four-dimensional space-time, 28–29, 31–32, 35–36, 41–42, 61–63, 95, 100–102, 107–14, 116–38, 219–21, 226–235, 269n28, 296n18; and geometry, 3, 5–8, 13–14, 22; and history, 20–22, 29–32, 34–37, 40–41, 165–67, 234–35, 237n9; and imitation, 42; on miracles, 61–65, 107, 243n14; moral authority of, 21–22, 28; and mythology, 41–42, 83, 96, 128–30, 178, 197–8, 204–5; and perception, 54–55; and politics, 2–3, 21–22, 28; and prophecy, 2–3, 34, 63–64, 163–66, 169–70, 234–35; and relief etching, 24–26, 71, 164–65, 222–23, 243n13; and symbols, 52–53. *See also* Contraries; Mundane Soul; Vortex
Blake, William, works by: *All Religions Are One*, 6–7, 215–16; *America*, 102–3, 234–35; *The Book of Urizen*, 5–6, 12–13, 42, 82–83, 85–94, 96–97, 99–102, 130, 132–34, 152–53, 172–73, 178, 202, 226, 228–32; "A Cradle Song," 56, 58, 67–69, 75–76, 78–79, 81, 108–10, 130, 136, 158–59, 171–72, 255n23; *The Deity from Whom Proceed the Nine Spheres*, 119–26; *Europe*, 114–16, 151–52; *Ezekiel's Wheels*, 209–10; *The Four Zoas*, 8–9, 103, 148, 158, 162–70, 172–73, 178, 228; *The French Revolution*, 102–4; illustrations to Milton's "On the Morning of Christ's Nativity," 113–16, 153–54, 228; *Innocence*, 26; *Jacob's Dream*, 209–10; *Jerusalem*, 12, 37–38, 72–73, 108–13, 127–28, 135, 142–43, 149–50, 158–59, 170–73, 176–82, 184–87, 216–17, 226, 229–32, 234–35, 262n22; *The*

Marriage of Heaven and Hell, 2–3, 21, 24–26, 55–56, 81–83, 178–79, 182–83, 187–88, 191–92, 216, 219–22, 225–27, 234–35, 296n20; *Milton*, 12–13, 29, 37, 42, 63–64, 70–71, 93, 96, 98–100, 108–11, 119–20, 124–45, 150–51, 165–66, 168–70, 172–93, 205–10, 226, 234–35, 237n9, 248n41, 268n21, 282n61, 289n9, 296n20; *Milton's Mysterious Dream*, 46–48, 209–10; *Newton*, 1–6, 9–12, 14–15, 35–36, 96–97; "On the Virginity of the Virgin Mary & Joanna Southcott," 282n61; *A Public Address to the Chalcographic Society*, 42–43, 165, 198–99, 224; Queen Katharine's Dream, 209–10; "Satan Calling up his Legions," 228; *Songs of Experience*, 81–83, 130; *Songs of Innocence*, 19, 45–46, 53–59, 63–64, 67–82, 130; *The Sun at His Eastern Gate*, 128–30; There Is No Natural Religion, 6–7, 215–16, 258n41; "The Tyger," 1–2, 33, 37–38, 81–83, 130; *Visions of the Daughters of Albion*, 17–32, 34, 97–98, 111, 128–34, 140–41, 143, 158, 223–35, 245n25; *The Vision of the Last Judgment*, 55, 137–38, 175–76, 219–21, 223–35, 296n18, 296n20; "You don't believe I wont attempt to make ye," 282n61
body, 26; and action, 150–51; and force, 52–53; and mind, 45–47, 59, 65–67, 96, 214–15, 223–26, 233; and perspective, 168–69; and soul, 150–51, 153–55; and space, 138–39
Bohr, Niels, 9–10, 37–38, 43, 52–53, 107, 241n31, 251nn62–63; "The Unity of Human Knowledge," 195–96

Boltzmann, Ludwig, 38–40
Borges, Jorge Luis, 30–31, 245n25
Boscovich, Roger Joseph, 9–11, 28–30, 33, 35–36, 45–46, 49–56, 81–82, 92–94, 210–12; *Theoria Naturalis Philosophiae*, 31–32
Bradley, James, 189–91
Bronowski, Jacob, 124, 127–28
Buddhism, 105–6, 145

calculus, 6–7, 9–10, 52–53, 197–98
Calvinism, 13–14, 18, 40–41, 126–27, 150–51, 202–4, 269n31, 274n6
Čapek, Milič, 52–53, 128–32, 134–35
Cary, Henry, 119–20
causation, causality, 1–6, 11–13, 97–98; and analogy, 33; emanative cause, 153–54; and human action, 179–80; knowledge of, 46; nonlinear, 196–97
Cavell, Stanley, 40
child, children, 59–61; imagery of, 75–76; and language, 72–75
Clarke, Samuel, 64
Coleridge, Samuel Taylor, 3–4, 6–7, 32, 52–53, 149–50, 159–60, 213–15, 285n81; "Essay on Method," 24–26, 37; *Biographia Literaria*, 66–67, 76–77; *Christabel*, 26–27; suspension of disbelief, 45–46
Collingwood, R. G., 186–87
community, 157–58
conscience. *See* morality
contingency, 10–11, 29–30, 88–89, 95, 156, 176–77, 278n40. *See also* events; history, historicism
continuity, 34, 54, 142–43, 178–79, 195, 208
Contraries, 12–13, 34–35, 37, 55–56, 59, 64–65, 69, 76–77, 80–89, 102–6, 162–63, 196–97, 216, 219–22, 227. *See also* Blake, William

Coppo di Marcovaldo, 119–20
Cornford, Francis, 8–9, 214
corpuscles, 8, 10, 49, 52–54, 180–82, 189–92, 232. *See also* atoms, atomic theory
cosmogenesis, 9–10, 12–13, 96–97, 100–102
Coste, Pierre, 93–94
Cudworth, Ralph, 148; *True Intellectual System of the Universe*, 214
Cumberland, George, 37–38
Currie, Mark, 186–87
custom. *See* habit

d'Alembert, Jean-Baptiste le Rond, 229
Damon, S. Foster, 20–21
Damrosch, Leopold, Jr., 41–42, 165–66
Dante, 107–8; *Commedia*, 187–88; *Paradiso*, 119–27, 136–37
Darwin, Erasmus, 33, 37, 97–98; on analogy, 36; *The Economy of Vegetation*, 102–4; *Loves of the Plants*, 22, 31–32; *Zoonomia*, 33
Davy, Humphry, 31–32, 238n14, 248n41; *Elements of Chemical Philosophy*, 37
deconstruction, 55–56, 71–73, 148, 202, 211–13
Derrida, Jacques, 2–3, 79–80
Descartes, René, 9–10, 13–14, 51, 147, 150–51, 154–55, 278n40, 291n20
design, 5–6, 9–10, 17–19, 22–23, 26–27, 35–38, 69–71, 75–79, 81, 89–90, 93–94, 126–32, 140–41, 158, 171–72, 180–82, 221–26, 233. *See also* pictorialism
Design, 1–2, 8, 33, 60–63, 72–73, 81, 85–86, 199–200, 204–5
Diderot, Denis, *Encyclopédie*, 229

dreams, 67–68, 169–72, 175–76, 187–88, 192, 207–12, 280n52

Eagleton, Terry, 234–35
Eaves, Morris, 20–21
education: childhood, 59–61; learning, 66–67, 69–70; and progress, 70–71
Einstein, Albert, 2, 4, 9–11, 14–15, 24–26, 28–32, 38–41, 43, 51, 54, 95, 174–75, 190–91, 238n10, 246n32, 246n35, 251n63. *See also* quantum theory; relativity
Eliot, T. S., 30–31, 245n26
emanation, emanationism, 2, 5, 34–35, 54–55, 76–77, 100–102, 113–17, 136–38, 144–45, 147–48, 152–54, 170–71, 190–91, 202, 214–15, 224–25, 233. *See also* light
emotion. *See* feelings
empiricism, 20, 29–32, 34–40; and associationism, 45–46; and hypothesis, 191–92, 195–96; and infinite, 41; and judgment, 234; and miracles, 61–63; pragmatic, 198–99; and psychology, 196–97
energy, 81–82, 85–86, 142–43, 214–16, 219–23; and Eternity, 97, 103–4; and gravity, 168–69, 174–75; and matter, 89–92
Enlightenment, 8, 10–14, 21, 54–55, 148–50, 196–97, 214–16; and analogy, 32; and educational reform, 70–71; and power of ideas, 29–30
Erdman, David V., 135–36, 163–64, 243n13, 252n1
eroticism, 24–27, 140–41, 221–22
esprit géométrique, 1–2, 154–55, 198–99
Essick, Robert N., 2–3, 243n13
Esterhammer, Angela, 2–3, 63–64, 86
Eternity, 12–14, 42, 85, 97, 99–100, 103, 106, 126–30, 133–34, 136–37,

142, 147, 153–54, 170–73, 175–78, 226–27, 233–34. *See also* Blake, William; infinity
Euclid, Euclidean method, 3, 6–10, 18, 37–38, 57, 69, 86, 99, 111, 116–23, 128–32, 136, 162, 166–67, 198–99, 221–22. *See also* geometry
events, 2–3, 13, 20–21, 29–30, 40; and apprehension, 206–7; and experience, 63–64; and relations, 196–97; and space-time, 95. *See also* history, historicism
experience, 18, 22–24, 27–28, 33, 35–36, 40–41, 144; and abstraction, 197–98; and events, 63; of ideas, 45–47; of language, 73–74; private, 192; and time, 175; and understanding, 195
eyes, 9–10, 18–19, 22–26, 46–48, 59–60, 85, 104, 118–19, 143, 169, 223–24, 243n13. *See also* vision, visualization

Faraday, Michael, 10
feelings, 56–57, 59, 62–65, 82–83, 141, 158–59, 187–88; vectoral character of, 97–98
feminism, 20–22
field theory, 10–11, 14–15, 20–21, 28–29, 33–35, 51–53, 147, 150–51, 153, 176–77, 241n31. *See also* relativity; Whitehead, Alfred North, "extensive continuum"
Flaxman, John, 35–38
Fletcher, Giles, *Christ's Triumph after Death*, 114–16
force, 13–15, 35–36, 45–46, 49–55, 79–82, 87, 92–97, 104, 168–69, 223; and matter, 212–13. *See also* gravity
form, formalism, 3, 29, 37–38, 41–42, 75–77, 85–86; and matter, 156–57; and narrative, 167–68; and pictorial representation, 107
fractals, 57, 59, 61–63, 76–77, 81–82, 145; and associations, 92–93
Fraser, Alexander Campbell, 94–95, 152–53, 206
Frye, Northrop, 20–21, 59, 105–6, 134–35, 192, 202–4; *Fearful Symmetry*, 32

Galileo (Galileo Galilei), *Dialogue concerning the Two Chief World Systems*, 99
Gamow, George, 180–82
Gardner, Stanley, 60–61
gender, 20–22
geometry, 2–3, 5–10, 13–15; cultural prestige of, 6–7; and Eternity, 100–102; Euclidean, 3, 6–7, 18, 37–38, 57, 69, 99, 107–8, 116–17, 119–23, 128–32, 136, 143–45, 148, 166–67, 221–22, 238n10; and experience, 18–19; non-Euclidean, 6–7, 9–10, 105, 119–20, 127–28, 198–99, 238n10, 296n13; and relativism, 20–21; and sense, 197–99; and signification, 162; and social criticism, 22; and space-time, 31–32, 107–8, 116–17, 119–24, 126–38, 174–75; and spheres, 111–23, 126–27, 133–34; and topology, 19, 106, 133–34, 140–41; of torus, 229. *See also* space, spatiality; time, temporality
Gilchrist, Alexander, 32, 127–28, 198–99
Glausser, Wayne, 214
Gödel, Kurt, 29–30, 99
Goldsmith, Steven, 2–3, 28
Gothic, 26–27
gravity, 1–4, 13–15, 28–29, 35–36, 49, 53–54, 87, 92–93, 102–3, 141,

gravity *(continued)*
    151–52, 164–65, 174–75, 183–85,
    212–14, 238n10. *See also* field
    theory; relativity
Griffiths, Devin S., 33, 36, 247n36
Gueroult, Martial, 154–55

habit, 63–64, 72–73, 86, 179–80,
    202, 208
Haggarty, Sarah, 3
Hamilton, Sir William, 52–53
Harper, George Mills, 147–48
Hartley, David, 4, 45–47, 53–54,
    60–61, 65–67, 71–82, 92–93;
    Observations on Man, 45–46, 69–70.
    See also associationism
Hayley, William, 144
Hazlitt, William, *Essay on Human
    Action*, 180
Hedrick, Lisa Landoe, 155–56
Heimann, P. M., 180–82
Heisenberg, Werner, 9–10, 43;
    uncertainty principle, 11–12
Heller-Roazen, Daniel, 69
Helmig, Christoph, 177–78
Henderson, Andrea, 14–15
Herbert, George, 149–50
Herschel, William, 103
Hinton, Charles Howard, 114–16,
    246n32
history, historicism, 10–11, 20–22, 28,
    163–65, 173–74; and ideas, 29–32;
    and time, 144–45, 224–25. *See also*
    time, temporality
Hoerner, Fredrick, 26
Hogarth, William, 37–38; *The Analysis
    of Beauty*, 117–18; *Satire on False
    Perspective*, 209–10
holenmerianism, 153–55
homogeneity, 57, 111, 129–30, 138,
    142–43, 153
Hooke, Robert, 10–11

horizon, 18–24, 42–43, 88–89,
    97–100, 102, 145, 205–6, 221–22,
    226–27, 233–34
Hume, David, 6–7, 29–30, 32–34,
    36–37, 46–47, 141, 160–61,
    198–99, 213–14, 258n46; *Dialogues
    Concerning Natural Religion*, 6–7,
    13–14, 18, 155; *Enquiry concerning
    Human Understanding*, 36; "Of
    Miracles," 60–65; *Treatise of Human
    Nature*, 208

Iamblichus, 148
idealism, 40–41, 43, 46–47, 52–55,
    57, 66–67, 91–92, 127–28, 143,
    155–56, 179–80, 195–97, 206,
    210–12, 214–15, 225–26, 232. *See
    also* realism
ideas, 29–30; and analogy, 32, 41–42;
    decomplex, 59–61, 65–74, 77–81,
    88; and sense experience, 41, 45–48,
    51. *See also* associationism
identity, identification, 22, 33, 41–42,
    52–53, 62–63, 65–68, 75–76, 102,
    135, 141–42, 192, 204–5, 208–9
illumination. *See* light
imagination, 24–26, 41–42, 63–64,
    175–76, 179–80, 205–6, 243n14;
    and action, 180; and creation,
    94–95; and reading, 56–57
imitation, 5, 26, 42, 87, 156–55,
    192–93
immanency, 3–4, 21, 55, 59, 67–68,
    86, 137–38, 144–45, 155–59,
    225–27, 233
incunabulus, 55–56, 255n23
induction, 33–37, 41, 54, 60–61,
    175–76
infinity, 12, 41, 49–51, 55–65,
    76–77, 85, 92–93, 104, 111–13,
    116–18, 128–34, 142–45, 189–90,
    199, 221–23, 226–28, 241n31;

infinitesimals, 7–8, 59–60, 106, 173–74, 188–90, 199. *See also* Eternity
inspissation, 114–16, 137–38, 140–41, 226, 228. *See also* More, Henry
intention, 30–31
introjection, 59
introspection, 85–86, 107, 251n62
irony, 26–28, 37–38, 45, 158, 168, 215–17, 234–35

Jacob, Alexander, 201–202, 210–12
James, William, 143
Jarvis, Simon, 79–80
Johnson, Joseph, 179–80, 187–88, 214–15, 252n1, 275n7
judgment, 21, 34–35, 91–92, 170–71, 233–34
justice, 21, 55, 270n36

Kant, Immanuel, 6–7, 12, 40–41, 97–98, 176–77, 195–96, 233–34
Katsos, Isidoros, 190–91
Klein bottle, Kleinian object, 19–20, 107–8, 111, 137–38, 143–44, 242n3
knowledge, 6–7, 10–11, 40, 46, 56–57, 98, 108–12, 142–45, 195–96, 199–200, 208–9
Koyré, Alexander, 150–51

Langen, Timothy, 30–31
language, 59–60, 69–70; experience of, 73–75; and narrative, 178; of nature, 72–73; philosophical, 72–73; and signification, 161–62; written, 73–74
Latini, Brunetto, 119–20
Latitudinarianism, 149–50, 180
Leibniz, Gottfried Wilhelm, 41–42, 49, 54, 95
Lewis, C. S., 107–8, 111–13, 267n15

Lewis, Matthew, 26–27
liberalism, 2–3, 21, 27–28, 104, 235, 274n6
light, 153–54, 180–82, 189–91, 238n10
limits, 41, 49–55, 63–65, 92–93, 98–99, 106, 156–57, 161–62, 165, 190–91, 198–99, 202
Locke, John, 4, 24–26, 37–38, 40–41, 46–47, 65, 85–86, 93–94, 152–53, 199–201, 208, 210–12; caverned mind, 17–18; *An Essay concerning Human Understanding*, 94–96; and sensation, 160–61, 188–90
Lord, Alfred, 70–71
Lucretius, 214, 268n21, 287n96. *See also* atoms, atomic theory

Mach, Ernst, 38–41, 195–97
Maguire, Muireann, 30–31
Marvell, Andrew, "To His Coy Mistress," 116–17
Marxism, 2–3
mathematics, 1–2, 6–8, 12–13, 19, 86, 92–93, 104, 119–24; and formalism, 3, 82; and realism, 28–30, 57; and senses, 198–99, 215–16; and space, 49–54, 155–56. *See also* geometry
matter, materialism, 2–5, 8–10, 17–20, 24–29, 35–36, 40–41, 85–86, 196–97, 210–12, 232–33; and creation, 93–97, 152–53, 210–12; and energy, 89–92; and force, 49–53, 212–13; and form, 156–57; and immaterialism, 81–83, 88–89, 93, 138–39, 147–48; and impressions, 45–48; and interpenetrability, 138–41; and mechanism, 3–4, 14–15, 25–26, 35–36, 40–41, 49, 64, 94–95, 103, 135, 139–40, 148, 153–55, 179–80, 195, 212–13, 222–23, 225–26, 228; and mind,

matter, materialism *(continued)*
201–2; and motion, 208; "nutshell" theory of, 53–54, 57, 93–94, 223; and perception, 54–55, 158–60, 180–82, 201–7, 212–13; and soul (spirit), 89–92, 95, 151–55, 213–14; and space, 152–53, 190–91, 210–12; threat of, 66–67. *See also* geometry; perception; senses, sensation; space, spatiality

Matthiessen, Hannes Ole, 118–19
Maturin, Charles, 26–27
Maxwell, James Clerk, 38–42
mechanism, mechanical philosophy, 3–4, 14–15, 25–26, 35–36, 40–41, 49, 64, 94–95, 103, 135, 139–40, 148, 153–55, 179–80, 195, 212–13, 222–23, 225–26, 228. *See also* Newton, Isaac
media, 5, 53–54; graphic, 2–3; mediation, 40, 70–71
Merleau-Ponty, Maurice, 19, 107, 137–38, 226
metaphysics, 8–9, 13–14, 17–18, 40, 43, 173–75; and induction, 34; and matter, 40–41; and mechanical philosophy, 147–48; and real world, 195–96; and relativity, 32
millennialism, 70–71, 148–50, 285n80
Milton, John, "l'Allegro," 128–30; *Comus*, 22; *Lycidas*, 80–81; "Methought I saw My Late Espoused Saint," 144–45; "On the Morning of Christ's Nativity," 113–16; *Paradise Lost*, 8, 22, 124, 126–27, 140–41
mimesis, 87. *See also* realism
miniatures, miniaturization, 59, 65–67
Minkowski, Hermann, 132, 246n32
Minkowski diagram, 19–20
miracles, 60–65
Möbius strip, 18–19

models, modeling, 1–2, 33, 38–42, 82, 92–93, 118–19, 143, 160–61, 187–88, 238n10
morality, 21–22, 28; and allegory, 37–38; and status quo, 27–28
More, Henry, 22, 34–35, 53–54, 114–16, 137–41, 147–55, 182, 188–89, 197, 201–4, 214, 224–26, 228, 237n9, 275n14
Mundane Soul, 8, 96, 148–49, 152, 158–59, 168–69, 171–72, 179–88, 201–5, 207–9, 226, 229–32, 280n49. *See also* Blake, William

Nagel, Thomas, 234
Nagy, Gregory, 178
names, naming, 86–87
narrative: and form, 167–68; and language, 178; and perspective, 168–69; and revision, 162–64, 169–73; and temporality, 86–89, 170–72
neoclassicalism, 35–38, 42, 130–32, 221–22
Neoplatonism, 2, 22, 28–29, 34–35, 53–54, 85–86, 88, 147–49, 156–57, 182, 210–12, 214–15, 243n14, 277n25
Newton, Isaac, 1–6, 8–11, 13–14, 54, 189–90, 210–12; and absolute space, 111, 135–36, 159–60, 166–67, 197, 229, 289n9; on aether, 3–4, 10–11, 28–29, 53, 180–82, 214; and creation of matter, 152–53; critique of, 106; *De gravitatione ed aequipondio fluidorum*, 7–8, 95–96, 262n21; on gravity, 3–4, 35–36, 53, 214, 238n10; and instantaneity, 224–25; on mechanics, 11–12, 14–15, 33; on molecules, 92–94; and Neoplatonism, 147–48; *Opticks*, 3–6, 10–11, 28–29, 53–54, 151–52,

223, 238n10; *Principia Mathematica*, 3–4, 183–84; Scholium to *Principia Mathematica*, 155, 184–85, 262n21
nominalism, 35–39, 100
nullibism, 147. *See also* Descartes, René; matter, materialism

objectivity, 29–34, 97–98, 164–68
observation, 3–4, 11–13, 34–42, 49, 195–96; and miracles, 64; and time, 175. *See also* perception
Oldenburg, Henry, 4
Oppenheimer, Robert, 37

Paine, Thomas, 2–3
Palter, Robert, 95
Panofsky, Erwin, 37–38, 57, 128–30
parody, 5, 8–9, 26–27, 60–61, 124, 131–33, 164–67
Pecheux, Mother M. Christopher, 113–14
perception, 14–24, 28–30, 42; and creation, 200–201; and dimensionality, 143; and environment, 52–55; expanded, 180–83, 188–90, 192, 204; of ideas, 45–48, 78–79; and immediacy, 97–98; and matter, 53–55, 82–83, 158–60, 180–82, 202–7; middle zone of, 198–99; narrowing of, 85–86; and objectivity, 31–32; and reading, 56–57; and senses, 189–90; and space, 197; and time, 172–73; translucency of, 232; and understanding, 43. *See also* senses, sensation
Percy, Thomas, *Reliques of Ancient English Poetry*, 78–79
performativity, 2–3, 11–12, 56–57, 63–64, 71–72
person, personhood, 69, 192–93. *See also* identity, identification

perspective, 2, 9–12, 19–20; change of, 188–89; Euclidean, 99; nonrelative, 88; and pictorialism, 37–38; rational, 57; and satire, 26–27
Peterson, Mark, 119–20, 123–24
Phillips, Adam, 192–93
philosophy: and analogy, 36–37; and geometry, 6–7, 127–28; and language, 72–73; mechanical, 147–53, 195–97; of mind, 25–26; and perception, 46, 52–53, 199–200, 210–13; and time, 175–76. *See also* idealism; matter, materialism; Neoplatonism; realism
physics: and analogy, 33–35; dualist, 13–15, 35–36, 43, 232–33; and experience, 195–96; and force, 45–46, 49–54, 92–94; and language, 72–73; and movement, 105–6, 111, 114–16; Newtonian, 1–4, 8–9, 14–15, 35–36, 87–88, 105–6, 148–49, 176–77; and perception, 41–43, 46–48, 154–55, 199–200; and space-time, 20–21, 120–23, 127–28, 137–45, 173–76, 189, 192, 210–12. *See also* atoms, atomic theory; metaphysics; relativity
physiology, 4, 13–14, 53–54, 60–61, 65–66; of experience, 141, 143, 166–67, 287n94; of ideas, 45–47, 185; and theology, 262n21
pictorialism, 2–3, 14–15, 37–38, 107, 136, 259n51
Planck, Max, 41, 246n35
plasticity, 107, 147–49, 151, 212–15
Plato, 5–10, 24–26, 111–13, 148–49, 206; Cave, 17–18; Demiurge, 5–6, 157–58, 162, 167–68, 170–71, 271n49, 292n39; *khora*, 5–6, 73–74, 159–62, 167–68, 278n39; *Meno*, 6–7; *Phaedrus*, 126–27; Receptacle, 5–12, 34–35, 73–74,

Plato *(continued)*
151–53, 155–62, 167–68, 170, 177–82, 214, 271n49; *Theaetatus*, 160–61; *Timaeus*, 5–6, 8–10, 12, 34–35, 151–52, 155–61, 167–68, 177–79, 190–91, 196–97, 262n21; World Soul, 157–58, 186–87, 214. *See also* Neoplatonism

Plotinus, 153–54, 159–61, 165–70, 280n47

poetry: and poet, 168–70; and revision, 162–64, 169–73; historicizing, 163–65. *See also* Blake, William, works by; prophecy

Pope, Alexander, 111–13; *Rape of the Lock*, 22

Popper, Karl, 195–96

Porterfield, William, 59–60

positivism, 40, 43, 195–96

Price, Richard, 47–48, 61–63

Priestley, F. E. L., 197, 214

Priestley, Joseph, 28–29, 31–33, 35–36, 45–54, 57, 59, 61–63, 93–94, 147, 149–50, 196–97, 210–14, 252n1, 274n6, 292n30, 294n46; *Disquisitions Relating to Matter and Spirit*, 13–14, 53–54, 93, 210–12, 214–15; *History and Present State of Discoveries Relating to Vision, Light, and Colours*, 59–60, 100; *History and Present State of Electricity*, 10–11, 70–71

Proclus, 34–35, 88, 148, 177–78, 190–91, 199

progress, 8, 11–12, 60–61, 185, 238n14; and educational reform, 70–71

prophecy, 2–3, 21, 30–31, 34, 63–64, 163–66, 169–70, 234–35

psychology, 34–35, 63–64, 196–97, 251n62; associationist, 45–46, 256n24; and characters, 62–63, 87; and mind, 52–54, 65–66, 80–82, 175–76; and sympathy, 141. *See also* associationism; empiricism

quantum theory, 40, 43, 52–53, 174–75, 195–96, 241n31, 251n62, 257n34; and pictorial representation, 107. *See also* atoms, atomic theory; field theory; physics; relativity

Quasha, George, 170

race, and slave trade, 20–22

Raine, Kathleen, 147–48

Rainich, George Yuri, 128–32

Randall, John Herman, Jr., 165–66, 280n47

reading, 45–47, 53–59, 63–64, 69–70; and attention, 171–72; and narration, 168; and participation, 67–69, 75–76, 78–79, 88–92; physiological effects of, 65–66; and self-reflexivity, 79–81; and song, 70–72

realism, 29–32, 37–38, 42–43, 57, 88–89, 155–57, 206, 232–33. *See also* idealism

reason: and analogy, 36–37; critique of, 41; and miracles, 63; and progress, 8; and rhetoric, 6–7

Reid, Jasper, 138–40, 150–51, 153–54

Reid, Thomas, 6–7, 9–10, 17–18, 66–67, 160–61, 200–201; *Essay towards a New Theory of Vision*, 119; *Inquiry into the Human Mind on the Principles of Common Sense*, 119

relations, 2, 12–15; and events, 196–97; and phenomena, 195–96; and space, 51, 95

relativism, 105–6, 233–34

relativity, 2, 4, 9–13, 32–36, 51, 99, 238n10; descriptivistic understanding of, 195; and space-time, 20–21,

28–29, 95, 134, 173–75; and three-dimensional representation, 128–32. *See also* atoms, atomic theory; field theory; quantum theory
relief etching, 24–26, 71, 164–65, 222–23
Remnant, Peter, 94
rhetoric, 6–7
Robb, A. A., 134
Romanticism: and analogy, 31–32; and dreams, 209–10; and materialism, 2, 261n1; and miracles, 63; and neoclassicism, 35–36, 42, 130–32; and oral culture, 70–71; politics of, 105; and space, 111. *See also* Blake, William
Rosen, Steven M., 19
Rosenblum, Robert, 35–36

satire, 26–27, 37–38, 40–41, 85–86. *See also* irony; parody
satisfaction, 97–98, 179–80
Schiff, Richard, 226
Schofield, Robert, 47–48, 214–15
Scholasticism, 46, 153–55
Schrödinger, Erwin, 241n31
science: and analogy, 32, 37–42; and appearances, 40; and Neoplatonism, 2, 11–12; and objectivity, 31–32; and observation, 42; and prophecy, 34; and scientist, 31–32, 35–36. *See also* geometry; philosophy; physics
self: and interconnectedness, 192–93; and mind, 169–70; and perception, 46–47, 169; and perspective, 88. *See also* subjects, subjectivity
sense, sensation, 9–10, 24–26, 31–33, 35–36, 38–43, 45–46, 189–90; and abstraction, 197–98; and awareness, 97–98; and consciousness, 223–26; and data, 195–96; and events, 206–7; expanded, 180–83, 188–90, 192, 204; and God, 199–204; and language, 161–62, 195; limits of, 199; and mathematics, 198–99, 215–16; and matter, 158–60, 201–7, 212–13; of physical world, 51–55; and reading, 56–57. *See also* associationism; perception
sensualism, 24–25, 27–28, 289n9
September Massacres of 1792, 27–28
Shakespeare, William, 116–17
Shaviro, Steven, 86
Shelley, Percy Bysshe, 246n35; *A Defence of Poetry*, 30–31
Simplicius, 162
skepticism, 28, 33–34, 36–37, 46, 64–65, 160–61
Smith, Adam, 180; *Theory of Moral Sentiments*, 141
Smith, Robert, *Compleat System of Opticks*, 100
social, sociality, 2–3, 86; and perception, 54–55; and sympathy, 141; transcendence of, 18–22, 26–27
soul, 13–14, 54–55, 185; and body, 32, 37, 59–60, 89–92, 137–41, 147–48, 150–55; and dreams, 208–9; and matter, 40, 89–92, 94–95, 167–68, 206, 213–15; and space, 40, 153–54, 197. *See also* Mundane Soul
space, spatiality: 5–15, 19, 28–29, 195; atomic, 199; Cartesian, 128–32; and creation, 96–97; dimensions of, 19–20, 28–32, 35–36, 41–42, 57, 66–67, 99–103, 107–8, 111–38, 228, 233, 246n32; and distance, 173–74, 189–90, 202; empty, 155–56; and extension, 201–2; and fractals, 57; impenetrable, 93–96; and location, 189–90; and matter, 49–51, 138–39, 140–41,

space, spatiality *(continued)*
  147, 150–53, 190–91, 210–12, 229; passive, 157; and perception, 53–54, 159–60, 172–73, 197; and relation, 51, 95; and scale, 92–93; and soul, 40, 153–54, 197; and time, 66–67, 128–30, 132, 142–43, 155–56, 165–67, 175–77, 226–27; and topology, 107–11; and Vortex concept, 164–67. *See also* time, temporality
speech, 2–3, 30–31, 79–80, 169–71, 281n53
spirit. *See* soul
spissitude, 137–41, 153–54, 272n61. *See also* More, Henry
Steel, Carlos, 177–78
Stengers, Isabelle, 97–98, 176–77
Stewart, Dugald, 49–51, 210–12
Stewart, Garrett, 73–74, 79–80, 102, 160–61
structuralism, 11–12, 20–21, 37, 106, 216–17
subjects, subjectivity, 28–29, 40, 195–97; and experience, 31–32, 46, 62–65, 192, 225–26; and relativism, 202, 210–12, 234; and time, 143–44, 175. *See also* self
Swift, Jonathan, *Modest Proposal*, 26–27
symbols, symbolism, 41–42, 52–53, 58–59, 79, 88–89, 97–98, 106, 111–13, 137–38, 165–66, 222–23, 232, 245n25, 250n59. *See also* allegory
symmetry, 1–2, 12–13, 81, 132
sympathy, 141, 180, 185, 187–88, 192–93
synesthesia, 45–46, 74–75
synthesis, 33, 36, 59–60 73–75, 77–81, 86, 88–89, 175, 185, 216–17. *See also* ideas, decomplex

Taylor, A. E., 8–9, 155–57
Taylor, Thomas, 34–35, 137–38, 147–48, 198–99, 202–4; *Elements of the True Arithmetic of Infinites*, 76–77
Test and Corporation Acts, 149–50
Thompson, E. P., 234–35
thresholds, 18–19, 78–79, 88–92, 172–73
time, temporality, 5–6, 10–13, 195; and change, 27–28, 34–35, 37–38, 63, 170, 173–76, 196–97, 224–27; and continuity, 34; dimensions of, 28–29, 31–32, 35–36, 57, 99–102, 111, 114–16, 134, 145, 246n32; and duration, 116–17, 141–45, 170, 172–73, 177–78, 186–87, 224–25, 229; and eternity, 99; and experience, 175; and history, 144–45; and millennialism, 148–49; and moment, 142–45, 207–8; and movement, 165, 175–76, 189; and narrative, 86–89, 170–72; and perception, 20, 34, 53–54, 172–73; and process, 26, 37–38; and space, 128–30, 1332, 166–67, 175–77, 226–27; and transcendence, 30–31. *See also* Eternity; infinity; space, spatiality
topology. *See under* geometry; space, spatiality
Townsend, Chris, 202–5
transcendence, 18–19, 89–92, 98, 116, 126–28, 144–45, 155–56, 170, 172–73, 176–77, 204–6, 233–34, 244n21
Tucker, Abraham, 144–45, 147–49, 201–2, 207–9, 243n13; *Light of Nature Pursued*, 179–92, 204
typology, 30–31, 226–27

Unitarianism, 149–50, 213–14
utopia, utopianism, 24–26, 28, 54–55, 66–67, 70–71, 86, 99, 149–50, 162–63, 248n41

vehicles, 139–41, 150–51, 170, 182, 187–92, 201–2, 204, 216–17; and motion, 189. *See also* More, Henry
vision, visualization, 19–24, 42, 46–48, 59–60, 85, 107, 113, 114–23, 143; dimensions of, 28–29; expanded, 204; and relativism, 235. *See also* perception; senses, sensation
voice, 169
Vortex, 42, 46–47, 59, 69, 81–82, 85–86, 97, 128–30, 133–38, 140–45, 163–67, 170, 172–73, 224–27, 229–33, 273n71. *See also* Blake, William

Watson, Bishop Richard, 36–37
Whewell, William, 31–32, 64
Whitehead, Alfred North, 20–21, 24–26, 29–30, 34–36, 66–67, 81, 86, 95, 97–98, 105–6, 127–28, 142, 144–45, 148–153, 155–58, 162–63, 166–67, 172–73, 205–6, 215–16, 225–26, 233–24, 245n25, 263nn23–24, 277n25, 291n20; "bifurcation of nature," 195–96, 265n4; "extensive continuum," 30, 157–58, 173–79, 199–200, 207–8, 224–25, 283n62; on God, 132, 184–85; on soul, 32, 59–60. *See also* relativity
Whyte, L. L., 51
Wilkins, John, 72–73
Wilson, Mona, 32
Wollstonecraft, Mary, 22
wonder, 24–26, 63–64
Wordsworth, William, 5–7, 78–79, 80–81, 111, 203–5, 226–27, 238n14, 248n41
Worrall, David, 102–4, 243n14

Yeats, W. B., 137–38, 271n56
Yolton, John, 152–53, 212–13
Young, Thomas, 10

www.ingramcontent.com/pod-product-compliance
Lightning Source LLC
Chambersburg PA
CBHW051558230426
43668CB00013B/1894